Marine Microbial Bioremediation

Editors

Anjana K Vala

Department of Life Sciences
Maharaja Krishnakumarsinhji Bhavnagar University
Bhavnagar, India

Dushyant R Dudhagara

Department of Life Sciences
Bhakta Kavi Narsinh Mehta University
Junagadh, Gujarat, India

Bharti P Dave

Department of Biosciences, School of Sciences
Indrashil University, Rajpur Kadi,
Mehsana Gujarat, India

CRC Press
Taylor & Francis Group
Boca Raton London New York

CRC Press is an imprint of the
Taylor & Francis Group, an **informa** business
A SCIENCE PUBLISHERS BOOK

The cover credit goes to authors of Chapter 2, Yi Wei et al.

First edition published 2022
by CRC Press
6000 Broken Sound Parkway NW, Suite 300, Boca Raton, FL 33487-2742

and by CRC Press
2 Park Square, Milton Park, Abingdon, Oxon, OX14 4RN

Library of Congress Cataloging-in-Publication Data

Names: Vala, Anjana K., 1973- editor. | Dudhagara, Dushyant R., 1987- editor. | Dave, Bharti P., 1956- editor.
Title: Marine microbial bioremediation / editors, Anjana K. Vala, Dushyant R. Dudhagara, Bharti P. Dave.
Description: First edition. | Boca Raton : CRC Press, Taylor & Francis Group, 2021. | "CRC Press is an imprint of the Taylor & Francis Group, an Informa Business." | Includes bibliographical references and index.
Identifiers: LCCN 2021004502 | ISBN 9780367425333 (hardcover)
Subjects: LCSH: Marine bioremediation. | Microbial biotechnology. | Marine pollution.
Classification: LCC TD192.5 .M27 2021 | DDC 628.1/683--dc23
LC record available at https://lccn.loc.gov/2021004502

ISBN: 978-0-367-42533-3 (hbk)
ISBN: 978-0-367-77003-7 (pbk)
ISBN: 978-1-003-00107-2 (ebk)

DOI: 10.1201/9781003001072

Typeset in Times
by TVH Scan

Preface

Microorganisms are omnipotent, hence have applications in diverse fields ranging from biotechnology, pharmaceuticals, green chemistry, environmental clean-up, biomass, and energy generation to nanobiotechnology. Microbial bioremediation has a big role to play as microbes can not only transform toxic heavy metals to non toxic form, they can also degrade highly recalcitrant compounds.

Marine ecosystem, the largest among all, is a reservoir of several bioresources. It also plays an important role in transportation, energy generation, extraction of precious materials, recreation, etc. All these activities pollute the marine environment to a great extent. Besides, the wastes generated in terrestrial environment ultimately find their way into the marine environment. Marine ecosystem thus serves as a sink for pollutants imposing severe health hazards to marine organisms and ultimately to human beings.

Marine microbes possess unique traits such as novel proteins, enzymes etc. that enable microbes to detoxify the marine environment. Bioremediation of various toxic pollutants by diverse marine microbes and the mechanisms involved therein are covered in this book. We believe the book would be useful to scientific community involved in research related to environmental microbiology and marine microbiology in particular. The book would also be of benefit to the under graduate and graduate student community and researchers at large.

We are happy and thankful to the contributors, the leading scientists with immense experience and expertise in the field. It is only their sincere efforts and inputs that made this task possible amidst the pandemic.

We thank Dr. Raju Primlani, Senior Editor, Science Publishers and his team for constant support, patience and assistance.

We are grateful to our respective university authorities (Maharaja Krishnakumarsinhji Bhavnagar University, Indrashil University and Bhaktakavi Narsinh Mehta University) for encouragement and support. Thanks are also due to our faculty and research colleagues, research scholars, family and friends.

Last but not the least, we (Anjana Vala and Prof Bharti Dave) are grateful to our mentor Prof H C Dube, Former Head, Department of Life Sciences, MK Bhavnagar University, who has always been a source of inspiration for taking up scientific endeavors and imbibing in us the spirit of writing books.

Thank you all.

Anjana K Vala
Dushyant Dudhagara
Prof Bharti P Dave

Contents

1

Degradation of Polycyclic Aromatic Hydrocarbons By Halophilic Aquatic Fungi

Elisabet Aranda[1], Gabriela Ángeles de Paz[1], María del Rayo Sánchez[2] and Ramón Alberto Batista-García[3,*]

INTRODUCTION: PAHS IN THE ENVIRONMENT

Polycyclic aromatic hydrocarbons (PAHs), also called polynuclear aromatic hydrocarbons or polyarenes, represent a well-known group of toxic chemicals widely distributed in the environment. They are formed by the fusion of two or more rings in linear, angular or clustered configuration. According to the number of fused benzene rings, they are classified as *small* PAHs, when they have less than up to six benzene rings (naphthalene, anthracene, phenanthrene, pyrene) and large PAHs (more than six benzene rings) (dibenzo [a, f] perylene, dibenzo [a, l] pentacene, benzocoronene, etc.). These organic hydrophobic molecules have a low solubility, which increase with the increase in the ring numbers.

The main source of PAHs are *petrogenic* or petroleum derived, *pyrogenic* or combustion derived and *biogenic* or organism derived – formed naturally through biological processes (Wang et al. 2014b) (Figure 1.1). The petrogenic source of PAHs results from geologic seepage, petroleum spills and petroleum refined product. The release of exhausts and vehicle emissions and organic materials such as creosote, coal tar, crankcase oil, and wood burning account for pyrogenic input. This pyrogenic source can be also natural, as it originated from volcanic eruptions and natural forest fires. In each process, the content on different structures will depend on the starting source and the temperature reached. Then, usually petrogenic processes lead to high percentages of small PAHs with three or fewer rings and pyrogenic processes lead to high percentages of high PAHs with four or more rings (Benner et al. 1990). However, most of the sea pollution is due to daily practices such as navigation, which implies the presence of low amounts of hydrocarbons in the seawater.

[1] Institute of Water Research, University of Granada. Ramón y Cajal, 4, Bldg. Fray Luis, 18071 Granada. Spain.

[2] Centro de Investigación en Biotecnología. Universidad Autónoma del Estado de Morelos. Cuernavaca, Morelos. Mexico.

[3] Centro de Investigación en Dinámica Celular, Instituto de Investigación en Ciencias Básicas y Aplicadas. Universidad Autónoma del Estado de Morelos. Cuernavaca, Morelos. Mexico.

[*] Corresponding author: rabg@uaem.mx

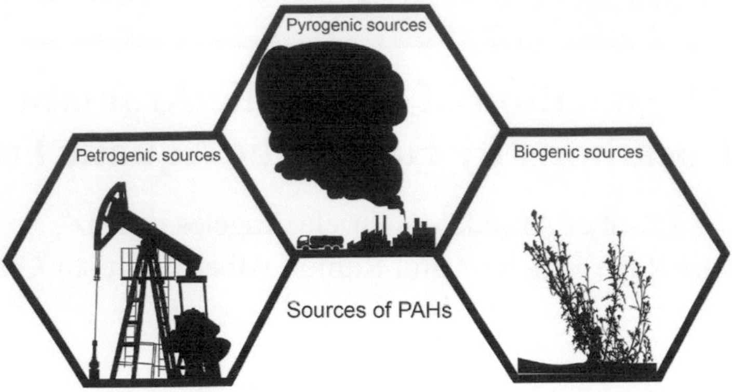

FIGURE 1.1 Sources of PAHs in the environment.

Distribution of PAHs

The distribution of PAHs depends on several factors, like chemical and physical properties of these molecules. PAHs are present in the atmosphere as vapors or adsorbed into airborne particulate matter and commonly derived from combustion and volatilization. According to the vapor pressure of different PAHs, they will be distributed differently. The atmosphere is the main dispersal source of PAHs (Manoli and Samara 1999).

Atmospheric PAHs are deposited in soils and sediments being bound to soil particles. Attached to them, they become mobiles, and the mobility will depend on the physical properties of the particles and their mobility in soils. Thus, it is not rare to find PAHs at relative high concentrations in remote areas due to their ability to be transported over long distances. Many hypersaline environments, including coastal lagoons, salt and soda lakes, salterns and seawater are polluted with PAHs compounds and some of these environments usually have a huge ecological, economic and scientific value.

Finally, they can be deposited in sediments and water bodies as the results of effluent from industrial outfalls, storm water from highways or urban runoff from storm drains (Sanders et al. 2002). In river sediments, the concentration is generally higher than in the surrounding water body because they tend to be attached into the organic matter. In water, they usually enter via atmospheric fallout. It has been estimated that 10-80% of PAHs present in the world oceans come from atmospheric sources (Moore and Ramamoorthy 2012). However, petroleum spillage, particularly in the surrounding of oil platforms, industrial discharges, atmospheric deposition, and urban run-off has a relevant role in coastal environments (Neff 1979). As a result, PAHs can be found in all the aquatic environments all over the world (Lima et al. 2005).

Toxicity of PAHs

PAHs are included in the list of priority pollutants by different regulation agencies, such as the US Environmental Protection Agency (USEPA), EU legislation concerning PAHs in food and environment, due to the carcinogenic, teratogenic and mutagenic properties (Luch 2009) (Table 1.1). https://ec.europa.eu/jrc/sites/jrcsh/files/Factsheet%20PAH_0.pdf. PAHs are lipophilic compounds, which can be readily absorbed in a variety of tissues. Particularly, PAH can covalently bind to proteins, RNA and DNA, this last covalent union being the most associated with carcinogenicity (Marston et al. 2001). In addition, some studies have reported their ability to interfere with hormone metabolizing enzymes of the thyroid glands, and their adverse effects on reproductive as well as immune system (Oostinngh et al. 2008).

TABLE 1.1 Names and structures of PAHs frequently monitored according to recommendations by the US Environmental Protection Agency (EPA), the EU Scientific Committee for Food (SCF) and the European Union (EU).

List	Common Name	Structure	List	Common Name	Structure
EPA	Naphtalene	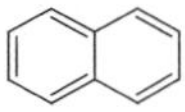	EPA, SCF, EU	Benzo [a] anthracene	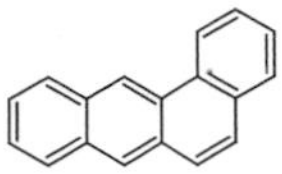
EPA	Anthracene	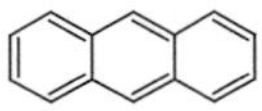	EPA, SCF, EU	Benzo [a] pyrene	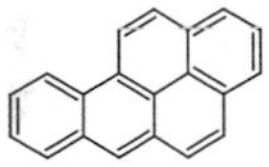
EPA	Phenantrene	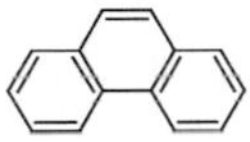	EPA, SCF, EU	Dibenzo [a, h] anthracene	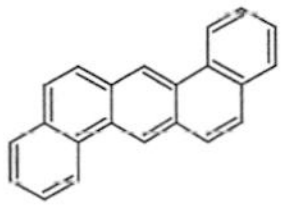
EPA	Acenaphtene	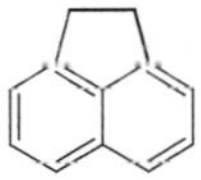	SCF, EU	Benzo [j] fluoranthene	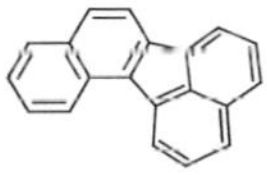
EPA	Acenaphthylene	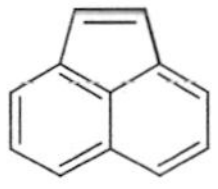	EPA, SCF, EU	Benzo [k] fluoranthene	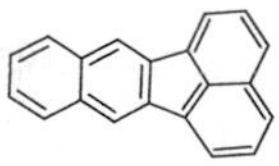
EPA	Fluorene	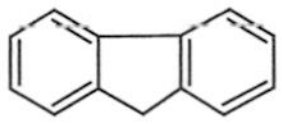	EU + SCF	Dibenzo [a, e] pyrene	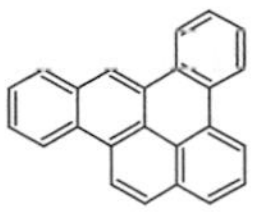
EPA	Fluoranthene	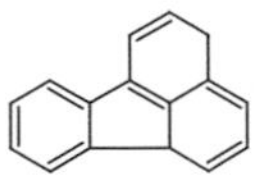	EU + SCF	Dibenzo [a,h] pyrene	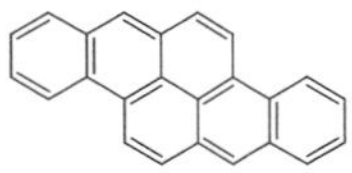
EPA	Pyrene	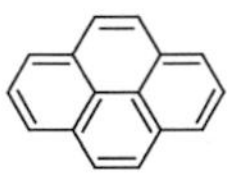	EU + SCF	Dibenzo [a, i] pyrene	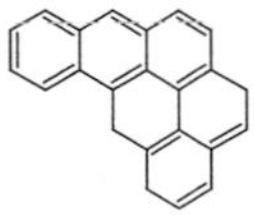
EPA, SCF, EU	Chrysene	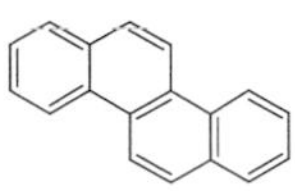	EPA, SCF, EU	Benzo [ghi] perylene	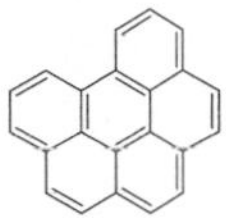

These compounds are resistant to degradation and they tend to be accumulated for long periods in the environment, causing adverse effects at all the tropic levels. In addition, they are susceptible to photochemical oxidation by UV-light and sunlight in the visible exposure, since the specific molecule makes them absorb the light under this wavelength. Several studies have shown the toxic effects of oxygenated PAHs (Arfsten et al. 1996, Lundstedt et al. 2007), concluding that the acute phototoxic effects of PAHs and nitro-, and oxo-substituted should be considered when conducting environmental risk assessments. PAHs also have an important effect on the microbial communities which are usually accompanied by the enrichment of PAH-tolerant bacteria and fungi (Wang et al. 2014a). Recent studies have shown how the presence of PAHs can accelerate the propagation of genes related to tolerance to pharmaceutical compounds (Wang et al. 2017), highlighting the changes that PAHs can exert on microbial communities.

HALOPHILIC FUNGI IN CONTAMINATED SITES: TAXONOMIC GROUPS

Halophilic aquatic-fungi have been shown to have an important role in biotechnological applications. Particularly, some studies have shown the ability of some halophilic aquatic fungi in degradation of PAH, using them as carbon and energy source for its metabolism (Uratani et al. 2014, Bonugli-Santos et al. 2015). These fungus have developed special phenotypic and molecular adaptations against low water activity and extreme conditions, such as the production of meristem-like structures (Zalar et al. 2005) and genes coding for different transport alkali metal ions (Arino et al. 2010, Gostinčar et al. 2011, Zajc et al. 2013).

There are many taxonomic studies which have described several species of halophilic terrestrial fungi with the ability to degrade PAHs (Márquez-Rocha et al. 2000, Di Gregorio et al. 2016, Pozdnyakova et al. 2016, Kadri et al. 2017). On the other hand, aquatic fungi have not been extensively explored nor employed for bioremediation processes (Harms et al. 2011).

Halophily is expressed in several groups of individual orders and groups of fungi taxonomically distant (Gunde-Cimerman et al. 2009). However, some of them have similar morphology and physiology (Carmichael and de Hoog 1980, Zalar et al. 1999, 2005). The halophily might be an evolutionary ancient trait, then, all of these fungi could have independently evolved in different unrelated groups and it could have adapted to halophilic situations (De Hoog et al. 2005).

Basidiomycota Division

Within the Basidiomycota division, there are three main orders with halophilic features, *Trichonosporales*, *Sporidiales* and *Wallemiales* (Cantrell et al. 2011). Additionally, some marine basidiomycetes in the genus *Marsimellus* have been intensively analysed, highlighting its biotechnological potential due to their ability to successfully degrade PAHs in saline conditions, such as *Marasmiellus ramealis* JK 314 (Lange et al. 1996) and *Marasimellus* sp. (Vieira et al. 2018).

The order *Wallemiales* is one of the most xerophilic fungal taxa. It comprises the genus *Wallemia* with three different species: *W. ichtyophaga, W. sebi* and *W. Muriae* (Zalar et al. 2005). To date, the entire genus *Wallemia* and the entire order *Wallemiales* are considered xerophilic (Gunde-Cimerman et al. 2009). Several taxonomic analyses have resolved this genus into three different species and they were all isolated from several hypersaline natural and anthropogenic sources (Kis-Papo et al. 2003, Zalar et al. 2005), showing a considerable molecular distance between them and having many distinguishable morphological characters, the xerophilic capability being one of the few phenotypic similarities (Zalar et al. 2005). Within these species, *Wallemia ichthyophaga* have been widely studied due to its extreme salt tolerance and osmolarity glycerol signalling pathway (Plemenitaš et al. 2014).

In addition, the predominant basidiomycetous yeast isolated from aquatic environment is the member of the genus *Rhodotorula, Rhodosporidium, Trichosporon, Cryptococcus* and *Sporobolomyces* (Nagahama et al. 2001). Species belonging to the genus *Rhodotorula* and *Trichosporon* have also shown a great adaptation to hypersaline environments (Butinar et al. 2005a).

Ascomycota Division

Ascomycota division also has several representative halophilic groups and a high resistance to low water activity. The most representative orders in this group are *Capnodiales, Dothideales* and *Eurotiales*, which involve the most dominating yeasts in the hypersaline water (Gunde-Cimerman et al. 2000). These fungi are commonly separated into two groups: i) the melanized black fungi, including meristematic yeast-like fungi, with phylogenetically related genus, and ii) the non-melanized fungi. Among them, black fungi are the most abundant (Musa et al. 2018).

The melanized black fungi involves different taxa and a diverse phylogeny. However, they have melanized cell walls and yeast-like multilateral division (Sterflinger 2006). The halophilic dominating species in this non-taxonomic group are *Hortea werneckii, Aurobasidium pullulan* and *Phaeotheca triangularis* (Zalar et al. 1999) and phylogenetically relevant *Cladosporium* species. Since many members of this group have been related with systemic mycosis, they became an important target in physiological analysis (Turk et al. 2004, Sterflinger 2006, Chowdhary et al. 2015).

Halophilic species belonging to the order *Capnodiales* usually have a very long genome with large repetitive sequences. This redundancy in their genome might be an advantage under stress conditions (Gostinčar et al. 2011, Plemenitaš et al. 2014). It was also suggested that many members in the taxa can be adapted to changes in the environment (Crous et al. 2009). *Cladosporium sphaerospermum (*Dugan et al. 2008, Zalar et al. 2007) and the genus *Phaeoteca* (Zalar et al. 1999) are frequently isolated from hypersaline water and correspond to the halophilic fungi group (Crous et al. 2009). Otherwise, the genus *Cladosporium* contains some opportunist fungi as well as non-pathogenic species (Bensch et al. 2012). They appear frequently isolated from salterns and salty lakes (Gunde-Cimerman et al. 2000). Furthermore, several species of *Cladosporium* have been isolated from contaminated soils and described as PAH degrader fungi (Giraud et al. 2001, Potin et al. 2004, Fayeulle et al. 2019). Additionally, recent studies demonstrated a better yield in PAH degradation whether they used aquatic hypersaline fungi such as *Cladosporium* sp. (Birolli et al. 2018).

Dothideales also contains halophilic melanized yeast. They share a similar morphology with a thick, melanized cell wall and, in this regard, an extremophilic ecotype (Sterflinger 1998). These microorganisms have resulted in the most dominant fungal group isolated in hypersaline waters. Moreover, their biomass increased with concentrations of salt in selective media (Gunde-Cimerman et al. 2000). In general, they are rarely related with pathogenicity, but *Hortea werneckii,* the most halophilic member in this group, is easily transmitted and is the main pathogenic agent responsible for *tinea nigra,* a superficial fungal infection (Bonifaz et al. 2008).

Aspergillus niger, A. sydowii, Eurotium amstelodami, Penicillium crhysogenum and species of *Emericella* are the most representative halophilic members in the order *Eurotiales.* They have been identified in hypersaline environments (Butinar et al. 2005b, Gunde-Cimerman et al. 2009) and showed halotolerance as a plesiomorphic trait shared by the entire order (Gunde-Cimerman et al. 2009). Species of *Aspergillus,* including its teleomorphs, and the genus *Emericella* comprise the most xerophilic fungi into this order (De Hoog et al. 2005, Lenassi et al. 2013). Some *Aspergillus* strains such as *A. niger* and *A. sclerotiorum* CBMAI 849 have shown degradation abilities of high concentrations of toxic pollutants, like PAHs (Wu et al. 2009, Passarini et al. 2011a). Furthermore, species like *A. destruens* and *A. sydowii* could remove over 90% of benzo-*a*-pyrene and phenanthrene successively at hypersaline conditions (1M NaCl) (González-Abradelo et al. 2019). Further investigations of the genetics and ecological roles of *Aspergillus* spp. and its teleomorph *Eurotium* spp. have to be performed, since some studies stated that they may belong to the most highly evolved microorganisms on Earth due to its extremophilic adaptations (Butinar et al. 2005c).

Finally, the order *Saccharomycetales* presents species of the genera *Pichia* which are frequently isolated in saline aquatic environments, the most common being *Pichia membranifaciens* (Soares et al. 1997) and *Yarrowia lipolytica* isolated from marine environment. They have shown their ability to use bromine contaminated hydrocarbons as a carbon source and to degrade it (Vatsal et al. 2011).

Many Ascomycota and Basidiomycota fungi were identified from very extreme aquatic environments, thanks to advances in "omics" analyses, but the information is still limited (Barone et al. 2019). Nevertheless, some studies made by RNA-sequencing and the eukaryotic metatranscriptomic analyses revealed the taxonomic composition in saline environments and showed the fact that they are metabolically active at the same time, especially the fungus belonging to the order *Malasseziales, Dothiedales* and *Sporodiales* (Stock et al. 2012, Bernhard et al. 2014, Pachiadaki et al. 2014, Edgcomb et al. 2016). This finding might ensure the success of industrial applications in environmental polluted fields (Gunde-Cimerman and Zalar 2014, Plemenitaš et al. 2014).

To sum up, halophilic fungi possess a great capacity to degrade pollutants due to catabolic enzymes and to extend their mycelial system, they all belong to different and relatively far phylogenetic groups, but they have an important biotechnological advantage such as their great adaptive abilities (Table 1.2). Additionally, members from different halophilic taxa have been proposed as model organisms for the study of halotolerance in eukaryotes instead of the well-known *Saccharomyces cerevisiae* (Gunde-Cimerman et al. 2009).

TABLE 1.2 Biodegradation of PAHs by aquatic halophilic fungus.

Division	Order	Strain	Pollutant	Reference
Ascomycota	Capnodiales	*Davidiella tassiana*	Anthracene	Aranda et al. 2017
Ascomycota	Debaryomyces	*Debaryomyces hansenii*	Phenol	Jiang et al. 2016
Ascomycota	Eurotiales	*Aspergillus flavus*	Anthracene	Aranda et al. 2017
Ascomycota	Eurotiales	*Aspergillus BAP14*	benzo-α-pyrene	Wu et al. 2009
Ascomycota	Eurotiales	*A. sclerotiorum*	benzo-α-pyrene	Passarini et al. 2011b
Ascomycota	Capnodiales	*Cladosporium* sp.	Anthracene	Birolli et al. 2018
Ascomycota	Saccharomycetales	*Candida lipolytica*	Petroleum compounds	Boguslawska-Wąs and Dąbrowski 2001
Basidiomycota	Sporidiales	*Rhodotorula mucilaginosa*	Petroleum compounds	Boguslawska-Wąs and Dąbrowski 2001
Ascomycota	Saccharomycetales	*Yarrowia lipolytica*	Bromoalkanes	Vatsal et al. 2011
Basidiomycota	Agaricales	*Marasimellus* sp.	Pyrene and BaP	Vieira et al. 2018

ADAPTIVE MECHANISMS OF HALOPHILIC FUNGI

Microorganisms that grow optimally with ≥% of NaCl are considered to be halophiles. According to their salt concentration requirements for growth, halophiles are classified as slight halophiles (0.2-0.85 M or 1-5% NaCl), moderate halophiles (0.85-3.4 M or 5-20% of NaCl), and extreme halophiles (3.4-5.1 or 20-30% of NaCl) (DasSarma and Arora 2001); among the three categories, some are obligate halophiles. However, halotolerant microorganisms do not require salt for growth but can tolerate, in some cases, extreme concentrations of salt (Dalmaso et al. 2015). Both halophiles and halotolerant can grow over a wide range of salinity (DasSarma and Arora 2001). In 2008, Dr. A. Oren proposed that a halophile is a microorganism that grows optimally at salt concentrations of 50 g/L (~0.85 M NaCl) or higher, and tolerates 100 g/L of salt (~1.7 M NaCl) at least, and that marine microorganism is not synonymous to halophilic microorganism.

The microorganisms living in hypersaline environments (where salt concentration is higher than in the sea, ~0.5 M of NaCl) are challenged by osmotic pressure and toxicity of Na^+ (Gunde-Cimerman and Zalar 2014). In salt non-adapted organisms, osmotic imbalance triggers the exit of water from the cells with a reduction of the turgor pressure and the dehydration of the cell, leaving a highly concentrated cytoplasm. Ionic stress, caused by entry of ions to the cell, caused the damage of membrane systems, as well as denaturation of proteins and consequently cell death (Petelenz-Kurdziel et al. 2011, Gunde-Cimerman and Zalar 2014). Nonetheless, halophiles and halotolerants are equipped with mechanisms to tolerate both ionic and osmotic conditions.

Two main adaptive strategies are used by these microorganisms to maintain the cytoplasm hyperosmotic:

1. The "*salt in*" strategy, mostly used by aerobic halophilic Archaeas (e.g. *Halobacteriaceae*) and Bacteria (e.g. *Salinibacter*), where the counterbalance of high-salt concentrations is through accumulation of other inorganic ions such as KCl to even up the extracellular molar concentrations (Oren et al. 2008, Weinisch et al. 2018). In this case, intracellular enzymatic machinery is well adapted to function in high ionic concentration. For instance, proteins from these salt-in strategists have elevated frequencies of acidic residues on the surfaces, leading to the organization of a hydrated salt network and the prevention of rigid folded conformations (Oren et al. 2008, Gunde-Cimerman et al. 2018).
2. The "*salt-out*" strategy uses the extrusion of Na^+, mainly through active transporters, and the accumulation of compatible solutes, either by synthesis or capture, that do not interfere with cellular processes, to avoid the Na^+ entry (Gunde-Cimerman et al. 2018). This strategy has been reported to be present in bacteria, algae and fungi.

Up to date, the knowledge regarding the salt-tolerance mechanisms of eukaryote microorganisms has been generated from *Saccharomyces cerevisiae*, a salt-tolerant yeast which is unable to grow in hypersaline environments. Oren and Gunde-Cimerman laboratories were pioneers in the fungal discovery from different hypersaline environments, including solar salterns, salt marshes, sea water and even in the Dead Sea in Israel and the Great Salt Lake in USA (Steiman et al. 1995, Gunde-Cimerman et al. 2000, Kis-Papo et al. 2014, Butinar et al. 2005a, b, c). Since then, two fungal models have been proposed to study haloadaptations: the extreme halotolerant yeast *Hortaea werneckii* and the obligate halophile *Wallemia ichthyophaga* (Gunde-Cimerman and Zalar 2014).

The data from genomic, transcriptomic and biochemistry approaches have revealed that these fungi have differences in the haloadaptation strategies compared with *S. cerevisiae* (Kogej et al. 2005, 2007, Lennasi et al. 2013, Zajc et al. 2014, Plemenitas et al. 2014). However, halophile and halotolerant fungi adaptations to hypersalinity are not fully unraveled and more comparative studies are needed.

Haloadaptations of Halophilic Fungi

Osmosensing and the Synthesis of Compatible Solutes

One of the main strategies used by prokaryotes and eukaryotes to deal with high-salt concentrations is the accumulation of compatible solutes to maintain low intracellular Na^+ (Bloomberg et al. 2000, Kogej et al. 2007, Plemenitas et al. 2014,

Gunde-Cimerman et al. 2018). Compatible solutes (osmolytes) are small highly water-soluble molecules that do not interfere with cellular metabolism, that are either *de novo* synthetized or transported from the environmental surroundings. Osmolytes can be categorized as: (i) zwitterionic solutes (modified aminoacids: ectoine, betaine, hidroxyectoine), (ii) noncharged solutes (sugars: trehalose, sucrose; modified sugars: α-glucosylglycerol, α-mannosylglyceramide; poly-alcohols: glycerol, mannitol, arabitol, etc.) and (iii) anionic solutes (aminoacids: L-α-glutamate, β-glutamate and modified aminoacids) (Roberts 2005, Khatibi et al. 2019). Model organisms such as *Saccharomyces cerevisiae,* a salt-tolerant yeast, and the halotolerant fungi *Aspergillus nidulans* have been reported to maintain positive turgor pressure at high salinity (≤1 M of NaCl) through the production and accumulation of glycerol, trehalose and other organic compatible solutes (Yale et al. 2001, Kogej et al. 2005, Plemenitas et al. 2014). In the extreme halotolerant *H. werneckii* (growing from 0% to 32% NaCl, and a broad optimum from 6 to 10% NaCl) and the obligate halophile *W. ichthyophaga* (that can live only in media with above 10% of NaCl and up to 32% NaCl), the strategy of compatible solutes is also present but osmolyte dynamics change according to salt concentration and growth-phase (Butinar et al. 2005c, Kogej et al. 2007, Kuncik et al. 2010, Zajc et al. 2013). For instance, at low salinities *H. werneckii* accumulates a mixture of glycerol>erythritol>arabitol>mannitol (0-10% NaCl), whereas at high salinity the main osmolytes are glycerol and erythritol (15-25% NaCl). This wide osmoadaptation response of *H. werneckii* contributes to its growth at a wide salinity range (Kogej et al. 2007). In *W. ichthyophaga,* the main polyol detected in different salt concentrations was glycerol (15-25%), except when growing in the lowest concentration of salt (10% NaCl) where arabitol mainly contributed to the total amount of polyols (Zajc et al. 2013). Other halotolerant yeast *Deburyomices hansenii* and the poly extrem tolerant yeast *Aureobasidium pullulans,* frequently isolated from marshes and from solar saltern water, which can tolerate up to 17% NaCl in the growth medium, accumulate glycerol depending on the growing phase and the salinity (DasSarma and Arora 2001, Kogej et al. 2005). In contrast, in *Aspergillus montevidensis* there is an upregulation on genes related to the synthesis of amino acids such as alanine, aspartate and proline in hypersaline conditions suggesting that this can be used as compatible solutes (Ding et al. 2019).

Changes in the osmolarity are sensed mainly, but not exclusively, by the **h**igh **o**smolarity **g**lycerol (HOG) pathway, and consequently the accumulation of compatible solutes. This short-term response to increased salinity has been well documented in the last years in *S. cerevisiae* and some Aspergilli (Saito and Posas 2010, Hawigara et al. 2016). HOG pathway belongs to the prototypical MAPK cascade consisting of three different modules: MAPK, MAPK kinase (MAPKK), and MAPK kinase kinase (MAPKKK). MAPKKK activation is followed by phosphorylation of MAPKK, which in turn phosphorylates MAPK (Hohmann et al. 2007). Shol and Sln1 are two plasmatic membrane osmosensors proteins that, upon hyperosmotic stress, stimulate HOG pathway, where the signals converge at the MAPK kinase Pbs2 (MAPKK), which phosphorylates its downstream MAP kinase Hogl (MAPK) (Widmann et al. 1999, Hohmann 2002, Plemenitas et al. 2014). Phosphorylated Hogl controls the production of glycerol or other solutes, activating the enzymes related to its production in the cytoplasm and activating the transcription of osmoresponsive genes in the nucleus (Proft et al. 2006, Plemenitas et al. 2014). Once the turgor is re-established, Hog1 is dephosphorylated (Saito and Posas 2012). The HOG pathway counterparts have been identified in *H. werneckii* and *W. ichthyophaga*, sharing most of the pathway elements with *S. cerevisiae;* however, there are some remarkable differences. Homologous proteins of the pathway such as Sho1, Ste20, and Ste11 have been identified in the genomes of *W. ichthyophaga* and *H. werneckii*, with two copies of each component in the latter (Lenassi et al. 2013, Zajc et al. 2013, Plemenitas et al. 2014, Gunde-Cimerman et al. 2018). For instance, two isoforms of Sho1 osmosensor are present in *H. werneckii* (HwSho1A and HwSho1B). Interestingly, both fully complement the function of the homologous *S. cerevisiae* Sho1 protein (Fettich et al. 2011). In contrast, a complementation study of *S. cerevisiae* with some of the components of HOG pathway from *W. ichthyophaga* show that there is poor interaction between WiSho1 (sensor) and WiPbs2 (MAPKK). Furthermore, some components from the osmosensing apparatus are absent from *W. ichthyophaga* genome, suggesting that the SHO1 branch components are not involved in HOG signaling in this halophilic fungus (Konte et al. 2016). It is well known that in *S. cerevisiae*, *S. pombe*, *D. hansenii*, *A. nidulans*, *A. orizae*, HOG pathway activation after an osmotic stress leads to Hog1 phosphorylation which allows its nuclear translocation to transcriptionally regulate osmoresponsive genes as well as the direct activation of key enzymes (glycerol-3-phosphate-dehydrogenase, glycerol-3-phosphatase, mannitol-phosphate-dehydrogenase, among others) (de Vries et al. 2003, Gori et al. 2005, Furukawa et al. 2005, Saito y Posas 2012). Instead, phosphorylated Hog1 is the basal state in *W. ichtyophaga* when growing in optimal conditions, while hypersalinity stimulation lead to Hog1 dephosphorylation (Plemenitas et al. 2014). Lately, transcriptomics from haloterant *Aspergillus salisburgensis* and the halophile and *Aspergillus sclerotialis* show that both branches of HOG pathway are present, but no biochemical studies related to phosphorylated HOG have been conducted in these fungi. Interestingly, in species of Aspergilli genus Hog1 gene is duplicated (*A. nidulans* and *A. fumigatus*, each with HogA/SakA and MpkC) or triplicated (*A. oryzae* and *A. niger*, each with HogA/SakA, MpkC, and MpkD), suggesting that the signal transduction in this genus might be more complex than in other filamentous fungi (Hagiwara et al. 2016); however, the functional role of the duplication in halophile fungi isolated from hypersaline environments is still not fully understood.

In addition to the biosynthesis of glycerol, fungi internalize extracellular glycerol through the glycerol/H+ symporter Stl1 and glycerol facilitator Fps1 during hyperosmotic conditions if there is availability (Ferreira et al. 2005). This symporter and facilitator are present in halotolerant *H. werneckii*, *D. hansenii*, *A. montevidensis*, *A. sclerotialis* and in the halophiles *W. ichthyophaga*, *A. salisburgensis*, *E. rubrum* (Lenassi et al. 2013, Zajc et al. 2013, Plemenitas et al. 2013, Kis-Papo et al. 2014, Ding et al. 2019, Tafer et al. 2019).

Ion Transporters

High concentrations of intracellular Na^+ cations are harmful to the cell functions, while K^+ cations are needed to compensate the negative charge from proteins as well as inorganic and organic negatively charged polyanions and to activate key metabolic processes. Intracellular homeostasis of these alkali metal cations is fundamental to maintain physiological processes' functioning. Salt-adapted fungi could have a hyperosmotic cytoplasm vs. the extracellular media to maintain the turgor pressure. Frequently, these keep a high ratio of K^+/Na^+ by K^+ accumulation and Na^+ extrusion. Extreme halotolerant *H. werneckii* and polyextreme *A. pullulans*, which maintain low amounts of intracellular Na^+ in hypersalinity conditions, are named "extruders" through the efficient efflux or compartmentalization of Na^+ cations (Kogej et al. 2005, Plemenitas et al. 2016). Others, such as the halotolerant *D. hansenii* and the obligate halophile *W. ichthyophaga,* allow relatively high internal concentration of sodium when growing in hypersalinity and consequently are named "includers" (Kogej et al. 2005, Zajc et al. 2014). To eliminate any superfluous Na^+ ions, the cells spend a great deal of energy on active transmembrane transport of ions. This is achieved through transporters that have higher affinity to K^+ than to Na^+ during influx, or transporters that facilitate the efflux or the compartmentalization of the toxic Na^+. In *S. cerevisiae*, six cytoplasmic transporters have been described: the potassium uptake system Trk1 and Trk2 and the Tok1 K^+ channels, the inorganic phosphate (Pi)-Na^+ symporter Pho89, both efflux systems Ena Na^+-ATPases, and the Nha1 Na^+/H^+ antiporter (reviewed by Ariño et al. 2010, Lenassi et al. 2013, Ariño et al. 2016). Additionally, other K^+ influx systems such as K^+-H^+ symporter (Hak), K^+-Na^+ P-type ATPase (Acu), and K^+ efflux channel Tok have been described in non-conventional yeasts and recently in other fungi (reviewed by Ramos et al. 2011, Benito et al. 2011). Additionally, fungi have intracellular cation/H^+ antiporters such as Vnx1, Nhx1 and Kha1 in the vacuoles, endosomes and Golgi apparatus, respectively (reviewed by Ariño et al. 2010).

In contrast, halotolerant yeast *H. werneckii* the obligate basidiomycete *W. ichthyophaga* increase the genes encoding metal cation transporters or express differentially transport related proteins. In the first case, the number of the alkali cation transporters are increased with duplications in the Nha1 Na^+/H^+ antiporters (*Hw*Nha1A and *Hw*Nha1B) and expansion of the genes of Trk K^+ channels up to 8 homologues (Lenassi et al. 2013). The genome of *W. ichthyophaga* shows that there are a low number of cation transporter genes with the exception of duplications on the intracellular Kha1 cation/H+ antiporter, the rare Acu K^+-Na^+ P-type ATPase (and only one presents the P-type ATPase domain) and Nha1 (Zajc et al. 2013, Lenassi et al. 2013). The transcriptomic analysis shows that the expression of these genes is low and independent of the salt concentration. These findings indicate that *W. ichthyophaga* cope with hypersalinity using other mechanisms to compensate the Na^+ intrusion but not through cation transporters enrichment, probably because the concentrations of salt, although extremely high, are relatively constant (Zajc et al. 2013, Plemenitas et al. 2014). *Eurotium rubrum,* in the presence of Dead Sea water, augmented the expression of zinc/iron permease, voltage-dependent calcium channels, and heavy-metal-transporting ATPase genes; *A. montevidensis* regulate 15 genes related to ion transport in the presence of 3 M of NaCl, but apparently there is no transporters gene enrichment (Kis-Papo et al. 2014, Ding et al. 2019).

A comparison of genomes of halophile *A. salisburgensis*, obligate halophile *W. ichthyophaga* and halotolerants *H. werneckii* and *A. sclerotialis* and *Eurotium rubrum*, show that there is an expansion in the Major Facilitator Superfamily (MFS) and the alkali cation transporters in the halotolerant fungi compared with the halophiles (Tafer et al. 2019). *A. salisburgensis* and *A. sclerotialis* genomes have all the alkali cation transporters with the exception of the high affinity potassium transporter (HAK1), the potassium antiporter (KHA1), and the outward-rectifier potassium channel TOK1 in the halophile, while in halotolerant the endosomal/prevacuolar sodium/hydrogen exchanger (NHX1) and HAK1 (Ariño et al. 2010, Tafer et al. 2019) are missing. However, transcriptomic analysis of the halotolerant *A. sclerotialis* growing in 20% vs. 5% of NaCl show that 96 genes were upregulated (in some cases 50 times), 14 genes that are related to transmembrane transport. In contrast, *A. salisburgensis* growing in the same conditions, shown a downregulated expression of only a pair of cation transporters (Tafer et al. 2019).

All of these evidences suggest that halophiles and halotolerants show differences in the presence of cation transporter genes or their expression according to their life styles.

Membrane Fluidity in Halophilic Fungi

Salt-adapted fungi needs to maintain a fluid membrane to guarantee optimal biochemical functioning of cellular processes, although it is necessary to remodel it when there is an ionic and an osmotic stress (Gostincar et al. 2011). Membrane fluidity is influenced by the length, branching and degree of saturation of fatty acids, the amount of sterols, and the proportion of types of lipids (Rodriguez-Vargas et al. 2007) that affects the functioning of membrane-attached proteins involved in ion homeostasis or compatible solute transport system (Turk et al. 2011, Dakat et al. 2014). Low membrane fluidity is observed when there is an increase in saturated lipids or sterols incorporated into the membrane. It has been proposed that the decrease in membrane fluidity could avoid the leaking of compatible solutes, particularly glycerol, through passive influx (Turk et al. 2003). For instance, it has been reported that halotolerant *Z. rouxii* and halophile *Phaeotheca triangularis* decrease the membrane fluidity when grown in hypersaline conditions through the change in the lipid composition or the amount of sterol present in the membrane (Hosono et al. 1992, Turk et al. 2004). Halotolerant *H. werneckii* and *A. pullulans* augment the membrane fluidity through the increase of fatty acid unsaturation (containing a high percentage of $C18{:}2^{\Delta 9,12}$) (Turk et al. 2007). A transcriptomic analysis of the halophile *A. montevidensis* ZYD4 showed the increase in the expression of 12 genes of unique locus related to lipid transport and metabolism, particularly the upregulation of a fatty acid synthase gene which could augment the biosynthesis of palmitate (C16:0, a saturated fatty acid) reducing the membrane fluidity (Ding et al. 2019). These results suggest that the adaptations in the membrane could be related to the lifestyle of the fungi, where the most of the halotolerant fungi have more flexible mechanisms to respond to the constant changes in their surroundings.

Morphological Adaptations of Halophilic Fungi

Halophilic fungi have adapted morphological ecotypes to live in hypersalinity (Kis-Papo et al. 2014, Kogej et al. 2006). For instance, *W. ichthyophaga* significantly grow in multicellular clumps, enhanced cell wall thickness and synthetize a thick cover of extracellular polysaccharides (Kralk-Kunkic et al. 2010); *H. werneckii* extensively accumulate melanin (Kogej et al. 2007), while the extremely halophilic *A. montevidensis* ZYD4 decrease the level of yellow pigments on colonies in the presence of high-salt levels (Liu et al. 2017). At high-salt conditions, halotolerant *A. salisburgensis* and halophile *E. rubrum* increased the expression of gene products involved in beta-glucan biosynthetic processes and chitin binding, suggesting that there is a modification in the cell wall (Kis-Papo et al. 2014, Tafer et al. 2019).

Hydrophobins

Hydrophobins are small amphipathic proteins with eight cysteine residues as a signature that are exclusively from pluricellular fungi. These proteins have been reported to be involved in the resistance of the spores to desiccation, to allow hyphae to grow in a hydrophobic aerial environment and the attachment of the fungi to different surfaces, because of their capacity to spontaneously assemble into amphipathic layers at hydrophobic : hydrophilic interfaces (Linder et al. 2015). Lately, an undescribed mechanism of salt-resistance has been proposed. The specific expression of hydrophobins depending on the NaCl concentration was observed in *W. ichthyophaga*. This fungus, although a basidiomycete, has a compact genome but a striking expansion of hydrophobin gene family (26) compared to other Wallemiales species such as *W. sebi* (12) (Zajc et al. 2013). *W. ichthyophaga* hydrophobins surprisingly contain a high proportion of acidic amino acids compared to homologs from other fungi, a signature of archaeal halophilic proteins (Madern et al. 2000, Zajc et al. 2013). Theoretically, if the acidic amino acids are exposed to the hydrophobin surface, they could bind salt and water, avoiding salt-induced changes in protein conformation (Siglioccolo et al. 2011). Other transcriptomic studies have reported the upregulation of hydrophobins transcripts such as ceratoulmin in the halotolerant *A. sclerotialis,* which was increased by a factor of 2641 among the 96 upregulated genes in hypersalinity but not in the halophile *A. salisburgensis* (Tafer et al. 2019). However, there is no experimental evidence that hydrophobins are needed to survive in hypersaline conditions.

Particularities

An analysis of genomes of halotolerant *Penicillium chrysogenum* and halophile *E. rubrum* shows that the proteins potentially coded in the genes have a high percentage of aspartic and glutamic acid. Particularly, hydrophobins from *W. ichthyophaga* have shown the same pattern compared to homologous hydrophobins from other fungi (Zajc et al. 2013). Interestingly, acidic amino acid enrichment is a signature of proteins adapted to hypersaline environments mainly from halophile prokaryotes (Kis-Papo et al. 2014, DasSarma and DasSarma 2015).

Surprisingly, the obligate halophile basidiomycete *W. ichthyophaga* has a compact genome, with a reduced number of protein-coding genes suggesting that this might be a characteristic of obligate extremophile (Zajc et al. 2013, Gostincar et al. 2019). However, in other halophile fungi, such tendency has not been confirmed since *A. salisburgensis* and *Eurotium rubrum* have an average genome of 26.2 Mb and 21.9 Mb, respectively, compared with a halotolerant such as *A. sclerotialis* (27.9 Mb) from the same genus (Kis-Papo et al. 2014, Tafer et al. 2019). However, it has been noticed that *A. salisburgensis* has a decrease by 27% in genome size and gene content compared with *A. sclerotiales*. Thereupon, fungi (*W. ichthyophaga* and *A. salisburgensis*) living in niches with extremely stable conditions have optimized its genome, keeping only the genes necessary for its survival (Zajc et al. 2013, Tafer et al. 2019). Among other adaptations that have been reported in the halophile *A. montevidensis* and *A. glaucus,* is the switch to asexual reproduction on hypersaline conditions (Liu et al. 2017, Ding et al. 2019).

In conclusion, although there are studies in the strategies that allow halotolerant and halophile fungi to cope with salt, there is still gaps to fill in the knowledge of the life style of these microorganisms, since many of the proposed mechanisms have not been tested experimentally.

DEGRADATION OF PAHS IN HYPERSALINE ENVIRONMENTS

PAHs are present in many hypersaline environments, including hypersaline wastewaters. The potential of halophilic microorganisms to remove pollutants under hypersaline (1 M NaCl) condition has been poorly explored, with really few studies conducted up to date. However, there is a growing interest in using biological treatments to remove PAHs from the environments, in particular from water bodies and wastewaters contaminated with xenobiotics, which many times present extreme conditions. Many of these polluted wastewaters exhibit serious restriction for life, but at the same time are colonized for a huge biodiversity of microorganisms.

Different industries such as petrochemical, pesticides, pharmaceutical and agro-food, among others, generate hypersaline wastewaters with high level of NaCl (10-120 mg/L) (revised in Lefebvre and Moletta 2006). Thus, the use of halophilic microorganisms emerges as a great alternative to restore negatively impacted ecosystems with aromatic compounds such as PAHs.

In this section, we will focus on halophilic fungi because they have been poorly investigated in relation with their biotechnological application as bioremediator agents to remove PAHs under hypersaline conditions. In addition, fungi are well known for their high adaptability to growth under low water activity conditions (Gostinčar et al. 2009). Due to this, they represent promissory agents for mycoremediation processes.

Figure 1.2 (figure and legend taken from González-Abradelo et al. 2019) summarizes the main published works describing the removal of PAHs by fungi in the presence of NaCl. However, the first study using halophilic fungi to remove PAHs from wastewater under hypersaline conditions was recently published by González-Abradelo et al. (2019). In this work, author demonstrated the ability of

Aspergillus sydowii and *Aspergillus destruens* to biodegrade benzo-α-pyrene and phenantrene, as well as pharmaceutical compounds from industry wastewater. Figure 1.3 (figure and legend taken from González-Abradelo et al. 2019) represents the experimental strategy followed in the previous work. Previously, *A. sydowii* was investigated in mycotreatments because it was capable of removing hydrocarbons under hyperalkaline conditions (pH=12) (Batista-García et al. 2017).

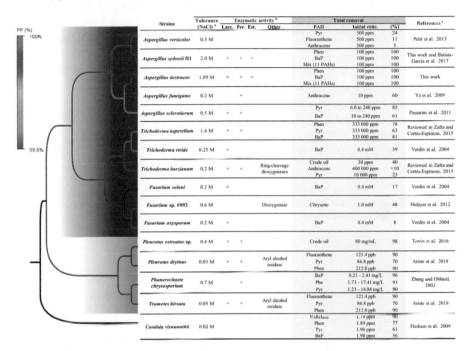

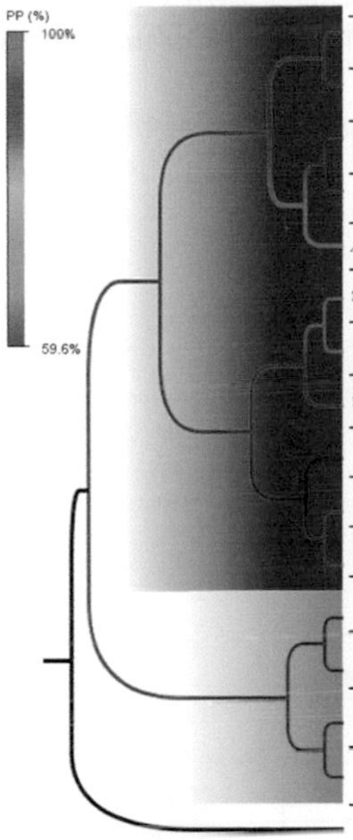

Strains	Tolerance (NaCl)[a]	Enzymatic activity[b] Lacc.	Per.	Est.	Other	Total removal PAH	Initial conc.	(%)	References[c]
Aspergillus versicolor	0.5 M					Pyr Fluoranthene Anthracene	500 ppm 500 ppm 500 ppm	24 11 5	Petit et al. 2013
Aspergillus sydowii H1	2.0 M	+	+	+		Phen BaP Mix (11 PAHs)	100 ppm 100 ppm 100 ppm	100 100 100	This work and Batista-García et al. 2017
Aspergillus destruens	1.89 M	+	+	+		Phen BaP Mix (11 PAHs)	100 ppm 100 ppm 100 ppm	100 100 100	This work
Aspergillus fumigatus	0.2 M		+			Anthracene	10 ppm	60	Ye et al. 2009
Aspergillus sclerotiorum	0.5 M	+	+			Pyr BaP	6.0 to 240 ppm 10 to 240 ppm	85 61	Passarini et al. 2011
Trichoderma asperellum	1.4 M	+	+			Phen Pyr BaP	333 000 ppm 333 000 ppm 333 000 ppm	78 63 81	Reviewed in Zafra and Cortés-Espinosa, 2015
Trichoderma viride	0.25 M	+				BaP	0.4 mM	39	Verdin et al. 2004
Trichoderma harzianum	0.2 M	+	+		Ring-cleavage dioxygenases	Crude oil Anthracene Pyr	30 ppm 400 000 ppm 10 000 ppm	40 <10 25	Reviewed in Zafra and Cortés-Espinosa, 2015
Fusarium solani	0.2 M	+				BaP	0.4 mM	17	Verdin et al. 2004
Fusarium sp. F092	0.6 M				Dioxygenase	Chrysene	1.0 mM	48	Hidayat et al. 2012
Fusarium oxysporum	0.2 M	+				BaP	0.4 mM	8	Verdin et al. 2004
Pleurotus ostreatus sp.	0.4 M	+	+			Crude oil	80 mg/mL	98	Torres et al. 2016
Pleurotus dryinus	0.03 M	+	+		Aryl alcohol oxidase	Fluoranthene Pyr Phen	121.4 ppb 86.8 ppb 212.0 ppb	90 70 90	Ariste et al. 2018
Phanerochaete chrysosporium	0.7 M		+			BaP Phe Pyr	0.21 - 2.41 mg/L 1.73 - 17.41 mg/L 1.23 - 14.88 mg/L	96 93 90	Zheng and Obbard, 2002
Trametes hirsuta	0.05 M	+	+		Aryl alcohol oxidase	Fluoranthene Pyr Phen	121.4 ppb 86.8 ppb 212.0 ppb	90 70 90	Ariste et al. 2018
Candida viswanathii	0.02 M					Naftalene Phen Pyr BaP	1.19 ppm 1.89 ppm 1.96 ppm 1.98 ppm	90 77 61 56	Hesham et al. 2009

FIGURE 1.2 Comparison of the PAHs removal using different fungal strains. The tolerance to NaCl reported for the same strain or other member of the species (References in Supplementary Material) is shown for illustrative purpose (a), as only *A. sydowii*, *A. destruens* and *Fusarium sp. F092* have been tested for degradation of PAHs under saline conditions. (b) Enzyme activities evidenced in cultures with PAHs. (c) Only the references related to PAH removal are shown. The phylogram of the fungal strains was obtained from the Internal Transcribed Spacer rDNA (ITS) marker alignment as described in Supplementary Materials (González-Abradelo et al. 2019). Ascomycete and Basidiomycete strains are shaded in blue and green, respectively. Lacc: laccase, Est: esterase, Per: peroxidase, BaP: benzo-α-pyrene, Phe: phenantrene, Pyr: Pyrene, ppm: parts per million, ppb: parts per billion. Figure and legend taken from González-Abradelo et al. (2019).

As Figure 1.2 shows, no related phylogenetic fungi are tolerant to NaCl and at the same time, they can remove high concentration of aromatic hydrocarbons. For example, *A. versicolor* is capable of removing phenanthrene, fluorene and anthracene under 0.5 M NaCl (Petit et al. 2013), while *Trichoderma viride* removes benzo-α-pyrene under 0.25 M NaCl (Verdin et al. 2004). *Trichoderma asperellum* exhibited high biodegradation levels (>60%) of benzo-α-pyrene, pyrene and phenanthrene in the presence of 1.4 M NaCl (Zafra and Cortéz-Espinosa 2015). Also note that all of these fungi produced active extracellular enzymes involved in PAH degradation under these hypersaline conditions. This fact is very important for downstream applications because these fungi do not only grow under hypersaline conditions, but they also produce halostable oxidative enzymes (Figure 1.2).

Other *Aspergillus* species have been studied in bioremediation strategies. For example, the marine strain *Aspergillus* sp. BAP14 was isolated from sediment and showed 60% biodegradation of benzo-α-pyrene in 12 days (Wu et al. 2019). Moreover, *A. sydowii* also isolated from marine habitats removed 30% of anthracene in 14 days (Birolli et al. 2017).

In conclusion, few works have investigated the potential of halophilic fungi for pollutant removal. Additional efforts are needed to explore and implement mycoremediation processes at field level. However, firstly, we need to produce vast amount of knowledge related with the molecular mechanisms involved in biotransformation of PAHs in the presence of high concentration of NaCl. Up to date, there is no omics approach to describe the metabolism of PAHs under halophilic conditions.

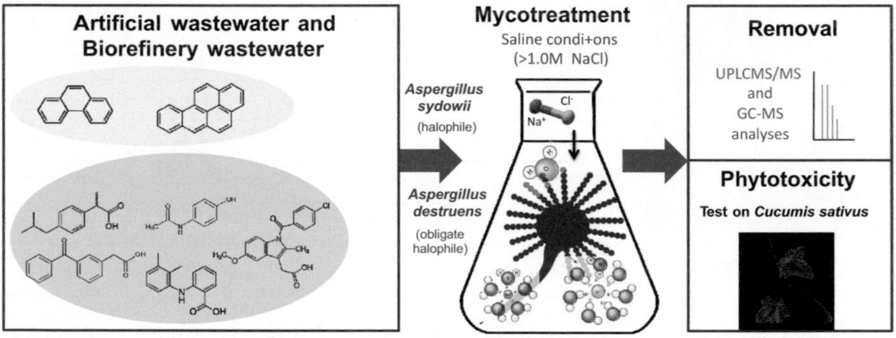

FIGURE 1.3 Experimental strategy fallowed by González-Abradelo et al. (2019). Figure and legend taken from González-Abradelo et al. (2019).

ACKNOWLEDGMENTS

EA gratefully thanks MINECO-ERDF for the Ramón y Cajal contract [RYC-2013-12481] and GAP thanks Consejo Nacional de Ciencia y Tecnología (CONACyT) for the pre-doctoral fellowship (CVU:772485).

REFERENCES

Aranda, E. 2016. Promising approaches towards biotransformation of polycyclic aromatic hydrocarbons with Ascomycota fungi. Curr. Opin. Biotechnol. 38: 1-8.

Aranda, E., P. Godoy, R. Reina, M. Badia-Fabregat, M. Rosell, E. Marco-Urrea, et al. 2017. Isolation of Ascomycota fungi with capability to transform PAHs: Insights into the biodegradation mechanisms of *Penicillium oxalicum*. Int. Biodeterior. Biodegradation. 122: 141-150.

Arfsten, D.P., D.J. Schaeffer and D.C. Mulveny. 1996. The effects of near ultraviolet radiation on the toxic effects of polycyclic aromatic hydrocarbons in animals and plants: a review. Ecotox. Environ. Safe. 33(1): 1-24.

Argumedo-Delira, R., A. Alarcón, R. Ferrera-Cerrato, J.J. Almaraz and J.J. Peña-Cabriales. 2012. Tolerance and growth of 11 *Trichoderma* strains to crude oil, naphthalene, phenanthrene and benzo[a]pyrene. J. Environ. Manage. 95: S291-S299.

Arino, J., J. Ramos and H. Sychrova. 2010. Alkali metal cation transport and homeostasis in yeasts. Microbiol. Mol. Biol. Rev. 74: 95-120.

Banks, M.K. and K.E. Schultz. 2005. Comparison of plants for germination toxicity tests in petroleum-contaminated soils. Water. Air. Soil Pollut. 167: 211-219.

Barone, G., S. Varrella, M. Tangherlini, E. Rastelli, A. Dell'Anno, R. Danovaro, et al. 2019. Marine fungi: Biotechnological perspectives from deep-hypersaline anoxic basins. Diversity 11: 113.

Batista-García, R.A., E. Balcázar-López, E. Miranda-Miranda, A. Sánchez-Reyes, L. Cuervo-Soto, D. Aceves-Zamudio, et al. 2014. Characterization of lignocellulolytic activities from a moderate halophile strain of *Aspergillus caesiellus* isolated from a sugarcane bagasse fermentation. PLoS One 9: e105893.

Batista-García, R.A., V.V. Kumar, A. Ariste, O.E. Tovar-Herrera, O. Savary, H. Peidro-Guzmán, et al. 2017. Simple screening protocol for identification of potential mycoremediation tools for the elimination of polycyclic aromatic hydrocarbons and phenols from hyperalkalophile industrial effluents. J. Environ. Manage. 198: 1-11

Benito B., B. García de blas, A. Fraile-Escanciano and A. Rodríguez-Navarro. 2011. Potassium and sodium uptake systems in fungi. The transporter diversity of Magnaporthe oryzae. Fungal Genet. Biol. 48: 812-822.

Benner Jr, B.A., N.P. Bryner, S.A. Wise, G.W. Mulholland, R.C. Lao and M.F. Fingas. 1990. Polycyclic aromatic hydrocarbon emissions from the combustion of crude oil on water. Environ. Sci. Technol. 24(9): 1418-1427.

Bensch, K., U. Braun, J.Z. Groenewald and P.W. Crous. 2012. The genus *Cladosporium*. Stud. Mycol. 72: 1-401.

Bernhard, J.M., K. Kormas, M.G. Pachiadaki, E. Rocke, D.J. Beaudoin, C. Morrison, et al. 2014. Benthic protists and fungi of Mediterranean deep hypsersaline anoxic basin redoxcline sediments. Front. Microl. 5: 1-13.

Birolli, W.G. de A., D. Santos, N. Alvarenga, A.C.F.S. Garcia, L.P.C. Romão and A.L.M. Porto. 2018. Biodegradation of anthracene and several PAHs by the marine-derived fungus *Cladosporium* sp. CBMAI 1237. Mar. Pollut. Bull. 2: 525-533.

Bisht, S., P. Pandey, B. Bhargava, S. Sharma, V. Kumar and D. Krishan. 2015. Bioremediation of polyaromatic hydrocarbons (PAHs) using rhizosphere technology. Brazilian J. Microbiol. 46: 7-21.

Blomberg, A. 2000. Metabolic surprises in *Saccharomyces cerevisiae* during adaptation to saline conditions: Questions, some answers and a model. FEMS microbiology letters. 182(1): 1-8.

Boguslawska-Wąs, E. and W. Dąbrowski. 2001. The seasonal variability of yeasts and yeast-like organisms in water and bottom sediment of the Szczecin Lagoon. Int. J. Hyg. Environ. Health. 2: 525-533.

Bonifaz, A., H. Badali, G.S. de Hoog, M. Cruz, J. Araiza, M.A. Cruz, et al. 2008. *Tinea nigra* by *Hortaea werneckii*, a report of 22 cases from Mexico. Stud. Mycol. 61: 77-82.

Bonugli-Santos, R.C., S.M.R. dos Vasconcelos, M.R.Z. Passarini, G.A.L. Vieira, V.C.P. Lopes, P.H. Mainardi, et al. 2015. Marine-derived fungi: Diversity of enzymes and biotechnological applications. Front. Microbiol. 6: 1-15.

Butinar, L., S. Santos, I. Spencer-Martins, A. Oren and N. Gunde-Cimerman. 2005a. Yeast diversity in hypersaline habitats. FEMS Microbiol. Lett. 244(2): 229-234.

Butinar, L., P. Zalar, J.C. Frisvad and N. Gunde-Cimerman. 2005b. The genus *Eurotium*-Members of indigenous fungal community in hypersaline waters of salterns. FEMS Microbiol. Ecol. 512: 155-166.

Butinar, L., S. Sonjak, P. Zalar, A. Plemenitaš and N. Gunde-Cimerman. 2005c. Melanized halophilic fungi are eukaryotic members of microbial communities in hypersaline waters of solar salterns. Bot. Marina. 48(1): 73-79.

Cantrell, S.A., J.C. Dianese, J. Fell, N. Gunde-Cimerman and P. Zalar. 2011. Unusual fungal niches. Mycologia. 6: 1161-1174.

Carmichael, J.W. and de G.S. Hoog. 1980. The Black Yeasts, II: *Moniliella* and *Allied* Genera. Mycologia. 5: 1056.

Carvalho, P.N., M.C.P. Basto, C.M.R. Almeida and H. Brix. 2014. A review of plant-pharmaceutical interactions: From uptake and effects in crop plants to phytoremediation in constructed wetlands. Environ. Sci. Pollut. Res. 21: 11729-11763.

Castillo-Carvajal, L.C., J.L. Sanz-Martín and B.E. Barragán-Huerta. 2014. Biodegradation of organic pollutants in saline wastewater by halophilic microorganisms: A review. Environ. Sci. Pollut. Res. 21: 9578-9588.

Chowdhary, A., J. Perfect and G.S. de Hoog. 2015. Black molds and melanized yeasts pathogenic to humans. Cold Spring Harb. Perspect. Med. 8: 1-21.

Crous, P.W., C.L. Schoch, K.D. Hyde, A.R. Wood, C. Gueidan, G.S. de Hoog, et al. 2009. Phylogenetic lineages in the *Capnodiales*. Stud. Mycol. 64: 17-47.

Cruz, A., M.I. Pariente, I. Vasiliadou, B. Padrino, D. Puyol, R. Molina, et al. 2017. Removal of pharmaceutical compounds from urban wastewater by an advanced bio-oxidation process based on fungi *Trametes versicolor* immobilized in a continuous RBC system. Environ. Sci. Pollut. Res. 25: 34884-34892.

Dary, O., G.P. Georghiou, E. Parsons and N. Pasteur. 1990. Microplate adaptation of Gomori's assay for quantitative determination of general esterase activity in single insects. J. Econ. Entomol. 83: 2187-2192.

DasSarma, S. and P. Arora. 2001. Halophiles. Encyclopedia of Life Science eLS. John Wiley & Sons, Ltd. https://doi.org/10.1038/npg.els.0000394.

DasSarma, S. and P. DasSarma. 2015. Halophiles and their enzymes: Negativity put to good use. Curr. Opin. Microbiol. 25: 120-126.

De Hoog, S., P. Zalar, B.G. Van Den Ende and N. Gunde-Cimerman. 2005. Relation of halotolerance to human-pathogenicity in the fungal tree of life: An overview of ecology and evolution under stress. pp. 371-395. *In*: N. Gunde-Cimerman, A. Oren, A. Plemenitaš. (eds.). Adaptation to Life at High Salt Concentrations in Archaea, Bacteria and Eukarya. Springer, Dordrecht.

De Vries, R.P., S.J. Flitter, P.J. Van De Vondervoort, M.K. Chaveroche, T. Fontaine, S. Fillinger, et al. 2003. Glycerol dehydrogenase, encoded by gldB is essential for osmotolerance in *Aspergillus nidulans*. Mol. Microbiol. 49(1): 131-141.

Di Gregorio, S., S. Becarelli, G. Siracusa, M. Ruffini-Castiglione, G. Petroni, G. Masini, et al. 2016. *Pleurotus ostreatus* spent mushroom substrate for the degradation of polycyclic aromatic hydrocarbons: The case study of a pilot dynamic biopile for the decontamination of a historically contaminated soil. J. Chem. Technol. Biotechnol. 91: 1654-1664.

Dugan, F.M., U. Braun, J.Z. Groenewald and P.W. Crous. 2008. Morphological plasticity in *Cladosporium sphaerospermum*. Persoonia: Mol. Phyl. Evol. Fungi 21: 9-16.

Edgcomb, V.P., M.G. Pachiadaki, P. Mara, K.A. Kormas, E.R. Leadbetter and J.M. Bernhard. 2016. Gene expression profiling of microbial activities and interactions in sediments under haloclines of *E. Mediterranean* deep hypersaline anoxic basins. ISME J. 11: 2643-2657.

Fayeulle, A., E. Veignie, R. Schroll, J.C. Munch and C. Rafin. 2019. PAH biodegradation by telluric saprotrophic fungi isolated from aged PAH-contaminated soils in mineral medium and historically contaminated soil microcosms. J. Soils Sediments. 19: 3056-3067.

Fettich, M., M. Lenassi, P. Veranič, N. Gunde-Cimerman and A. Plemenitaš. 2011. Identification and characterization of putative osmosensors, HwSho1A and HwSho1B, from the extremely halotolerant black yeast *Hortaea werneckii*. Fungal Genet. Biol. 48(5): 475-484.

Floch, C., E. Alarcon-Gutiérrez and S. Criquet. 2007. ABTS assay of phenol oxidase activity in soil. J. Microbiol. Methods 71: 319-324.

Giraud, F., P. Guiraud, M. Kadri, G. Blake and R. Steiman. 2001. Biodegradation of anthracene and fluoranthene by fungi isolated from an experimental constructed wetland for wastewater treatment. Water Res. 17: 4126-4136.

González-Abradelo, D., Y. Pérez-Llano, H. Peidro-Guzmán, M. Sánchez-Carbente, J.L. del R. Folch-Mallol, E. Aranda, et al. 2019. First demonstration that ascomycetous halophilic fungi *Aspergillus sydowii* and *Aspergillus destruens* are useful in xenobiotic mycoremediation under high salinity conditions. Bioresour. Technol. 279: 287-296.

Gori, K., H.D. Mortensen, N. Arneborg and L. Jespersen. 2005. Expression of the GPD1 and GPP2 orthologues and glycerol retention during growth of *Debaryomyces hansenii* at high NaCl concentrations. Yeast 22(15): 1213-1222.

Gostinčar, C., M. Grube, S. De Hoog, P. Zalar and N. Gunde-Cimerman. 2009. Extremotolerance in fungi: Evolution on the edge. FEMS Microbiol. Ecol. 71(1): 2-11.

Gostinčar, C., M. Lenassi, N. Gunde-Cimerman and A. Plemenitaš. 2011. Fungal adaptation to extremely high salt concentrations. Adv. Appl. Microbiol. 77: 71-96.

Gunde-Cimerman, N., P. Zalar, S. de Hoog and A. Plemenitaš. 2000. Hypersaline waters in salterns–natural ecological niches for halophilic black yeasts. FEMS Microbiol. Ecol. 32(3): 235-240.

Gunde-Cimerman, N., J. Ramos and A. Plemenitaš. 2009. Halotolerant and halophilic fungi. Mycol. Res. 11311: 1231-1241.

Gunde-Cimerman, N. and P. Zalar. 2014. Extremely halotolerant and halophilic fungi inhabit brine in solar salterns around the globe. Food Technol. Biotechnol. 2: 170-179.

Gunde-Cimerman, N., A. Plemenitaš and A. Oren. 2018. Strategies of adaptation of microorganisms of the three domains of life to high salt concentrations. FEMS Microbiol. Rev. 42(3): 353-375.

Hachi, M., A. Chergui, A.R. Yeddou, A. Selatnia and H. Cabana. 2017. Removal of acetaminophen and carbamazepine in single and binary systems with immobilized laccase from *Trametes hirsuta*. Biocatal. Biotransformation 35: 51-62.

Harms, H., D. Schlosser and L.Y. Wick. 2011. Untapped potential: Exploiting fungi in bioremediation of hazardous chemicals. Nat. Rev. Microbiol. 9: 177-192.

Haroune, L., S. Saibi, J.P. Bellenger and H. Cabana. 2014. Evaluation of the efficiency of *Trametes hirsuta* for the removal of multiple pharmaceutical compounds under low concentrations relevant to the environment. Bioresour. Technol. 171: 199-202.

He, Z., H. Xiao, L. Tang, H. Min and Z. Lu. 2013. Biodegradation of di-n-butyl phthalate by a stable bacterial consortium, HD-1, enriched from activated sludge. Bioresour. Technol. 128: 526-532.

Hernández-López, E.L., L. Perezgasga, A. Huerta-Saquero, R. Mouriño-Pérez and R. Vazquez-Duhalt. 2016. Biotransformation of petroleum asphaltenes and high molecular weight polycyclic aromatic hydrocarbons by *Neosartorya fischeri*. Environ. Sci. Pollut. Res. 23(11): 10773-10784.

Hesham, A.E., S.A. Alamri, S. Khan, M.E. Mahmoud and H.M. Mahmoud. 2009. Isolation and molecular genetic characterization of a yeast strain able to degrade petroleum polycyclic aromatic hydrocarbons. African J. Biotechnol. 8: 2218-2223.

Huang, J., G. Ning, F. Li and G.D. Sheng. 2015. Biotransformation of 2,4-dinitrotoluene by obligate marine *Shewanella marisflavi* EP1 under anaerobic conditions. Bioresour. Technol. 180: 200-206.

Hussain, N., T. Abbasi and S.A. Abbasi. 2015. Vermicomposting eliminates the toxicity of Lantana (Lantana camara) and turns it into a plant friendly organic fertilizer. J. Hazard. Mater. 298: 46-57.

Jiang, Y., Y. Shang, K. Yang and H. Wang. 2016. Phenol degradation by halophilic fungal isolate JS4 and evaluation of its tolerance of heavy metals. Appl. Microbiol. Biotechnol. 4: 1883-1890.

Kadri, T., T. Rouissi, S. Kaur-Brar, M. Cledon, S. Sarma and M. Verma. 2017. Biodegradation of polycyclic aromatic hydrocarbons PAHs by fungal enzymes: A review. J. Environ. Sci. 1: 52-74.

Kaur, N., T.E. Erickson, A.S. Ball and M.H. Ryan. 2017. A review of germination and early growth as a proxy for plant fitness under petrogenic contamination — knowledge gaps and recommendations. Sci. Total Environ. 603-604: 728-744.

Khatibi, S.M.H., F.Z. Vahed, S. Sharifi, M. Ardalan, M.M. Shoja and S.Z. Vahed. 2019. Osmolytes resist against harsh osmolarity: Something old something new. Biochimie. 158: 156-164.

Kirk, J.L., J.N. Klironomos, H. Lee and J.T. Trevors. 2002. Phytotoxicity assay to assess plant species for phytoremediation of petroleum-contaminated soil. Bioremediat. J. 6: 57-63.

Kis-Papo, T., V. Kirzhner, S.P. Wasser and E. Nevo. 2003. Evolution of genomic diversity and sex at extreme environments: Fungal life under hypersaline Dead Sea stress. Proc. Natl. Acad. Sci. U.S.A. 100: 14970-14975.

Kis-Papo, T., A.R. Weig, R. Riley, D. Peršoh, A. Salamov, H. Sun, et al. 2014. Genomic adaptations of the halophilic Dead Sea filamentous fungus *Eurotium rubrum*. Nat. Commun. 5: 3745.

Konte, T., U. Terpitz and A. Plemenitaš. 2016. Reconstruction of the high-osmolarity glycerol (HOG) signaling pathway from the halophilic fungus *Wallemia ichthyophaga* in *Saccharomyces cerevisiae*. Frontiers Microbiol. 7: 901.

Kuncic K., M.T. Kogej, D. Drobne and N. Gunde-Cimerman. 2010. Morphological response of the halophilic fungal genus *Wallemia* to high salinity. Appl. Environ. Microbiol. 76: 329-337.

Lange, B., S. Kremer, H. Anke and O. Sterner. 1996. Metabolism of pyrene by basidiomycetous fungi of the genera *Crinipellis*, *Marasmius* and *Marasmiellus*. Can. J. Microbiol. 42: 1179-1183.

Lee, H., Y.S. Choi, M.J. Kim, N.Y. Huh, G.H. Kim, Y.W. Lim, et al. 2010. Degrading ability of oligocyclic aromates by phanerochaete sordida selected via screening of white rot fungi. Folia Microbiol. 55: 447-453.

Lee, H., Y. Jang, Y.S. Choi, M.J. Kim, J. Lee, H. Lee, et al. 2014. Biotechnological procedures to select white rot fungi for the degradation of PAHs. J. Microbiol. Methods. 97: 56-62.

Lefebvre, O. and R. Moletta. 2006. Treatment of organic pollution in industrial saline wastewater: A literature review. Water Res. 40: 3671-3682.

Lenassi, M., C. Gostinčar, S. Jackman, M. Turk, I. Sadowski, C. Nislow, et al. 2013. Whole genome duplication and enrichment of metal cation transporters revealed by de novo genome sequencing of extremely halotolerant black yeast *Hortaea werneckii*. PLoS One 8: 1-18.

Liao, W.L., D.H. Tseng, Y.C. Tsai and S.C. Chang. 1997. Microbial removal of polycyclic aromatic hydrocarbons by *Phanerochaete chrysosporium*. Water Sci. Technol. 35: 255-264.

Lima, A.L.C., J.W. Farrington and C.M. Reddy. 2005. Combustion-derived polycyclic aromatic hydrocarbons in the environment: A review. Environ. Forensics 6(2): 109-131.

Liu, F., G.G. Ying, R. Tao, J.L. Zhao, J.F. Yang and L.F. Zhao. 2009. Effects of six selected antibiotics on plant growth and soil microbial and enzymatic activities. Environ. Pollut. 157: 1636-1642.

Liu, S.H., G.M. Zeng, Q.Y. Niu, Y. Liu, L. Zhou, L.H. Jiang, et al. 2017. Bioremediation mechanisms of combined pollution of PAHs and heavy metals by bacteria and fungi: A mini review. Bioresour. Technol. 224: 25-33.

Luch, A. 2009. On the impact of the molecule structure in chemical carcinogenesis. *In*: A. Luch. (ed.). Molecular, Clinical and Environmental Toxicology. Experientia Supplementum, vol 99. Birkhäuser Basel. https://doi.org/10.1007/978-3-7643-8336-7_6.

Lundstedt, S., P.A.White, C.L. Lemieux, K.D. Lynes, I.B. Lambert, L. Öberg, et al. 2007. Sources, fate and toxic hazards of oxygenated polycyclic aromatic hydrocarbons (PAHs) at PAH-contaminated sites. AMBIO 36(6): 475-486.

Luo, Y., J. Liang, G. Zeng, M. Chen, D. Mo, G. Li, et al. 2018. Seed germination test for toxicity evaluation of compost: Its roles, problems and prospects. Waste Manag. 71: 109-114.

Maila, M.P. and T.E. Cloete. 2005. The use of biological activities to monitor the removal of fuel contaminants: Perspective for monitoring hydrocarbon contamination: A review. Int. Biodeterior. Biodegrad. 55: 1-8.

Manoli, E. and C. Samara. 1999. Polycyclic aromatic hydrocarbons in natural waters: Sources, occurrence and analysis. TrAC Trac-Trends Anal. Chem. 18(6): 417-428.

Márquez-Rocha, F.J., V.Z. Hernández-Rodríguez and R. Vázquez-Duhalt. 2000. Biodegradation of soil-adsorbed polycyclic aromatic hydrocarbons by the white rot fungus *Pleurotus ostreatus*. Biotechnol. Lett. 22: 469-472.

Marston, C.P., C. Pereira, J. Ferguson, K. Fischer, O. Hedstrom, W.M. Dashwood, et al. 2001. Effect of a complex environmental mixture from coal tar containing polycyclic aromatic hydrocarbons (PAH) on the tumor initiation, PAH-DNA binding and metabolic activation of carcinogenic PAH in mouse epidermis. Carcinogenesis 22: 1077-1086.

Martínez-Martínez, M., I. Lores, C. Peña-García, R. Bargiela, D. Reyes-Duarte, M.E. Guazzaroni, et al. 2014. Biochemical studies on a versatile esterase that is most catalytically active with polyaromatic esters. Microb. Biotechnol. 7: 184-191.

Migliore, L., S. Cozzolino and M. Fiori. 2003. Phytotoxicity to and uptake of enrofloxacin in crop plants. Chemosphere 52: 1233-1244.

Moore, J.W. and S. Ramamoorthy. 2012. Organic Chemicals in Natural Waters: Applied Monitoring and Impact Assessment. Springer Science and Business Media. Springer Science and Business Media, New York.

Musa, H., F.H. Kasim, A.A. Nagoor Gunny and S.C.B. Gopinath. 2018. Salt-adapted moulds and yeasts: Potentials in industrial and environmental biotechnology. Process Biochem. 69: 33-44.

Nagahama, T., M. Hamamoto, T. Nakase, H. Takami and K. Horikoshi. 2001. Distribution and identification of red yeasts in deep-sea environments around the northwest Pacific Ocean. Anton. Leeuw. Int. J. G. 80: 101-110.

Neff, J.M. 1979. Polycyclic aromatic hydrocarbons in the aquatic environment. Biol. Conserv.; (United Kingdom). 18: 1.

Olicón-Hernández, D.R., J. González-López and E. Aranda. 2017. Overview on the biochemical potential of filamentous fungi to degrade pharmaceutical compounds. Front. Microbiol. 8: 1-17.

Oostingh, G.J., M. Schmittner, A.K. Ehart, U. Tischler and A. Duschl. 2008. A high-throughput screening method based on stably transformed human cells was used to determine the immunotoxic effects of fluoranthene and other PAHs. Toxicol. *In Vitro*. 22: 1301-1310.

Oren, A. 2008. Microbial life at high salt concentrations: Phylogenetic and metabolic diversity. Saline Syst. 4(1): 2.

Pachiadaki, M.G., M.M. Yakimov, V. Lacono, E. Leadbetter and V. Edgcomb. 2014. Unveiling microbial activities along the halocline of Thetis, a deep-sea hypersaline anoxic basin. ISME J. 8: 2478-2489.

Passarini, M.R.Z., M.V.N. Rodrigues, M. da Silva and L.D. Sette. 2011a. Marine-derived filamentous fungi and their potential application for polycyclic aromatic hydrocarbon bioremediation. Mar. Pollut. Bull. 62: 364-370.

Passarini, M.R.Z., L.D. Sette and M.V.N. Rodrigues. 2011b. Improved extraction method to evaluate the degradation of selected PAHs by marine fungi grown in fermentative medium. J. Braz. Chem. Soc. 22: 564-570.

Petelenz-Kurdziel, E., E. Eriksson, M. Smedh, C. Beck, S. Hohmann and M. Goksör. 2011. Quantification of cell volume changes upon hyperosmotic stress in *Saccharomyces cerevisiae*. Integr. Biol. 3(11): 1120-1126.

Petit K., J. Colina, F. Yegres, H. Moran and N. Richard-Yegres. 2013. Biodegradation of polycyclic aromatic hydrocarbons (PAHs) by fungi isolated from water contaminated with oil, white rot and aloe from *Aloe vera*. Rev. QuímicaViva 3: 287-304.

Plemenitaš, A., M. Lenassi, T. Konte, A. Kejžar, J. Zajc, C. Gostinčar, et al. 2014. Adaptation to high salt concentrations in halotolerant/halophilic fungi: A molecular perspective. Front. Microbiol. 5: 1-12.

Potin, O., C. Rafin and E. Veignie. 2004. Bioremediation of an aged polycyclic aromatic hydrocarbons PAHs-contaminated soil by filamentous fungi isolated from the soil. Int. Biodeterior. Biodegrad. 541: 45-52.

Pozdnyakova, N.N., M.P. Chernyshova, V.S. Grinev, E.O. Landesman, O.V. Koroleva and O.V. Turkovskaya. 2016. Degradation of fluorene and fluoranthene by the basidiomycete *Pleurotus ostreatus*. Prikl Biokhim Mikrobiol. 52: 590-598.

Pozdnyakova, N., E. Dubrovskaya, M. Chernyshova, O. Makarov, S. Golubev, S. Balandina, et al. 2018. The degradation of three-ringed polycyclic aromatic hydrocarbons by wood-inhabiting fungus *Pleurotus ostreatus* and soil-inhabiting fungus *Agaricus bisporus*. Fungal Biol. 122(5): 363-372.

Prista C, M.C. Loureiro-Dias, V. Montiel, R. Garcia and J. Ramos. 2005. Mechanisms underlying the halotolerant way of *Debaryomyces hansenii*. FEMS Yeast Res. 5: 693-701.

Ramos J, J. Arino and H. Sychrova. 2011. Alkali-metal-cation influx and efflux systems in nonconventional yeast species. FEMS Microbiol. Lett. 317: 1-8.

Roberts, M.F. 2005. Organic compatible solutes of halotolerant and halophilic microorganisms. Saline Syst. 1(1): 5.

Rončević, S., J. Spasojević, S. Maletić, J.M Jazić, M.K. Isakovski, J. Agbaba, et al. 2016. Assessment of the bioavailability and phytotoxicity of sediment spiked with polycyclic aromatic hydrocarbons. Environ. Sci. Pollut. Res. 23: 3239-3246.

Sanders, M., S. Sivertsen and G. Scott. 2002. Origin and distribution of polycyclic aromatic hydrocarbons in surficial sediments from the Savannah River. Arch. Environ. Contam. Toxicol. 43(4): 0438-0448.

Sharma, A., S.B. Singh, R. Sharma, P. Chaudhary, A.K. Pandey, R. Ansari, et al. 2016. Enhanced biodegradation of PAHs by microbial consortium with different amendment and their fate in *in situ* condition. J. Environ. Manage. 181: 728-736.

Sklenář, F., Z. Jurjević, P. Zalar, J.C. Frisvad, C.M.Visagie, M. Kolařík, et al. 2017. Phylogeny of xerophilic aspergilli (subgenus *Aspergillus*) and taxonomic revision of section Restricti. Stud. Mycol. 88: 161-236.

Soares, C.A.G., M. Maury, F.C. Pagnocca, F.V. Araujo, L.C. Mendonça-Hagler and A.N. Hagler. 1997. Ascomycetous yeasts from tropical intertidal dark mud of southeast Brazilian estuaries. J. Gen. Appl. Microbiol. 43: 265-272.

Steiman, R., P. Guiraud, L. Sage, F. Seigle-Murandi and J.L. Lafond. 1995. Mycoflora of soil around the Dead Sea I—Ascomycetes (including *Aspergillus* and *Penicillium*), Basidiomycetes, Zygomycetes. Sys. App. Microbiol. 18(2): 310-317.

Sterflinger, K. 1998. Temperature and NaCl-tolerance of rock-inhabiting meristematic fungi. Anton. Leeuw. Int. J. G. 74: 271-281.

Sterflinger, K. 2006. Black yeasts and meristematic fungi: Ecology, diversity and identification. pp. 501-514. *In*: C.A. Rosa, P. Gábor. (eds.). Biodiversity and Ecophysiology of Yeasts. Springer, Berlin, Heidelberg.

Stock, A., H.W. Breiner, M. Pachiadaki, V. Edgcomb, S. Filker, V. La Cono, et al. 2012. Microbial eukaryote life in the new hypersaline deep-sea basin Thetis. Extremophiles. 16: 21-34.

Teijon, G., L. Candela, K. Tamoh, A. Molina-Díaz and A.R. Fernández-Alba. 2010. Occurrence of emerging contaminants, priority substances (2008/105/CE) and heavy metals in treated wastewater and groundwater at Depurbaix facility (Barcelona, Spain). Sci. Total Environ. 408: 3584-3595.

Tirado-Torres, D., M. Gayosso-Canales, Y. Marmolejo-Santillán, C. Romo-Gómez and O. Acevedo-Sandoval. 2016. Removal of polycyclic aromatic hydrocarbons by *Pleurotus ostreatus* sp. ATCC38540 in liquid medium. Acad. J. Sci. Res. 4: 376-379.

Turk, M., L. Méjanelle, M. Šentjurc, J.O. Grimalt, N. Gunde-Cimerman and A. Plemenitaš. 2004. Salt-induced changes in lipid composition and membrane fluidity of halophilic yeast-like melanized fungi. Extremophiles. 8: 53-61.

Turk, M., Z. Abramović, A. Plemenitaš and N. Gunde-Cimerman. 2007. Salt stress and plasma-membrane fluidity in selected extremophilic yeasts and yeast-like fungi. FEMS Yeast Res. 7(4): 550-557.

Uratani, J.M., R. Kumaraswamy and J. Rodríguez. 2014. A systematic strain selection approach for halotolerant and halophilic bioprocess development: A review. Extremophiles 18: 629-639.

Uribe-Alvarez, C., M. Ayala, L. Perezgasga, L. Naranjo, H. Urbina and R. Vazquez-Duhalt. 2011. First evidence of mineralization of petroleum asphaltenes by a strain of *Neosartorya fischeri*. Microb. Biotechnol. 4: 663-672.

van Asperen, K. 1962. A study of housefly esterases by means of a sensitive colorimetric method. J. Insect Physiol. 8: 401-416. doi: 10.1016/0022-1910(62)90074-4.

Vatsal, A., S.S. Zinjarde and A.R. Kumar. 2011. Growth of a tropical marine yeast *Yarrowia lipolytica* NCIM 3589 on bromoalkanes: Relevance of cell size and cell surface properties. Yeast. 28: 721-732.

Verdin, A., A.L.H. Sahraoui and R. Durand. 2004. Degradation of benzo[a]pyrene by mitosporic fungi and extracellular oxidative enzymes. Int. Biodeterior. Biodegradation 53: 65-70.

Vieira, G.A.L., M.J. Magrini, R.C. Bonugli-Santos, M.V.N. Rodrigues and L.D. Sette. 2018. Polycyclic aromatic hydrocarbons degradation by marine-derived basidiomycetes: Optimization of the degradation process. Braz. J. Microbiol. 494: 749-756.

Wang, J., J. Wang, Z. Zhao, J. Chen, H. Lu, G. Liu, et al. 2017. PAHs accelerate the propagation of antibiotic resistance genes in coastal water microbial community. Environ. Pollut. 231: 1145-1152.

Wang, W. and K. Freemark. 1995. The use of plants for environmental monitoring and assessment. Ecotoxicol. Environ. Saf. 30(3): 289-301.

Wang, W., R. Zhong, D. Shan and Z. Shao. 2014a. Indigenous oil-degrading bacteria in crude oil-contaminated seawater of the Yellow sea, China. Appl. Microbiol. Biotechnol. 98(16): 7253-7269.

Wang, Z., C. Yang, J.L. Parrott, R.A. Frank, Z. Yang, C.E. Brown, et al. 2014b. Forensic source differentiation of petrogenic, pyrogenic and biogenic hydrocarbons in Canadian oil sands environmental samples. J. Hazard. Mater. 271: 166-177.

Wu, Y.R., T.T. He, J.S. Lun, K. Maskaoui, T.W. Huang and Z. Hu. 2009. Removal of benzo[a]pyrene by a fungus *Aspergillus* sp. BAP14. World J. Microbiol. Biotechnol. 25: 1395-1401.

Ye, J., H. Yin, J. Qiang, H. Peng and H. Qin. 2011. Biodegradation of anthracene by *Aspergillus fumigatus*. J. Hazard. Mater. 185: 174-181.

Zafra, G. and D.V. Cortes-Espinosa. 2015. Biodegradation of polycyclic aromatic hydrocarbons by *Trichoderma* species: A mini review. Environ. Sci. Pollut. Res. Int. 22(24): 19426-19433.

Zajc, J., Y. Liu, W. Dai, Z. Yang, J. Hu, C. Gostinčar, et al. 2013. Genome and transcriptome sequencing of the halophilic fungus *Wallemia ichthyophaga*: Haloadaptations present and absent. BMC Genomics 14: 617.

Zajc, J., S. Džeroski, D. Kocev, A. Oren, S. Sonjak, R. Tkavc, et al. 2014. Chaophilic or chaotolerant fungi: A new category of extremophiles? Front. Microbiol. 5: 708.

Zalar, P., G.S. de Hoog and N. Gunde-Cimerman. 1999. *Trimmatostroma salinum*, a new species from hypersaline water. Stud. Mycol. 43: 57-62.

Zalar, P., G.S. de Hoog, H.J. Schroers, J.M. Frank and N. Gunde-Cimerman. 2005. Taxonomy and phylogeny of the xerophilic genus *Wallemia* (*Wallemiomycetes* and *Wallemiales*, cl. et ord. nov.). Anton. Leeuw. 87(4): 311-328.

Zalar, P., G.S. de Hoog, H.J. Schroers, P.W. Crous, J.Z. Groenewald and N. Gunde-Cimerman. 2007. Phylogeny and ecology of the ubiquitous saprobe *Cladosporium sphaerospermum*, with descriptions of seven new species from hypersaline environments. Stud. Mycol. 58: 157-183.

2

Applications of Marine-Derived Fungi: Biocontrol, Cell Wall Degradation and Soil Remediation

Yi Wei[1], Ziyu Zhang[1], Zhugang Li[2], Yu Li[3] and Shi-Hong Zhang[1,*]

INTRODUCTION

Chemical fertilizers and pesticides are double-edged swords for agriculture. On the one hand, chemical fertilizers and chemical pesticides play important roles in ensuring stable, high yields of agricultural products and promoting growth of the agricultural economy (Wang et al. 1996, Jin and Zhou 2018), and indeed, according to statistical data from UN food and agriculture Organization (FAO) and Ministry of Agriculture and Rural Affairs of the People's Republic of China (MOA), fertilizer contributes nearly 50% to the increase of world crop production, and pesticide use saves about 40% of the world's total crop production (MOA 2015, FAO 2015). On the other hand, fertilizers and pesticides used unscientifically in agriculture, however, lead to unwanted consequences, such as degraded soil fertility, excessive pesticide residues, and agricultural non-point source pollution. Particularly, excessive use of pesticides and fertilizers influences the safety of the ecological environment and agricultural production, and further threatens human health and sustainable agricultural development (Xing and Zhu 2000, Fischer et al. 2010). Therefore, a safe alternative to fertilizers and pesticides is becoming increasingly urgent.

Biofertilizer, commonly defined as a substance which includes or involves living organisms or microorganisms and is helpful in promoting the growth of the plant's root system and germination of seeds, is the solution to stop the destruction of soil structure, water quality and flora. The main characteristics of biofertilizer are natural and organic, and on which organic agriculture mainly depends. Biofertilizer makes nutrients that are naturally abundant in soil or atmosphere, usable for plants and acts as supplements to agrochemicals; in addition, biofertilizers as essential components of organic farming play vital role in maintaining long term soil fertility and sustainability by fixing atmospheric dinitrogen, mobilizing fixed macro and

[1] Plant Protection College, Shenyang Agricultural University, Shenyang 110866, P. R. of China.
[2] Institute of Tillage and Cultivation for Crops, Heilongjiang Academy of Agricultural Sciences.
[3] Institute of Mycology, Jilin Agricultural University, Changchun 130062, P. R. of China.
[*] Corresponding author: 1357491337@qq.com; zhang_sh@jlu.edu.cn (S.H.Z.)

micro nutrients or convert insoluble phosphorous in the soil into available forms for plants, thereby increasing their efficiency and availability. It is due to the eco-friendly, easy to apply, non-toxic and cost-effective nature that biofertilizer has emerged as a highly potent alternative to chemical fertilizers. To modern agriculture, biofertilizers containing beneficial microorganisms or their metabolites are the best alternatives for replacing the conventional chemical-based fertilizers.

Biopesticides are defined as the directly used bioactive organisms or bioactive substances produced by organisms as pesticides, as well as synthetic pesticides with the same structure as natural compounds. Biopesticides include microbial pesticides, botanical pesticides, biochemical pesticides, genetically modified organisms and natural enemy organisms. According to the food and Agriculture Organization of the United Nations, biological pesticides are generally natural or genetic modifiers, mainly including microbial pesticides (fungi, bacteria, insect viruses, native animals, or genetically modified microorganisms), and biochemical pesticides (pheromones, hormones, plant regulators, insect growth regulators). Microbial pesticide is considered as the most promising alternative to chemical pesticides for its eco-friendliness and economy. Similar to microbial fertilizers, beneficial microorganisms or their metabolites are the best alternatives for replacing the conventional chemical-based pesticides.

No matter microbial fertilizer or microbial pesticide, they are inseparable from the role of microorganism. Studies on microorganisms isolated from land soils have covered a lot of the applied microorganisms. Some beneficial microorganisms including bacteria, actinomycetes, yeasts and filamentous fungi have been intensively studied and even used in organic fertilizer production and biological control of crop pests and diseases. Marine fungi, one of the major decomposers of marine environment, are found to produce many enzymes and novel biomolecules. Compared with terrestrial fungi, marine fungi have been studied relatively few and late; however, microorganisms in marine habitats have been described to produce compounds with properties that differ from those found in non-saline habitats (Torres et al. 2019); in addition, marine fungi have been widely concerned about metabolites, because of their special biological activity and metabolic mechanism. For example, Farha and Hatha (2019) explored bioprospecting potentials such as antimicrobial, anticancer and enzymatic activities of the marine sediment-derived fungus *Penicillium* sp. ArCSPf that was isolated from continental slope of Eastern Arabian Sea.

Marine fungi have been classified as obligate or facultative: obligate marine fungi are those that grow and sporulate exclusively in a marine or related habitat, whereas facultative marine fungi are those from freshwater or terrestrial origin that are able to grow and sporulate in marine environments (Kohlmeyer and Volkmann-Kohlmeyer 2003, Li and Wang 2009). In this chapter, the term "marine-derived fungi" is frequently used because most of the fungi isolated from marine samples are not because of obligate or facultative marine microorganisms. Considering that the distribution of fungi in the marine environment and the fact that a great diversity of fungi has been recovered from along coastlines, such as mangrove, sand, beach, river, and estuarine habitats, marine-derived fungi, sometimes, are used to denote terrestrial fungi carried toward marine environments by floods and winds (Mejanelle et al. 2000). Anyway, halotolerant or halophilic are the common characteristics, so marine-derived fungi actually belong to halotolerant or halophilic fungi.

Bioprocess is currently established as a useful tool, including the production of renewable raw materials, and the biodegradation of environmental contaminants (Gavrilescua and Chisti 2005). In organic fertilizer production from straw degradation, high activity and high thermal stability cellulose hydrolytic enzymes are essential, and in fungal disease biocontrol process, cell wall hydrolytic enzymes such as chitinases, which inhibit or destroy disease development are also important mechanisms for biocontrol. All these cell wall degradations are related to bioprocess in which certain enzyme is required (Liu et al. 2004). Fungi from marine environments are thoroughly adapted to surviving and growing under harsh conditions (Burton et al. 2002). Such habitat-related characteristics are desirable features from a general biotechnological perspective and are of key importance to exploit a microorganism's enzymatic potential (Trincone 2011, 2010). Indeed, many marine-derived fungi contain novel enzymes showing optimal activities at extreme values of salt concentrations, pH, and temperature, compared to enzymes isolated from terrestrial origins (Antranikian et al. 2005, Dionisi et al. 2012, Ferrer et al. 2012). In addition, some reports about mangrove fungi revealed that most marine fungi were able to produce enzymes like chitinase and insecticidal metabolites with high bioactivity for plant disease biocontrol (Xiao et al. 2009, Swe et al. 2009).

Microbial application for amelioration of saline soils is gaining popularity due to its better amelioration and reduction in economic and environmental costs. Within the last few decades, a series of marine halophilic and alkaliphilic fungi have been identified. This chapter is focused on the isolation and characterizations of marine-derived halotolerant or halophilic fungi, and their roles in straw degradation and fungal disease biocontrol by mechanism of cellulase or chitinase activities; the application of these marine fungi or their metabolisms in saline affected or chemical contaminated soil remediation is also highlighted.

Identification and Halotolerant or Halophilic Characteristics of Marine-Derived Fungi

Ancient Earth was covered in a global ocean (Burnham and Berry 2017). Studies on the microbial communities of deep subsurface sediments have indicated the presence of Bacteria and Archaea throughout the sediment column. From the biological and environmental evolution perspective, fungi are one of the earliest eukaryotes to colonize the ancient Earth (Horodyski and Knauth 1994). Indeed, microbial eukaryotes could also be present in deep-sea subsurface sediments; in addition, fungi are the most consistently detected eukaryotes in the marine sedimentary subsurface (Edgcomb et al. 2011). Most marine-derived fungi belong to halotolerant fungi which live in saline environments but do not necessarily require certain concentrations of salt; the rest of the marine-derived fungi are classed as halophilic fungi because these fungi require salt concentrations of at least 0.3 M (sodium salt, e.g. NaCl) to grow optimally, and even they thrive in high-salt environments. In halotolerant fungi, salinity can directly affect sporulation and growth of fungi: at higher salinities (>5%), there tends to be increased sporulation with more chlamydospores observed, an inhibition of conidiogenesis, and fewer hyphae (Mulder et al. 1989, Mahdy et al. 1996, Mulder and El-Hendawy 1999, Mandeel 2006). On the other hand, halophilic fungi do not always have to be in saline habitats; thus, there is no need to make a strict distinction between halotolerant and halophilic fungi (Arakaki et al. 2013).

There is no identical cultivation method for isolation of marine-derived fungi. Salinity is an important condition and can be achieved using sterilized seawater. The greatest diversity can be recovered from seawater, sediments or other floating matters using poor and rich media. Some information related to the isolation of marine-derived fungi can be found in literatures by Vrijmoed (2000) and Nakagiri (2012). Early researches have focused on the isolation and identification of *Eubacteria*, *Archaea* and *Algae* due to these prokaryotes' growth under salt stress and populated saline ecosystems (Oren 2002). Over the last two decades, marine fungi have been discovered accordingly in the saline environments such as in the Dead Sea, Atlantic Ocean, China Sea (Grishkan et al. 2003, Nazareth et al. 2012), and the solar salterns near to seacoast (Cantrell et al. 2006, Nayak et al. 2012). A large number of studies on biodiversity and physiology have focused on the characterization of halophilic fungi present in the sea related saline and hypersaline ecosystems, among which species of *Ascomycetes*, as well as some *Basidiomycetes*, have been described in detail (Gunde-Cimerman et al. 2000, Butinar et al. 2005a, b, Zalar et al. 2005a, b, Evans et al. 2013, Gunde-Cimerman and Zalar 2014, Zajc et al. 2014a, b, Gonçalves et al. 2017). In general, fungal communities in hypersaline environments are dominated by *Aspergillus*, *Penicillium* and some of their teleomorphic genera (formerly Eurotium, Emericella, and Eupenicillium). Other genera such as Alternaria, Cladosporium, Fusarium, Chaetomium, Wallemia and Hortaea were also reported (Gunde-Cimerman and Zalar 2001, Mandeel 2006, Moubasher et al. 2018). Some new species were also described from hypersaline environments including three species of the genus Wallemia (Zalar et al. 2005a, b), twelve species of the genus Cladosporium (Zalar et al. 2007, Schubert et al. 2007), two species of the genus Emericella (Zalar et al. 2008) and three species of the genus Gymnoascus (Zhou et al. 2016). Similar to the communities observed in hypersaline environments (Buchalo et al. 1998, 2000, Gunde-Cimerman et al. 2000, Butinar et al. 2005a, b, Kis-Papo et al. 2003, 2014, Gunde-Cimerman and Zalar 2014), partial hypersaline fungal communities dominated by *Aspergillus* and *Penicillium* species, with melanized dematiaceous forms are also observed in the soils of coastal zone or even inland lands (Moubasher et al. 1990, Grum-Grzhimaylo et al. 2016), suggesting their common origin and similar properties.

The Dead Sea, a typical high-salt habitat for microorganisms, contains 340 g/L of dissolved salt; a variety of filamentous fungi has been isolated from the Dead Sea by the Nevo group. *Gymnascella marismortui* is a remarkable salt-tolerant fungus that has been isolated from the surface water down to a depth of 300 m in the Dead Sea (Buchalo et al. 1998). *G. marismortui* grows optimally at NaCl concentrations between 0.5 and 2 M (Buchalo et al. 1998, 2000), suggesting that it is adapted to high-salt conditions and requires high salt concentrations. Among 476 fungal isolates from the Dead Sea, *Aspergillus terreus, Aspergillus sydowii, Aspergillus versicolor, Eurotium herbariorum, Penicillium westlingii, Cladosporium cladosporoides* and *Cladosporium sphaerospermum* were isolated consistently and probably form the stable core of the fungal community. In another study, approximately 43% of fungal isolates from the Dead Sea were found to belong to the genera *Eurotium* and *Aspergillus* (Yan et al. 2005).

Recently, Chamekh et al. (2019) identified 136 isolates from the soil of the Great Sebkha of Oran located in northwestern Algeria at 12 km from the Mediterranean Sea in Algeria, whose salt concentration is estimated at more than 100 $g.L^{-1}$ of dissolved salts. Most of the isolates are halotolerant but all of them can still grow on PDA medium without NaCl with the exception of *Wallemia sp.* H15 and the two strains of *Gymnoascus halophilus* H19 and H20 that are obligatorily halophilic, indicating the dominant flora of halotolerant fungi. 74% of the strains could grow at 12.5% NaCl and 5 strains (*A. subramanianii* strain A1, *Aspergillus* sp. strain A4, *P. vinaceum* and the two strains of *G. halophilus*) at 17.5%. The only strain that could grow at 20% was *Wallemia sp.* The optimum growth of most strains is 2.5% or 5% NaCl. The concentration of 10% is optimal for the growth of *G. halophilus.*

Collectively, the large diversity of the fungal species has been reported to inhabit high salt environments, and most of them can be regarded either as halotolerant or as extremely halotolerant (Table 2.1). Halotolerant fungi can grow without NaCl added to the medium but tolerate up to saturated NaCl levels (30%) (Gunde-Cimerman et al. 2000). Up till today, only *Wallemia ichthyophaga, Wallemia muriae, Phialosimplex salinarum, Aspergillus baarnensis, Aspergillus salisburgensis, Aspergillus atacamensis* and *Gymnoascus halophilus* are obligate halophilic fungi that strictly require NaCl from 5 to 10% (Piñar et al. 2016, Chamekh et al. 2019). Actually, several terrestrial soil strains such as *Gymnascella marismortui* (Buchalo et al. 1998), *Trichosporium* spp. (Elmeleigy et al. 2010), *Aspergillus unguis* (Nazareth et al. 2012) and *Aspergillus penicillioides* (Nazareth and Gonsalves 2014) have been previously reported to be obligate halophiles according to their minimum saline requirement.

Aspergillus penicillioides are commonly found in saline habitats, suggesting that the species are extensively adaptable to varied environments. Among 39 tested isolates of *A. penicillioides*, most strains had a minimum salt requirement of 5% for growth; one strain grew only on media supplemented with at least 10% solar salt (Nazareth and Gonsalves 2014). Interestingly, this species does not reproduce sexually (Tamura et al. 1999, Gostinčar et al. 2010, Gostinčar and Turk 2012); considering the genetic stability, this species is environmentally safe and has significant promise in environmental remediation applications.

TABLE 2.1 The representative marine-derived or related fungi and the characteristics.

Species/Strain	Source	$[Na^+]$ Range	pH Range	Reference
Wallemia sp.	Great Sebkha of Algeria	2.5-20%	NR	Chamekh et al. 2019
Aspergillus subramanianii A1	Great Sebkha of Algeria	0-17.5%	NR	Chamekh et al. 2019
Aspergillus sp. strain A4	Great Sebkha of Algeria	0-17.5%	NR	Chamekh et al. 2019
Penicillium vinaceum	Great Sebkha of Algeria	1-17.5%	NR	Chamekh et al. 2019
Gymnoascus halophilus S1-2	Great Sebkha of Algeria	2.5-17.5%	NR	Chamekh et al. 2019
Trichodermalixii IG127	Italy sea	1%	8.0	Pasqualetti et al. 2019
Clonostachys rosea IG119	Italy sea	3.8%	8.0	Pasqualetti et al. 2019
Aspergillus salisburgensis, Aspergillus atacamensis	Iquique, Chile	10-25%	NR	Martinelli et al. 2014
Aspergillus penicillioides	Mangroves of Goa, India	10-30%	NR	Nazareth and Gonsalves 2014
Aspergillus nidulans KK-99	Punjab, India	0-25%	4-10	Taneja et al. 2002
Eurotium herbariorum	Dead Sea, Israel	2-31%	7-9	Butinar et al. 2005a, b
Gymnascella marismortui	Dead Sea, Israel	5-30%	NR	Buchalo et al. 1998
Sodiomyces sp., *Acrostalagmus luteoalbus, Emericellopsis alkaline, Thielavia* sp., *Alternaria sect. Soda*	Russia, Mongolia, Kazakhstan, Kenya, Tanzania, Armenia	NR	8.5-11	Grum -Grzhimaylo et al. 2016
Hortaea werneckii	Ljubljana	5-31%	NR	Gunde-Cimerma-n et al. 2000
Aureobasidium pullulans	Amsterdam, Netherlands	0-17%	NR	Sterflinger et al. 1999
Myrothecium sp. IMER1	Wuhan, China	0-5%	8-10	Zhang et al. 2007
Myrothecium sp. GS-17	Gansu, China	NR	8-10	Liu et al. 2013
Cladosporium cladosporioides PXP-49	Hainan, China	0-20%	5-9	Xu et al. 2011
Wallemia sebi PXP-89	Hainan, China	0-20%	5-9	Peng et al. 2011a
Penicillium chrysogenum PXP-55	Hainan, China	0-20%	5-9	Peng et al. 2011b

Note: NR, no report.

Some halophilic fungi, such as *A. niger* and *C. cladosporoides*, have been indirectly isolated from sand and mud near the shore of salty aquatic bodies or from inflowing fresh water from floods and springs (Kis-Papo et al. 2003, 2014, Grum-Grzhimaylo et al. 2016). In terms of evolution and origin, these terrestrial halotolerant fungi should have come from nearby oceans. We also isolated the halophilic fungus *A. glaucus* CCHA from air-dried wild vegetation from the surface periphery of a solar salt field (Liu et al. 2011); this species shows extreme salt tolerance, with a salinity range of 5 to 32% (NaCl) required for growth (Liu et al. 2011). To our surprise, *A. glaucus* CCHA survives in solutions with a broad pH range of 2.0-11.5, indicating that it is a haloalkaliphilic fungus. Further investigation indicated that increasing the pH value (> 8.0) can induce *A. glaucus* CCHA to produce a variety of organic acids, including citric acid, oxalic acid and malic acid. In addition, *A. glaucus* CCHA shows resistance to aridity, heavy metal ions, and high temperature. The extremophilic nature of *A. glaucus* CCHA suggests that it has great promise in soil remediation applications.

China has remarkable biodiversity and many typical hypersaline environments. Research aimed at isolating and characterizing halotolerant or halophilic fungi from seas has progressed rapidly in China. A series of promising halophilic fungi, including *A. glaucus* CCHA, have been reported. Three marine-derived isolates were collected in Wenchang, Hainan Province, China, and identified as extremely halotolerant fungi: *Wallemia sebi* PXP-89 (Peng et al. 2011a), *P. chrysogenum* PXP-55 (Peng et al. 2011b), and *Cladosporium cladosporioides* PXP-49 (Xu et al. 2011). According to Xiao et al. (2009), 188 marine-derived fungi were collected from the sediment in Zhoushan Sea area, the mangrove at Yunxiao Country and Jiulongjang estuary in Fujian Province, China, of which, the ethyl acetate extract of strain 164 exhibited strong lethal effect on nematode *Rhabditis* sp. In another research, 31 nematode-trapping fungi belonging to three genera, *Arthrobotrys*, *Monacrosporium*, and *Dactylella*, were also recorded from mangrove habitat of Hong Kong (Swe et al. 2009).

The South China Sea covers a vast area. Zhang et al. (2013) investigated the diversity of fungal communities in nine different deep-sea sediment samples of the South China Sea by culture-dependent methods followed by analysis of fungal internal transcribed spacer sequences. 13 of 27 identified species were isolated from sediments of deep-sea environments for the first report. Moreover, these ITS sequences of six isolates shared 84-92% similarity with their closest matches in GenBank, which suggested that they might be novel phylotypes of genera *Ajellomyces*, *Podosordaria*, *Torula*, and *Xylaria*. The work focusing on isolating halotolerant/halophilic fungi is being carried out in our laboratory, and actually a series of halotolerant or halophilic species have been recently identified based on specimens collected from the South China Sea, Hainan, China (Table 2.2). All of these environments are suitable for extremophilic fungi and other microorganisms; therefore, isolating and identifying extremophilic fungi in China could lead to the development of promising new methods for modern agriculture.

Biocontrol Mechanisms of Marine-Derived Fungi

According to previous studies and statistics, approximately 35% of crop yield is lost to diseases in the field, and about 70-80% of plant diseases are caused by fungal pathogens. The main control measures thus are pointed to fungal diseases, and consequently, crops completely depend on the use of fungicides and pesticides to reduce yield loss, which leads to a proliferation of chemical fungicides. However, as frequently mentioned above, these chemicals have brought about a series of environmental and health problems as well as the development of fungicide resistance by pathogens; therefore, biocontrol as an alternative approach emerges as the times require to control plant diseases.

Chitin, glucan and protein comprise the cell walls of many fungi, including some yeasts, and make up the structural frameworks of nematode as well as of plant pests (Fig. 2.1). Many microbial genomes possess different genes encoding chitinolytic enzymes, which clearly reflects the importance of chitin of fungi. Fungal cell wall hydrolysis is the main weapon to break through the first barrier, to realize parasitism, to further complete antagonism and ultimately to achieve the goal for biological control (Fig. 2.1). Thus, the secretion of hydrolytic enzymes, such as chitinases, glucanases, and proteases (Gruber and Seidl-Seiboth 2012, Kubicek et al. 2011, Mandujano-Gonzalez et al. 2016) plays a critical role in cell wall degradation of fungi (Seidl et al. 2009).

Trichoderma spp, the typical mycoparasite fungi which have been successfully and widely used in agricultural practice, antagonizes many phytopathogenic fungi through many mechanisms including mycoparasitism, antagonism, competition, and induced systemic resistance in related plants. Mycoparasitism is the most important biocontrol mechanism adopted by *Trichoderma* or other mycoparasites against plant pathogens. It involves tropical growth of biocontrol agent towards the target organism and then sets up interactions between mycelia of both mycoparasitic fungus and host fungus. Mycoparasitic fungus hypha coils around host fungus and finally causes dissolution of target pathogen hyphal cell wall or membrane by the activity of corresponding enzymes. Chitinases are mainly studied in *Trichoderma, Oenicillium, Lecanicillium, Neurospora, Mucor, Metarhizium, Beauveria, Lycoperdon, Thermomyces* and *Aspergillus*, and Trichoderma chitinases have been extensively studied (Krause et al. 2000, Haki and Rakshit 2003, Kristensen et al. 2008, Sarkar et al. 2010, Trincone 2010, 2011, Hamid et al. 2013).

The use of cell wall degrading enzymes of *Trichoderma* is a promising alternative for inhibiting food storage diseases. In addition to chitinases, other enzymes such as aspartic protease P6281 secreted by the fungus *Trichoderma harzianum* have been verified to be important in mycoparasitism on phytopathogenic fungi (Deng et al. 2019). The recombinant P6281 (rP6281) expressed in *Pichia pastoris* showed high activity of 321.8 U/ml. Maximum activity was observed at pH 2.5 and 40°C, and the enzyme was stable in the pH range of 2.5-6.0. rP6281 significantly inhibited spore germination and growth of plant and animal pathogenic fungi such as *Botrytis cinerea, Mucorcircinelloides, Aspergillus fumigatus, Aspergillus flavus, Rhizoctonia solani*, and *Candida albicans*. Transmission electron microscopy revealed that rP6281 efficiently damages the cell wall of *Botrytis cinerea*. In addition, the protease significantly inhibited the development of grey mold that causes rotting of apple, orange, and cucumber, indicating that rP6281 may be developed as an effective anti-mold agent for fruit storage.

The biological control efficacy is determined by the activity and stability of enzymes secreted from biocontrol agents. In previous practice of biological control of diseases, the short validity period of biological control agents and the difficulty of field inoculation have led to poor control effect, which surely is an important reason for limiting the wide application of biological control. Till now, most biocontrol fungi were isolated from terrestrial soils, the common but not extreme environment that is not naturally conducive to the evolution of special metabolic pathways and the establishment of special enzymatic systems. Marine-derived microorganisms, the halotolerant or halophilic organisms living in high saline environment most likely contain series of enzymes with stable and specific activities. Actually, the specific conditions, such as salinity and pH extremes in marine environment make the marine fungal enzymes superior to homologous enzymes from terrestrial fungi (Booth and Kenkel 1986, Jones 2000, Gomes et al. 2008, Madhu et al. 2009, Pang et al. 2011, Intriago 2012, Passarini et al. 2013, Rämä et al. 2014). Therefore, the first task is to screen biocontrol fungi from marine-derived fungi, which is also the important strategy for solving above-mentioned defects of biocontrol agents.

Zhang et al. (2013) investigated the diversity of fungal communities in nine different deep-sea sediment samples of the South China Sea. Out of 27 identified species, 13 species isolated from sediments of deep-sea environments were first reported, which has been described as above. Recently, 28 fungal strains have been isolated from different natural marine substrates of East Sector of the Tyrrhenian Sea and plate screened for their production of chitinolytic activity (Pasqualetti et al. 2019). The two apparent best producers are *Trichodermalixii* IG127 and *Clonostachys rosea* IG119. IG127 appeared to be a slight halotolerant fungus, while *C. rosea* IG119 clearly proved to be a halophilic marine fungus (i.e. pH 8.0, salinity 38%). It is interesting to note that the maximum activity of chitinase is 3.3 U/Lh (Pasqualetti et al. 2019).

In the last four years, our laboratory also explored antimicrobial activities of the marine-derived fungal isolates from epicontinental sea sediment samples of the South China Sea (Table 2.2). We identified a total of 669 isolates, among which 421 isolates showed certain antifungal activity, indicating that most marine fungi have antifungal ability. A fungal isolate with high antifungal activity against many more plant pathogens was identified as *Trichoderma* spp. S58. Chitinase enzyme of S58 strain was purified and characterized. The S58 strain chitinase exhibited highest stability and activity. The crude as well as purified enzyme showed significant antimycotic activity against agricultural pathogens such as *Aspergillus flavus, Aspergillus niger, Aspergillus fumigatus, Magnaporthe oryzae* and *Fusarium oxysporum*. The high chitinolytic and antifungal activity indicates marine fungi are important biocontrol agents.

TABLE 2.2 Cell wall enzymatic activities of marine-derived fungi from South China Sea.

Species/Strain	Cellulase	Chitinase	Protease
***Trichoderma sp.* S58**	1.7	5.3	2.1
***Trichoderma sp.* S74**	2.4	4.2	3.3
***Gymnoascus halophilus* S33**	ND	1.2	5.1
***Gymnoascus halophilus* S144**	0.1	3.1	4.6
***Fusarium oxysporum* S352**	6.2	2.2	0.2
***Fusarium equiseti* S79**	4.5	3.2	0.1
***Alternaria sp.* S402**	6.2	ND	0.5
***Pleospora sp.* S371**	3.7	0.21	3.1
***Chaetomium sp.* S58**	4.4	ND	5.1
***Cladosporium sp.* S76**	1.3	1.4	0.4
***Penicillium sp.* S9**	4.5	2.6	3.9
***Chrysosporium sp.* S280**	0.6	3.5	ND
***Aspergillus sp.* S311**	4.8	4.7	ND
***Aspergillus terreus* S82**	5.1	1.1	0.6
***Aspergillus subramanianii* S47**	4.7	ND	3.4
***Lecanicillium sp.* S21**	2.5	3.6	1.8

Note: Protease activity detected according to Deng et al. (2018); cellulase activities detected according to Li et al. (2018); Chitinase detected according to Qin et al. (2008). Sea water or sea sediment samples were collected from the South China Sea by Wei Y and Zhang S-H. ND: the enzyme activity was not detected

After degradation of cell wall, the growth and development of fungal pathogen will be inhibited to a certain extent. Antibiotic produced by marine-derived fungi has cytotoxicity, which is another important mechanism for biocontrol. Microorganisms in marine or other saline habitats have been described to produce secondary metabolites with properties that differ from those found in non-saline habitats (Torres et al. 2019). *P. chrysogenum* SCSIO 41001 is a deep-sea-derived fungus (Chen et al. 2017). 15 new compounds, including 5 novel and 10 known compounds, have been identified from a deep-sea-derived fungus (Chen et al. 2017). Marine fungi from the sponge *Grantia compressa* proved to be an outstanding source of fungal diversity. Marine-derived fungi isolated from *G. compressa* were capable of producing many different metabolites; in particular, the compounds isolated from *Eurotium chevalieri* MUT 2316 showed promising bioactivity against well-known and emerging pathogens (Bovio et al. 2019). Farha et al. (2019) first explored the bioprospect potential of the sediment-derived fungus *Penicillium* sp. ArCSPf from continental slope of eastern Arabian Sea for the production of therapeutically active compounds. The fungal strain exhibited amylase, gelatinase, phytase, lipase and pectinase activity. The active fraction obtained from the ethyl acetate extract column fractionation of fungus showed antibacterial activity. Marine fungi, one of the major decomposers of marine environment, are found to produce potential enzymes and novel biomolecules. The above studies explored bioprospecting potentials such as antimicrobial, anticancer and enzymatic activities; we believe that these compounds should at most likely inhibit development of plant diseases.

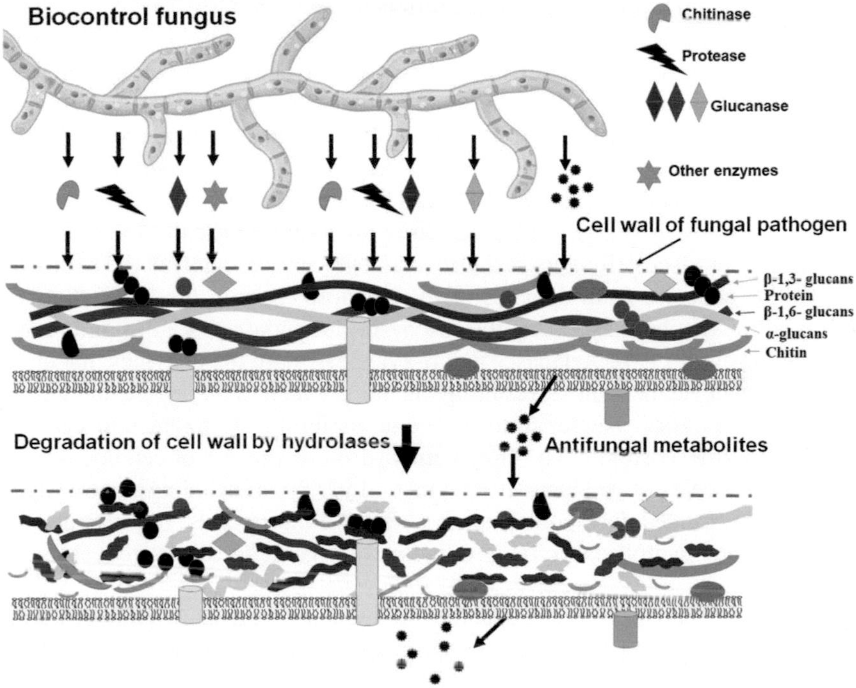

FIGURE 2.1 Biocontrol mechanism of marine-derived fungus. Cell wall of fungal pathogen was degraded by hydroenzymes (chitinase, cellulase and protease) of biocontrol fungus from marine environments.

Plant Cell Wall Degradation Enzymes of Marine-Derived Fungi

Marine habitats have usually been regarded as a source of microorganisms that possess robust proteins that help enable them to survive in harsh conditions. Fungi appear to dominate eukaryotic life in the buried marine subsurface of marine environments (Edgcomb et al. 2011). Fungi have high species diversity (Pachiadaki et al. 2016), are known to play key roles in decomposition of organic matter, represent a significant portion of the biomass in marine environment, and play important roles in biogeochemical cycles and food webs (Bass et al. 2007, Gadd 2007). Thus, marine-derived fungi and their cell wall degrading system play important roles in the recycling of agricultural wastes such as straw.

Plant cell wall is mainly composed of cellulose, hemicellulose and lignin. Cellulose is a homo-polymer of glucose subunits (cellobiose) with a crystalline structure; hemicellulose is a heteropolymer of pentose sugars with an amorphous structure, whereas lignin is a highly crystalline and rigid component of biomass. Cellulose and hemicellulose typically comprise two-thirds of the dry mass and varies with the type of straw. These three components of biomass can be converted to various valuable products such as straw organic fertilizer and man-made culture medium (soilless culture) through fungal fermentation.

Hydrolytic enzymes involve cellulose, hemicellulose, lignin, and pectin, which can degrade plant organic matter (e.g. maize, wheat or rice straw) (Castillo and Demoulin 1997, Santos et al. 2004, Batista-García et al. 2014, Li et al. 2018, Wei and Zhang 2018). The enzymatic hydrolysis of cellulose, particularly hydrogen-bonded and ordered crystalline regions, is a very structure. The hydrolytic system generally includes endoglucanases (EC 3.2.1.4), exoglucanases (EC 3.2.1.91), and β-glucosidases (EC 3.2.1.21). Endoglucanases randomly attack the internal chain of cellulose to produce cell oligosaccharides. Exoglucanases catalyze the hydrolysis of crystalline cellulose from the ends of the cellulose chain to produce cellobiose, which is ultimately hydrolyzed to glucose by β-glucosidases (Béguin and Aubert 1994, Tomme et al. 1995). Xylanases and β-xylosidases are the enzymes that attack the backbone of hemicellulose resulting in the production of xylose monomers. Lignin acts as physical barrier limiting the accessibility of enzymes to cellulose and hemicellulose substrates. The enzymatic hydrolysis process of cellulose is very slow. Multiple scales of products of enzymatic hydrolysis including nanocellulose are going to be formed during the microbe-cellulose interactions. Improvements in fungal enzymatic hydrolysis for the production of organic fertilizer from straw are necessary.

Much information is available on the production of cellulase enzymes using different substrates, and a variety of optimized cellulases have been identified from terrestrial fungi such as *Phanerochaete chrysosporium*, *Tricoderma reesei*, *Aspergillus niger*, *Penicillium oxalicum*, and *Gracibacillus* species (Szabo et al. 1996, Sukumaran et al. 2009, Yu et al. 2015, Huang et al. 2015b). The deep sea is an extreme habitat. Cellulases from marine bacteria have been reported a lot; however, much less information is available on marine fungi associated with cell wall degradation.

Trichoderma reesei and *Penicillium janthinellum* are known to be excellent cellulase producers, but their cellulases are not stable under alkaline conditions (Mernitz et al. 1996, Wang et al. 2005, Qin et al. 2008). *Aspergillus niger*, one of the most efficiently identified cellulose-degrading microorganisms, secretes large amounts of different cellulases during fermentation (Schuster et al. 2002). Endoglucanase B (EGLB), encoded by the endoglucanase gene (GenBank GQ292753) of *Aspergillus niger* BCRC31494, has been used in the fermentation industry because of its alkaline and thermal tolerance (Li et al. 2012a). EGLB is a member of glycosyl hydrolase family 5 of the cellulase superfamily. When the recombinant EGLB cDNA was expressed in *Pichia pastoris*, a purified protein of 51 kDa in size was obtained. The enzyme was specific for substrates with β-1,3 and β-1,4 linkages, and it exhibited optimal activity at 70°C and pH 4 (Li et al. 2012a). Interestingly, the relative activity of recombinant EGLB at pH 9 was significantly better than that of wild-type EGLB. The advantages of endoglucanase *EGLB*, particularly its tolerance to a broad range of pH values, indicate that this enzyme is promising as a means of genetically improving fungi for haloalkaline soil remediation.

The soft rot ascomycetes fungus *T. reesei* is utilized for industrial production of secreted enzymes, especially lignocellulose degrading enzymes. *T. reesei* uses several different enzymes for the degradation of plant cell wall derived material, including nine characterized cellulases, 15 characterized hemicellulases and at least 42 genes predicted to encode cellulolytic or hemicellulolytic activities (Häkkinen et al. 2014). The family 7 cellobiohydrolase (Cel7A) of *T. reesei* is comprised of a 36-amino acid CBM, a linker domain with O-glycan, and a large catalytic domain with N-linked glycan and a 50 Å tunnel for processing cellulose chains. The possibility of controlled hydrolysis of microcrystalline cellulose by *T. reesei* has been analyzed. The penetration of fungus into the ordered regions of microcrystalline cellulose during incubation resulted in reduced crystallinity of nanocellulose prepared by microbial hydrolysis compared to that of acid hydrolysis (Satyamurthy et al. 2011). However, in comparison to the fungal hydrolysis system, the anaerobic bacteria consortium is much more efficient in hydrolyzing microcrystalline cellulose to produce nanocellulose in a span of 7 days with a maximum yield of 12.3%, and nanocellulose prepared by this process has a bimodal particle size distribution (43 ± 13 and 119 ± 9 nm) (Satyamurthy and Vigneshwaran 2013). Thus, more efficient fungal hydrolyzing system must be strengthened.

Based on an analysis of the genomic sequence of haloalkaliphilic fungus *A. glaucus* CCHA, we found that *A. glaucus* CCHA expresses only one gene belonging to the GH5 family, AgCel5A. The open reading frame of *Agcel5A* consists of 1509 base pairs that encode a polypeptide of 502 amino acids. AgCel5A has 4 potential N-glycosylation sites and 3 O-glycosylation sites, which indicates high similarity to the characterized GH5 β-glucosidases from *Aspergillus niger* (65%) and *Trichoderma reesei* (31%). AgCel5A was cloned and heterologously expressed in *Pichia pastoris* GS115. Recombinant AgCel5A exhibited maximal activity at pH 5.0. AgCel5A is much more stable than PdCel5C from *Penicillium decumbens* (Liu et al. 2013); it retains more than 70% of its maximum activity at pH 8.0-10.0. In addition, AgCel5A exhibited stable degradation activity under high-salt (NaCl) conditions. In the presence of 4 M NaCl, AgCel5A retained 90% activity even after 4 h of pre-incubation. Interestingly, the activity of AgCel5A increased as the NaCl concentration was increased. The high resistance of AgCel5A to saline and alkaline conditions suggests that the *AgCel5A* gene is an ideal candidate for genetic improvement of soil fungi and industrial applications (Zhang et al. 2016, Li et al. 2018). As this fungus is being exploited for saline-alkaline soil remediation, nanocellulose production using this fungus is being analyzed in our lab. In a preliminary simulation test, multi-scale and type nanoparticles were detected when *A. glaucus* CCHA and its fermentation filtrate were added in the saline soil mixed with physically grated corn stalks (particle size: 0.5-2 mm). The complex nanoparticles should be MIMI nanoparticles as described above.

About 20 strains were isolated from sea mud samples collected in the East China Sea and then screened for their capacity to produce lignin-degrading enzymes (Chen et al. 2011). Out of the 20 strains, a moderately halo-tolerant endophytic fungus, *Pestalotiopsis sp.*, which was rarely reported as ligninolytic enzyme producer in the literature, had a great potential to secrete a considerable amount of laccase. The production of both cellulase and laccase by *Pestalotiopsis sp.* J63 was investigated under submerged fermentation and solid state fermentation with various lignocellulosic by-products as substrates. The result indicated that J63 produced 0.11 U/ml cellulase when alkaline-pretreated sugarcane bagasse was used as growth substrate under submerged fermentation.

The components of a natural medium were optimized to produce cellulase from a marine *Aspergillus niger* under solid state fermentation conditions by response surface methodology. Eichhornia crassipes and natural seawater were used as a major substrate and a source of mineral salts, respectively (Xue et al. 2012). Mineral salts of natural seawater could increase cellulase production. Raw corn cob and raw rice straw showed a significant positive effect on cellulase production. Incubation for 96 h in the natural medium increased the biomass to the maximum. The cellulase production was 17.80 U/g the dry weight of substrates after incubation for 144 h (Xue et al. 2012). The marine fungus and natural medium provide a practical method for environmentally friendly production of cellulase and utilizing straw resource.

Recently, following screening of fourteen fungi isolated from the deep-sea sponge *Stelletta normani* sampled at a depth of 751 metres, Batista-García et al. (2017) identified three halotolerant strains (*Cadophora* sp. TS2, *Emericellopsis* sp. TS11 and *Pseudogymnoascus* sp. TS12) which displayed high CMCase and xylanase activities. These three fungi displayed psychrotolerant and halotolerant growth on CMC and xylan as sole carbon sources, with optimal growth rates at 20°C. They produced CMCase and xylanase activities, which displayed optimal temperature and pH values between 50-70°C and pH 5-8 respectively, together with good thermostability and halotolerance. In solid-state fermentations, TS2, TS11 and TS12 produced CMCases, xylanases and peroxidase/phenol oxidases when grown on corn stover and wheat straw. Given the biochemical characteristics of these cell wall enzymes, it is likely that they may prove useful in future biomass conversion strategies involving cellulose materials.

Mycoremediation and Health Maintenance for Chemical-Contaminated Soils

Fungi can actively or passively absorb, collect, or immobilize inorganic metallic ions from their close environment. Some inorganic metallic ions are beneficial nutrient elements and necessary for organisms, while others such as heavy metal ions are not, and sometimes these ions are even harmful to organisms, especially when they are excessively accumulated. On the other hand, fungi, when exposed to stress environments like saline soils, have the ability of producing extracellular metabolites that serve as protective factors for their own survival. Some researches clearly demonstrated the relevance of the toxic metal ion resistance and the formation of nanoparticles. During the synthesis of metal nanoparticles by a fungus, the fungal mycelium is exposed to the metal salt solution, which prompts the fungus to produce enzymes and metabolites for its own survival. In this process, the toxic metal ions are reduced to the non-toxic metallic solid nanoparticles through the catalytic effect of the extracellular enzyme and metabolites of the fungus (Vahabi et al. 2011). It is because fungi play a crucial role in synthesis of metal nanoparticles that they are being considered as useful tools for producing nanoparticles today; in addition, more and more studies on synthesis of metal nanoparticles by fungi have been reported.

Nanoparticles are divided into intracellular and extracellular nanoparticles according to their accumulation locations. In the *Verticillium* fungus, the silver or gold nanoparticles were found to be deposited and bounded to the surface of the cytoplasmic membrane, suggesting the intracellular synthesis of gold or silver nanoparticles (Mukherjee et al. 2001, Sastry et al. 2003). From then on, the various intracellular metal nanoparticles were successively reported in the fungus *Phoma sp.* 32883 (Chen et al. 2003), *Trichothecium spp.* (Ahmad et al. 2005), *Verticillium luteoalbum* (Gericke and Pinches 2006), *Penicillium chrysogenum* (Sheikhloo and Salouti 2011), *Fusarium oxysporum* f. sp. *lycopersici* (Riddin et al. 2006), and *Aspergillus flavus* (Vala et al. 2014). As a matter of fact, there are much more researches focusing on the synthesis of extracellular nanoparticles. Extracellular nanoparticles can be divided into several sub-types on account of the capping components such as the proteins (enzymes), polypeptides, organic acids, or other metabolites secreted from the cells.

The extracellular nanoparticles have been extensively studied in various fungi. The silver nanoparticles were synthesized outside the cells of the salt tolerant yeast strain (MKY3), when MKY3 were treated with 1 mM soluble silver (Kowshik et al. 2003). Then biosynthesis of crystalline silver or gold nanoparticles was also reported by extremophilic yeasts (Mourato et al. 2011, Namasivayam et al. 2011). However, much more extracellular nanoparticles were described in filamentous fungi, for example, spherical, rod, square, pentagonal and hexagonal shape nanoparticles formed by the *Alternaria* Spices in 1 mM concentration of chloroaurate solution (Dhanasekar et al. 2015), the smaller spherical nanoparticles formed by *Aspergillus tamarii* PFL2, and the larger spherical nanoparticles formed by *Aspergillus niger* PFR6 and *Penicillium ochrochloron* PFR8 (Devi and Joshi 2015). The formation of these nanoparticles is mostly associated with heavy metal ions, and the fungal producer should have the characteristics of salt tolerance or resistance, which have been embodied in filamentous fungi such as *F. oxysporum* (Durán et al. 2005, Kumar et al. 2007a, Namasivayam et al. 2011), *F. acuminatum* (Ingle et al. 2008), *F. solani* (Ingle et al. 2009), *F. semitectum* (Basavaraja et al. 2007), *Trichoderma asperellum* (Mukherjee et al. 2008), *A. flavus* (Jain et al. 2011), *A. niger* (Gade et al. 2008), *Phoma glomerata* (Birla et al. 2009), *A. clavatus* (Verma et al. 2010), *Aspergillus* sp. (Pavani et al. 2012), *Trichoderma viride* (Fayaz et al. 2009), *Pestalotia* sp. (Raheman et al. 2011), *A. terreus* (Li et al. 2012b), *Trichophyton rubrum*, *Trichophyton mentagrophytes* and *Microsporum canisetc* (Moazeni et al. 2012), *Helminthosporium tetramera* (Shelar and Chavan 2014), and *P. gardeniae* (Rai et al. 2015).

In general, multiple stressful factors exist in saline soils and saline-alkaline soils that mostly involve heavy metal salts. Even if there is no heavy metal salt, other stressors such as sodium salt toxicity, lack of organic nutrition and drought will seriously affect the survival of organisms in saline environments, particularly in soda-affected soil, which is more serious than any other saline soils and leads to many negative effects on soil organic matter decomposition, acid-base imbalance and uptake of available nutrients (Rietz and Haynes 2003, Karlen et al. 2008). Therefore, in order to survive soil organisms must endure or resist many more stress factors beyond heavy metal ions. The synthesis of a kind of integrated-nanoparticles induced by soil structure and composition occurs, which involves not only salts or other soil components but also metabolites such as proteins, enzymes, polypeptides, organic acids, or other metabolites secreted from soil related microbial cells. In order to make it easy to express and remember, here we term MIMI nanoparticles as the microbial metabolites integrated nanoparticles.

As mentioned above, microbes are able to secrete considerable amounts of proteins, peptides, organic acids and any other organic metabolites, which might directly or indirectly affect the formation or/and substantial mass productivity of metal or non-metal nanoparticles. Both nitrate reductase and sulfite reductase are the main enzymes that directly impact the metal nanoparticles through reducing metal ions (Kumar et al. 2007b, Gholami-Shabani et al. 2014). However, in saline-alkali soils, the hydrolytic enzymes and organic acids that are beneficial to the soil health are important for the synthesis of MIMI nanoparticles.

Fungi like *A. glaucus* CCHA have the ability to secrete considerable amounts of organic acids. Furthermore, with the increase of salinity and pH value, the organic acids secreted from fungi gradually increase (Wei and Zhang 2018). Several organic acids, such as gallic acid, gluconic acid, citric acid, itaconic acid, kojic acid, and malic acid have been detected in the fermentation filtrate of *A. glaucus* CCHA while treated with 5% $NaHCO_3$ (Chen et al. 2018, unpublished data). Interestingly, molecular crosslinking can occur within gallic acids or gluconic acids under high salt and pH conditions (Ohno et al. 2001, Sanae et al. 2003). Both gallic acids and gluconic acid are efficient metal ion masking agents because the cross-linked organic acids tend to be further aggregated into nanoparticles. We are convinced that the cross-linked organic acids-assisted nanoparticles contribute to the synthesis of MIMI nanoparticles, despite the lack of direct evidence currently.

The biotic synthesized nanoparticles are metal related, and these metals mostly involve silver, gold, cadmium sulfite, copper oxide, platinum and zinc oxide. Some filamentous fungi secreted a series of enzymes under stress circumstances. Metal-associated enzymes, reducing cofactors, and organic materials have significant roles as reducing agents, which help in providing natural capping to synthesize nanoparticles, thereby preventing the aggregation of nanoparticles and helping them to remain stable for a long time, thus providing additional stability (Alani et al. 2012, Birla et al. 2009, Kumar et al. 2007c, Narayanan and Sakthivel 2010). The mechanism of extracellular production of metal nanoparticles by fungi is mostly found to involve the action of oxidoreductases. These enzymes could reduce the toxicity in the method of metal nanoparticle synthesis by reduction of the metal ions

(Gholami-Shabani et al. 2013, 2014, Huang et al. 2015a). Among the oxidoreductases, both nitrate reductase and sulfite reductase are intensively studied over the past decades. A number of researchers supported oxidoreductase for cell-free synthesis of metal nanoparticles (Kumar et al. 2007b, Gholami-Shabani et al. 2013, 2014, Prasad et al. 2014). The nitrate reductase secreted from the fungi is responsible for the reduction of metal ions and the subsequent synthesis of metal nanoparticles. When nitrate reductase is used, the color of the mixture turned reddish from white when tested with fungal filtrate, demonstrating the existence of nitrate reductase.

The beneficial effect of microbial application on saline-alkaline soil has been reported by Sahin and colleagues (2011). In the study, suspensions of three fungal isolates (*Aspergillus* spp. FS 9, 11 and *Alternaria* spp. FS 8) and two bacterial strains (*Bacillus subtilis* OSU 142 and *Bacillus megaterium* M3) at 10^4 spore/ml and 10^9 CFU/ml, respectively, were mixed with leaching water and applied to the soil columns in the Igdir plain of northeastern Turkey (Sahin et al. 2011). Gypsum is an economical alternative for replacing sodium with calcium in remediating saline-alkali soils (Gharaibeh et al. 2009, Oad et al. 2002). In the experimental process, gypsum was applied for the saline-alkali soil pretreatment, and the microorganisms are not halotolerant or halophilic (Aslantas et al. 2007, Turan et al. 2006); thus, the final results they obtained should not just be out of the function of microbes. Anyway, this study gives us an enlightened example of mycoremediation of saline-alkali soil by using haloalkaliphilic fungi. Actually, in our lab, we got a similar result in mycoremediation salt-affected soil by using salt tolerant fungal group (Fig. 2.2).

FIGURE 2.2 Mycoremediation of salt-affected soil using amendments supplemented with saline tolerant fungi (*Aspergillus glaucus* CCHA, *Aspergillus terreus*, *Trichoderma* sp. S58, and *Penicillium westlingii* (ratio = 12:9:4:4)). The area on the right received soil amendments mixed with haloalkaliphilic fungi, but the area on the left received salt-sensitive isolates. The experiment was conducted in salt-affected soil in Dajinghao Village, which is located in the Great Bend of Yellow River of Inner Mongolia, China. The properties of the saline soil before organic amendments were heavy salt soil (at a soil depth of 0-20 cm, NaCl = 9 g/kg, pH=9.3, measured by Shi and Zhu (2016). The photos were taken in 2019.

Organisms at simultaneous high salt concentration and high pH value require special adaptive mechanisms, which during the course of evolution would be both facilitative and essential for life supporting processes. Few researches focus on how haloalkaliphilic fungi cope with extremes of salt and pH value. We assume that haloalkaliphilic fungi adopt comprehensive strategies to survive the extreme environment; in other words, under saline-alkali conditions, soil fungi must possess certain mechanisms to alleviate the influence or damage of both salt and alkali. In terms of soil effects, only reducing soil soluble salt and regulating the pH value of soil solution can achieve the purpose of restoring saline-alkali soils.

1. Soil fungi have the ability to accumulate cation contents in cells. Saline soil will be improved with the accumulation of salt cation contents into fungal cells. *Hortaea werneckii*, the black yeast-like fungus isolated from hypersaline waters of salterns as their natural ecological niche, has been previously defined as halophilic fungus (Butinar et al. 2005a, b). *H. werneckii* cells were grown in liquid media at different salinities, ranging from 0 to 25% NaCl. The measurements of cation contents in cells grown at constant salt concentration have shown that the amounts of K^+ and Na^+ in *H. werneckii* were changing according to the NaCl concentration of the medium. When *H. werneckii* was grown in a medium without added NaCl, it accumulated a very low amount of Na^+, but with the increasing NaCl concentration of the medium, the amounts of the Na^+ content increased and in the end reached a higher value (Kogej et al. 2005).
2. Soil fungi produce different organic acid patterns (Scervino et al. 2010). The released organic acids allow the formation of organic mineral complexes (Richardson et al. 2001); on the other hand, with the release of organic acids, protons are produced that contribute to the acidification of the alkali soil solution.

   The saline-alkali soils and most cultivated soils are deficient in available forms of phosphorus. The release of these organic acids and other compounds in the rhizosphere by these microorganisms may be important in the solubilization of various inorganic phosphorus compounds (Scervino et al. 2010). In spite of this, based on the principle of acid-base neutralization, the organic acids also adjust the pH value of soil solution to a lower level.

   The reactions of the citric acid cycle are carried out by 8 enzymes that completely oxidize acetate, in the form of acetyl-CoA, into two molecules each of carbon dioxide and water. Organic acids citrate, iso-citrate, succinate, fumarate, malate and oxaloacetate are produced during each turn of the cycle. The high pH tolerance of *A. glaucus* has led to its utilization as an organic production strain (Barnes and Weitzman 1986). When *A. glaucus* CCHA, *A. terreus* S108, and *A. niger* S211 were cultured in an alkaline medium, key enzymes (e.g. citrate synthase, isocitrate dehydrogenase, succinyl-CoA synthetase, malate dehydrogenase) of the citric acid cycle were significantly up-regulated, suggesting that these genes contribute to the high pH tolerance of *A. glaucus* (Wei et al. 2013, Liu 2014, Zhou 2016, Wei and Zhang 2018).

Accordingly, alkali resistance might be improved in all these saline-alkali resistance fungi by overexpressing enzymes involved in the citric acid cycle. Actually, a series of the corresponding organic acids like citric acid, itaconic acid, and malic acid have been detected; in addition, both gallic acid and gluconic acid, which are associated with MIMI nanoparticles, have been investigated as well.

3. The roles played by most of the soil fungi (non-mycorrhizal fungi) at the cellular level to tolerate soil salt ions are probably similar to some of the strategies employed by ectomycorrhizal fungus, namely, binding to extracellular materials (Tam 1995, Aggangan et al. 2010, Gomes et al. 2018) or sequestration in the vacuolar compartment (Blaudez et al. 2000). We clearly observed that *A. glaucus* CCHA mycelia covered around the roots of rice like a net, but the mycelia did not penetrate and extend to rice cortex. We think this should also be an effective protection to the plants living under saline soil conditions.
4. Halophilic and alkaliphilic fungi are of biotechnological interest, as they produce extremozymes, which are useful in medical and environmental field because of their ability to remain active under the severe saline and alkaline conditions (Tiquia-Arashiro and Rodrigues 2016). The enzymes secreted by haloalkaliphilic fungi possess the bioreduction effect on salt ions of soil. This bioreduction of metal particles by certain biomasses is regarded as an organism's survival mechanism against toxic metal ions and occurs via an active or passive process or a combination of both (Ibrahim et al. 2001, Durán et al. 2005). Correlation between soil properties and soil enzymes from fungi or other microorganisms has been substantiated, and these enzyme activities are now widely used as important indicators of soil quality and soil biological activities (Rietz and Haynes 2003). As described above, high salt and pH value induce the secretion of organic acids. Similar to this case, cellulases and other so-called soil enzymes are also induced with the increasing of salt concentration and pH value. When hydrolytic enzymes are secreted into soil solution, soil properties will be improved accordingly. Take cellulases for instance, on one hand, soil cellulases can enhance the organic matters by degrading cellulose, on the other hand, cellulases in salt soils or salt solutions have been detected to form biotical nanoparticles (Riddin et al. 2006, Tiquia-Arashiro and Rodrigues 2016, Mohite et al. 2017). Recently, nanoparticles of varying size (10-300 nm) and shape (hexagons, pentagons, circles, squares, rectangles) were produced at extracellular levels by *A. glaucus* CCHA in our lab (unpublished data), further indicating that the formation of nanoparticles by haloalkaliphilic fungus is associated with saline-alkali soil remediation.

Overall, soil fungi buffer salinity and alkalinity by absorbing and/or constraining salt ions, secreting organic acids and/or macromolecule degradation enzymes, generating nanoparticles, and providing biomass; all of these effects of fungi reduce plant stress. Therefore, haloalkaliphilic fungi are excellent biological resources for soil mycoremediation.

Remarks and Prospects

Saline soil remediation by using salt-tolerant or resistant plants is one of the most effective methods because of its significant ecological, environmental and economic effects (Ilyas et al. 1993, Ghaly 2002, Nouri et al. 2017), but the method requires persistent management to produce meaningful changes in soil characteristics. Application of organic matter conditioners, which can ameliorate and increase the fertility of saline soils, is an alternative soil remediation technique (Wang et al. 2014).

Fertile soil is a vital complex that involves numerous species and immense biomass; soil organisms, particularly soil fungi, have significant effects on the soil ecosystem. The biosynthesis of fungal nanoparticles is beneficial to saline soil's remediation and health, and microbe-nanoparticles integrated technology (MiNIT) is easy, low-cost, nontoxic, and green and organic for saline soil remediation.

Soil inhabitant fungi build a metabolic bridge between insoluble organic matter and soil nutrients by producing cellulose degradation enzymes such as cellulase, as well as performing other biological processes. However, saline-alkaline soils generally lack fungi, which ordinarily play important roles in degrading insoluble organic matter such as crop straw into soluble and easily absorbed nutrients; therefore, applying organic matter supplemented with fermentation fungi to saline-alkali soil is a feasible strategy for soil remediation.

Haloalkaliphilic fungi are excellent biological candidates for soil mycoremediation, but to date very few species with both abilities to produce effective soil enzymes and to grow in saline-alkaline environments have been reported. Our lab currently isolated a series of halotolerant/philic fungi. Microbe isolation and identification are time-consuming and tedious tasks, and their applications require more laborious field experiments. Based on the relationship between the remediation effect of saline alkali soil and the production capacity of nanoparticles, and considering the fact that nitrate reductase and sulfite reductase are main enzymes for synthesizing nanoparticles, we optimized a simple and quick method for identifying saline remediation fungi. Generally, with the increase of NaCl concentration, the expression levels of the nitrate reductase and sulfite reductase genes are upregulated, and which is positively correlated with the yield of nanoparticles. Through designing degenerate primers of both genes and real-time PCR method, we can obtain the target fungal isolates.

On the other hand, to get better remediation effect, natural soil fungi require to be genetically modified at their degradation ability or saline-alkali resistance. Generally, several enzymes involved in salt and/or alkali resistance, such as the alkaline-stable endoglucanases B from *Aspergillus niger* BCRC31494 (Li et al. 2012a), the alkaline xylanase from *Aspergillus nidulans* KK-99 (Taneja et al. 2002), and the bilirubin oxidase from *Myrothecium sp.* IMER1 (Zhang et al. 2007), are highly abundant in fungi found in saline-alkali soil, but such fungi usually have a relatively low capacity for cellulose degradation, whereas fungi found in fertile soil show opposite characteristics. Thus, genetically modified strategy to create novel haloalkaliphilic fungi with high cellulase activity is a good choice. Fungi endowed with high resistance to saline-alkali environments or other beneficial genes would be promising candidates for saline-alkali soil remediation.

A series of salt and/or alkali resistance (or tolerance) genes have been characterized to provide a list of candidate genes to be applied in efforts to genetically improve soil fungi (Fang et al. 2014, Zhang 2016). In order to enhance the cellulose degradation ability of haloalkaliphilic fungi, additional cellulases with salt and alkali stability must be identified. Using cellulases with salt and alkali tolerance, two strategies can be employed to obtain saline/alkaline resistant fungi with enhanced enzyme secretion. Indeed, natural strains remain the first choice for soil remediation; therefore, isolating and screening suitable strains from extreme natural environments is still an important long-term task.

Haloalkaliphilic fungus *Aspergillus glaucus* CCHA, a fungal species with extreme tolerance to saline and alkaline conditions, has significant potential value in industrial and agricultural applications. Our group has been assessing the potential of *Aspergillus glaucus* CCHA in the mycoremediation of saline-alkaline soil in the Songnen plain of northeastern China (one of the three most famous saline and alkaline lands in the world) for three years (Shi and Zhu 2016). This study primarily indicates that the applied amendments mixed with haloalkaliphilic fungi significantly encourage steady growth and yield of rice in comparison to that achieved in the control plot.

ACKNOWLEDGEMENTS

The related work in our lab was partially supported by grants from the National Natural Science Foundation of China (grant nos. 31671972 and 31670141) and a project of the Ministry of Science and Technology of China (grant no. 2016YFD0300703). The authors would like to thank the Zhang Lab members, who provided the photographic pictures taken in their spare time. The authors are also grateful to former labmates Dr. Zheng-Qun LI, Dr. Yang SHI and Mr. Sen-Lin ZHANG, who contributed to fungal isolation and field trials, as well as to collaborators Ms. Ze-Juan Feng and Mr. Run-Zhi TAO who provided encouragement and assistance in promoting our scientific and technological achievements regarding saline-alkali soil mycoremediation using haloalkaliphilic fungi.

REFERENCES

Aggangan, N.S., H.K. Moon and S.H. Han. 2010. Growth response of Acacia mangium Willd. seedlings to arbuscular mycorrhizal fungi and four isolates of the ectomycorrhizal fungus *Pisolithus tinctorius* (Pers.) Coker and Couch. New Forests 39: 215-230.

Ahmadm, A., S. Senapati, M.I. Khan, R. Kumar and M. Sastry. 2005. Extra-/intracellular biosynthesis of gold nanoparticles by an alkalotolerant fungus *Trichothecium* sp. J. Biomed. Nanotechnol. 1: 47-53.

Alani, F., M. Moo-Young and W. Anderson. 2012. Biosynthesis of silver nanoparticles by a new strain of *Streptomyces* sp. compared with *Aspergillus fumigatus*. World J. Microbiol. Biotechnol. 28: 1081-1086.

Antranikian, G., C.E. Vorgias and C. Bertoldo. 2005. Extreme environments as a resource for fungi and novel biocatalysts. Adv. Biochem. Eng. Biotech. 96: 219-262

Arakaki R., D. Monteiro, R. Boscolo and E. Gomes. 2013. Halotolerance, ligninase production and herbicide degradation ability of basidiomicetes strains. Braz. J. Microbiol. 44: 1207-1214.

Aslantas, R., R. Cakmakci and F. Sahin. 2007. Effect of plant growth promoting rhizobacteria on young apple tree growth and fruit yield under orchard conditions. Sc. Hortic. 111: 371-377.

Barnes, S.J. and P.D. Weitzman. 1986. Organization of citric acid cycle enzymes into a multienzyme cluster. FEBS Lett. 201: 267-270.

Basavaraja, S., S.D. Balaji, A. Lagashetty, A.H. Rajasab and A. Venkataraman. 2007. Extracellular biosynthesis of silver nanoparticles using the fungus *Fusarium semitectum*. Mater. Res. Bull. 43: 1164-1170.

Bass, D., A. Howe, N. Brown, H. Barton, M. Demidova, H. Michelle, et al. 2007. Yeast forms dominate fungal diversity in the deep oceans. Proc. R. Soc. Lond. B. Biol. Sci. 274: 3069-3077.

Batista-García, R.A., E. Balcázar-López, E. Miranda-Miranda, A. Sánchez-Reyes, L. Cuervo-Soto, D. Aceves-Zamudio, et al. 2014. Characterization of lignocellulolytic activities from a moderate halophile strain of *Aspergillus caesiellus* isolated from a sugarcane bagasse fermentation. PLoS One 9(8): e105893. http://doi.org/10.1371/journal.pone.0105893.

Batista-García, R.A., T. Sutton, S.A. Jackson, O.E. Tovar-Herrera, E. Balcázar-López, M.D. Sánchez-Carbente, et al. 2017. Characterization of lignocellulolytic activities from fungi isolated from the deep-sea sponge *Stelletta normani*. PLoS One 12(3): e0173750. doi: 10.1371/journal.pone.0173750.

Béguin, P. and J.P. Aubert. 1994. The biological degradation of cellulose. FEMS Microbiol. Rev. 13: 25-58.

Birla, S.S., V.V. Tiwari, A.K. Gade, A.P. Ingle, A.P. Yadav and M.K. Rai. 2009. Fabrication of silver nanoparticles by *Phoma glomerata* and its combined effect against *Escherchia coli, Pseudomonas aeruginosa* and *Staphylococcus aureus*. Lett. App. Microbiol. 48: 173-179.

Blaudez, D., B. Botton and M. Chalot. 2000. Cadmium uptake and subcellular compartmentation in the ectomycorrhizal fungus *Paxillus involutus*. Microbiology 146: 1109-1117.

Booth, T. and N. Kenkel. 1986. Ecological studies of lignicolous marine fungi: A distribution model based on ordination and classification. pp. 297-310. *In*: S.T. Moss. (ed.). The Biology of Marine Fungi. Cambridge University Press, Cambridge, London, UK.

Bovio, E., L. Garzoli, A. Poli, A. Luganini, P. Villa, R. Musumeci, et al. 2019. Marine Fungi from the Sponge *Grantia compressa*: Biodiversity, chemodiversity and biotechnological potential. Mar. Drugs. 17(4). pii: E220. doi: 10.3390/md17040220.

Buchalo, A.S., E. Nevo, S.P. Wasser, A. Oren and H.P. Molitoris. 1998. Fungal life in the extremely hypersaline water of the Dead Sea: First records. Proc. R. Soc. London B. 265: 1461-1465.

Buchalo, A.S., E. Nevo, S.P. Wasser, A. Oren, H.P. Molitoris and P.A. Volz. 2000. Fungi discovered in the Dead Sea. Mycol. Res. News 104: 132-133.

Burnham, A.D. and A.J. Berry. 2017. Formation of Hadean granites by melting of igneous crust. Nat. Geosci. 10(6): 457-461.

Burton, S.G., D.A. Cowan and J.M. Woodley. 2002. The search for the ideal biocatalyst. Nat. Biotechnol. 20(1): 37-45. https://doi.org/10.1038/nbt0102-37.

Butinar, L., S. Sonjak, P. Zalar, A. Plemenitasˇ and N. Gunde-Cimerman. 2005a. Melanized halophilic fungi are eukaryotic members of microbial communities in hypersaline waters of solar salterns. Botanica Marina 48: 73-79.

Butinar, L., P. Zalar, J.C. Frisvad and N. Gunde–Cimerman. 2005b. The genus Eurotium – members of indigenous fungal community in hypersaline waters of salterns. FEMS Microbiol. Ecol. 51: 155-166.

Cantrell, S.A., L. Casillas-Martınez and M. Molina. 2006. Characterization of fungi from hypersaline environments of solar salterns using morphological and molecular techniques. Mycol. Res. 110: 962-970.

Castillo, G. and V. Demoulin. 1997. NaCl salinity and temperature effects on growth of three wood-rotting basidiomycetes from a Papua New Guinea coastal forest. Mycol. Res. 101: 341-344.

Chamekh, R., F. Deniel, C. Donot, J.L. Jany, P. Nodet and L. Belabid. 2019. Isolation, identification and enzymatic activity of halotolerant and halophilic fungi from the great sebkha of oran in northwestern of Algeria. Mycobiology 47(2): 230-241. doi: 10.1080/12298093.2019.1623979.

Chen, H.Y., D.S. Xue, X.Y. Feng and S.J. Yao. 2011. Screening and production of ligninolytic enzyme by a marine-derived fungal *Pestalotiopsis* sp. J63. Appl. Biochem. Biotechnol. 165(7-8):1754-1769. doi: 10.1007/s12010-011-9392-y.

Chen, J.C., Z.H. Lin and X.X. Ma. 2003. Evidence of the production of silver nanoparticles via pretreatment of *Phoma* sp. 32883 with silver nitrate. Lett. App. Microbiol. 37: 105-108.

Chen, S., J. Wang, Z. Wang, X. Lin, B. Zhao, K. Kaliaperumal, et al. 2017. Structurally diverse secondary metabolites from a deep-sea-derived fungus *Penicillium chrysogenum* SCSIO 41001 and their biological evaluation. Fitoterapia 117: 71-78. doi: 10.1016/j.fitote.2017.01.005.

Deng, J.J., W.Q. Huang, Z.W. Li, D.L. Lu, Y. Zhang and X.C. Luo. 2018. Biocontrol activity of recombinant aspartic protease from *Trichoderma harzianum* against pathogenic fungi. Enzyme Microb. Technol. 112: 35-42. doi: 10.1016/j.enzmictec.2018.02.002.

Devi, L.S. and S.R. Joshi. 2015. Ultrastructures of silver nanoparticles biosynthesized using endophytic fungi. J. Microsco. Ultrastruc. 3: 29-37.

Dhanasekar, N.N., G. Rahul, K.B. Narayanan, G. Raman and N. Sakthivel. 2015. Green chemistry approach for the synthesis of gold nanoparticles using the fungus *Alternaria* sp. J. Microbiol. Biotechnol. 25(7):1129-1135. doi:10.4014/jmb.1410.10036.

Dionisi, H.M., M. Lozada and N.L. Olivera. 2012. Bioprospection of marine fungi: biotechnological applications and methods. Rev. Argent. Microbiol. 44: 46-90.

Durán, N., P.D. Marcato, O.L. Alves, G.I. De Souza and E. Esposito. 2005. Mechanistic aspects of biosynthesis of silver nanoparticles by several *Fusarium oxysporum* strains. J. Nanobiotechnology 3(1): 8. doi: 10.1186/1477-3155-3-8.

Edgcomb, V.P., D. Beaudoin, R. Gast, J.F. Biddle and A. Teske. 2011. Marine subsurface eukaryotes: the fungal majority. Environ. Microbiol. 13(1): 172-183. doi: 10.1111/j.1462-2920.2010.02318.x.

Elmeleigy, M.A., E.N. Hoseiny, S.A. Ahmed and A.M. Alhoseiny. 2010. Isolation, identification, morphogenesis and ultrastructure of obligate halophilic fungi. J. Appl. Sci. Environ. Sanit. 5: 201-202.

Evans, S., R.W. Hansen and M.A. Schneegurt. 2013. Isolation and characterization of halotolerant soil fungi from the great salt plains of Oklahoma. Cryptogam Mycol. 34: 329-341. doi: 10.7872/crym.v34.iss4.2013.329

Fang, J., X. Han, L. Xie, M. Liu, G. Qiao, J. Jiang, et al. 2014. Isolation of salt stress-related genes from *Aspergillus glaucus* CCHA by random overexpression in *Escherichia coli*. Sci. World J. 39: 620959.

Farha, A.K. and A.M. Hatha. 2019. Bioprospecting potential and secondary metabolite profile of a novel sediment-derived fungus *Penicillium* sp. ArCSPf from continental slope of Eastern Arabian Sea. Mycology 10(2):109-117. doi: 10.1080/21501203.2019.1572034.

Fayaz, A.M., K. Balaji, M. Girilal, R. Yadav, P.T. Kalaichelvan and R. Venketesan. 2009. Biogenic synthesis of silver nanoparticles and their synergistic effect with antibiotics: A study against gram-positive and gram-negative bacteria. Nanomedicine 6: 103-109.

Ferrer, M., O.V. Golyshina, A. Beloqui and P.N. Golyshin. 2012. Mining enzymes from extreme environments. Curr. Opin. Microbiol. 10: 207-214.

Fischer, G., W. Winiwarter, T. Ermolieva, G.Y. Cao, H. Qui, Z. Klimont, et al. 2010. Integrated modeling framework for assessment and mitigation of nitrogen pollution from agriculture: Concept and case study for China. Agric. Ecosyst. Environ. 136(1-2): 116-124.

Gadd, G.M. 2007. Geomycology: biogeochemical transformations of rocks, minerals, metals and radionuclides by fungi, bioweathering and bioremediation. Mycol. Res. 111: 3-49.

Gade, A.K., P. Bonde, A.P. Ingle, P.D. Marcato, N. Durán and M.K. Rai. 2008. Exploitation of *Aspergillus niger* for synthesis of silver nanoparticles. J. Biobased Mater. Bioener. 2: 243-247.

Gavrilescua, M. and Y. Chisti. 2005. Biotechnology: A sustainable alternative for chemical industry. Biotechnol. Adv. 23(7-8): 471-499. https://doi. org/10.1016/j. biotechadv.2005.03.004.

Gericke, M. and A. Pinches. 2006. Microbial production of gold nanoparticles. Gold Bull. 39: 22-28.

Ghaly, F.M. 2002. Role of natural vegetation in improving salt affected soil in northern Egypt. Soil Till. Res. 64: 173-178.

Gharaibeh, M.A., N.I. Eltaif and O.F. Shunnar. 2009. Leaching and reclamation of calcareous saline-sodic soil by moderately saline and moderate-SAR water using gypsum and calcium chloride. J. Plant Nutr. Soil Sci. 172: 713-719.

Gholami-Shabani, M., A. Akbarzadeh, M. Mortazavi and M.K. Emzadeh. 2013. Evaluation of the antibacterial properties of silver nanoparticles synthesized with *Fusarium oxysporum* and *Escherichia coli*. Int. J. Life Sci. Biotechnol. Pharma Res. 2: 333-348.

Gholami-Shabani, M., A. Akbarzadeh, D. Norouzian, A. Amini, Z. Gholami-Shabani, A. Imani, et al. 2014. Antimicrobial activity and physical characterization of silver nanoparticles green synthesized using nitrate reductase from *Fusarium oxysporum*. Appl. Biochem. Biotechnol. 172: 4084-4098. doi: 10.1007/s12010-014-0809-2.

Gomes, D.N.F., M.A.Q. Cavalcanti, M.J.S. Fernandes, D.M.M. Lima and J.Z.O. Passavante. 2008. Filamentous fungi isolated from sand and water of "Bairro Novo" and "Casa Caiada" beaches, Olinda, Pernambuco, Brazil. Braz. J. Biol. 68: 577-582. doi: 10.1590/ s1519-69842008000300016.

Gomes, E.C.Q., V.M. Godinho, D.A.S. Silva, M.T.R. de Paula, G.A. Vitoreli, C.L. Zani, et al. 2018. Cultivable fungi present in Antarctic soils: Taxonomy, phylogeny, diversity and bioprospecting of antiparasitic and herbicidal metabolites. Extremophiles https://doi.o rg/10.1007/s00792-018-1003-1.

Gonçalves, V.N., G.A. Vitoreli, G.C.A. de Menezes, C.R.B. Mendes, E.R. Secchi, C.A. Rosa, et al. 2017. Taxonomy, phylogeny and ecology of cultivable fungi present in seawater gradients across the Northern Antarctica Peninsula. Extremophiles 21: 1005. https://doi.org/10.1007/s00792-017-0959-6.

Gostinčar, C., M. Grube, S. De Hoog, P. Zalar and N. Gunde–Cimerman. 2010. Extremotolerance in fungi: Evolution on the edge. FEMS Microbiol. Ecol. 71: 2-11.

Gostinčar, C. and M. Turk. 2012. Extremotolerant fungi as genetic resources for biotechnology. Bioengineered 3: 293-297

Grishkan, I., E. Nevo and S.P. Wasser. 2003. Soil micromycete diversity in the hypersaline Dead Sea coastal area, Israel. Mycol. Progress 2: 19-28.

Gruber, S. and V. Seidl-Seiboth. 2012. Self versus non-self: fungal cell wall degradation in *Trichoderma*. Microbiology 158: 26-34.

Grum-Grzhimaylo, A.A., M.L. Georgieva, S.A. Bondarenko, A.J.M. Debets and E.N. Bilanenko. 2016. On the diversity of fungi from soda soils. Fungal Divers. 76: 27-74.

Gunde-Cimerman, N., P. Zalar, G.S. de Hoog and A. Plemenitaš. 2000. Hypersaline waters in salterns: natural ecological niches for halophilic black yeasts. FEMS Microbiol. Ecol. 32: 235-240.

Gunde-Cimerman, N. and P. Zalar. 2001. Extremely halotolerant and halophilic fungi inhabit brine in solar salterns around the globe. Food Technol. Biotechnol. 4: 170-179.

Gunde-Cimerman, N. and P. Zalar. 2014. Extremely halotolerant and halophilic fungi inhabit brine in solar salterns around the globe. Food Technol. Biotechnol. 52: 170-179.

Haki, G.D. and S.K. Rakshit. 2003. Developments in industrially important thermostable enzymes: A review. Bioresour. Technol. 89: 17-34.

Häkkinen, M., M.J. Valkonen, A. Westerholm-Parvinen, N. Aro, M. Arvas and M. Vitikainen. 2014. Screening of candidate regulators for cellulase and hemicellulose production in *Trichoderma reesei* and identification of a factor essential for cellulase production. Biotechnol. Biofuels 7(14): 1-21.

Hamid, R., M.A. Khan, M. Ahmad, M.M. Ahmad, M.Z. Abdin, J. Musarrat, et al. 2013. Chitinases: An update. J. Pharm. Bioallied. Sci. 5: 21-29.

Horodyski, R.J. and P.L. Knauth. 1994. Life on land in the Precambrian. Science 263(5146): 494-498.

Huang, J., L. Lin, D. Sun, H. Chen, D. Yang and Q. Li. 2015a. Bio-inspired synthesis of metal nanoparticles and application. Chem. Soc. Rev. 44: 6330-6374.

Huang, Y., X. Qin, X.M. Luo, Q. Nong, Q. Yang, Z. Zhang, et al. 2015b. Efficient enzymatic hydrolysis and simultaneous saccharification and fermentation of sugarcane bagasse pulp for ethanol production by cellulase from *Penicillium oxalicum* EU2106 and thermotolerant *Saccharomyces cerevisiae* ZM1-5. Biomass Bioenergy 77: 53-63. http://dx.doi.org/10.1016/j.biombioe.2015.03.020.

Ibrahim, Z., A. Ahmad and B. Baba. 2001. Bioaccumulation of silver and the isolation of metal-binding protein from Pseudomonas diminuta. Brazil Arch. Biol. Tech. 44: 223-225.

Ilyas, M., R.W. Miller and R.H. Qureshi. 1993. Hydraulic conductivity of saline-sodic soil after gypsum application and cropping. Soil Sci. Soc. Am. J. 57: 1580-1585.

Ingle, A., A. Gade, S. Pierrat, C. Sonnichsen and M. Rai. 2008. Mycosynthesis of silver nanoparticles using the fungus Fusarium acuminatum and its activity against some human pathogenic bacteria. Curr. Nanosci. 4: 141-144.

Ingle, A., M.K. Rai, A. Gade and M. Bawaskar. 2009. Fusarium solani: a novel biological agent for the extracellular synthesis of silver nanoparticles. J. Nanopart. Res. 11: 2079-2085.

Intriago, P. 2012. Marine microorganisms: Perspectives for getting involved in cellulosic ethanol. AMB Express 2: 46. doi: 10.1186/2191-0855-2-46.

Jain, N., A. Bhargava, S. Majumdar, J.C. Tarafdar and J. Panwar. 2011. Extracellular biosynthesis and characterization of silver nanoparticles using *Aspergillus flavus* NJP08: A mechanism perspective. Nanoscale 3: 635-641.

Jin, S. and F. Zhou. 2018. Zero growth of chemical fertilizer and pesticide use: China's objectives, progress and challenges. J. Resour. Ecol. 9(1): 50-58. doi: 10.5814/j.issn.1674-764x.2018.01.006.

Jones, E.B.G. 2000. Marine fungi: Some factors influencing biodiversity. Fungal Divers. 4: 53-73.

Karlen, D.L., S.S. Andrews, B.J. Wienhold and T.M. Zobeck. 2008. Soil quality assessment: Past, present and future. J. Integr. Biosci. 6: 3-14.

Kis-Papo, T., I. Grishkan, A. Oren, S.P. Wasser and E. Nevo. 2003. Survival of filamentous fungi in hypersaline Dead Sea water. Microb. Ecol. 45: 183-190.

Kis-Papo, T., A.R. Weig, R. Riley, D. Peršoh, A. Salamov, H. Sun, et al. 2014. Genomic adaptations of the halophilic Dead Sea filamentous fungus *Eurotium rubrum*. Nat. Commun. 5. doi: 10.1038/ncomms4745.

Kogej, T., J. Ramos, A. Plemenitas and N. Gunde-Cimerma. 2005. The halophilic fungus *Hortaea werneckii* and the halotolerant fungus *Aureobasidium pullulans* maintain low intracellular cation concentrations in hypersaline environments. Appl. Environ. Microbiol. 71: 6600-6605.

Kohlmeyer, J. and E. Kohlmeyer. 1980. Marine mycology, the higher fungi. Marine Ecology 1(1): 103-104.

Kowshik, M., S. Ashtaputre, S. Kharrazi, W. Vogel, J. Urban, S.K. Kulkarni, et al. 2003. Extracellular synthesis of silver nanoparticles by a silver-tolerant yeast strain MKY3. Nanotechnology 14: 95-100.

Krause, A., J. Stoye and M. Vingron. 2000. The SYSTERS protein sequence cluster set. Nucleic Acids Rcs. 28: 270-272.

Kristensen, J.B., R.L. Meyer, B.S. Laursen, S. Shipovskov, F. Besenbacher and C.H. Poulsen. 2008. Antifouling enzymes and the biochemistry of marine settlement. Biotechnol. Adv. 26: 471-481. doi: 10.1016/j.biotechadv.2008.05.005.

Kubicek, C.P., A. Herrera-Estrella, V. Seidl-Seiboth, D.A. Martinez, I.S. Druzhinina, M. Thon, et al. 2011. Comparative genome sequence analysis underscores mycoparasitism as the ancestral life style of *Trichoderma*. Genome Biol. 12: R40.

Kumar, A.S., A.A. Ansari, A. Ahmad and M.I. Khan. 2007a. Extracellular biosynthesis of CdS quantum dots by the fungus Fusarium oxysporum. J. Biomed. Nanotechnol. 3: 190-194.

Kumar, S.A., M.K. Abyaneh, S.W. Gosavi, S.K. Kulkarni, A. Ahmad and M.I. Khan. 2007b. Sulfite reductase mediated synthesis of gold nanoparticles capped with phytochelatin. Biotechnol. Appl. Biochem. 47: 191-195. doi:10.1042/BA20060205.

Kumar, S.A., M.K. Abyaneh, S.W. Gosavi, S.K. Kulkarni, R. Pasricha, A. Ahmad, et al. 2007c. Nitrate reductase-mediated synthesis of silver nanoparticles from $AgNO_3$. Biotechnol. Lett. 29: 439-445.

Li, C.H., H.R. Wang and T.R. Yan. 2012a. Cloning, purification and characterization of a heat- and alkaline-stable endoglucanase B from *Aspergillus niger* BCRC31494. Molecules 17: 9774-9789.

Li, G., D. He, Y. Qian, B. Guan, S. Gao, Y. Cui, et al. 2012b. Fungus mediated green synthesis of silver nanoparticles using *Aspergillus terreus*. Int. J. Mol. Sci. 13: 466-476.

Li, Q. and G. Wang. 2009. Diversity of fungal isolates from three Hawaiian marine sponges. Microbiol. Res. 164(2): 233-241.

Li, Z.Q., X. Pei, Z.Y. Zhang, Y. Wei, Y.Y. Song, L.N. Chen, et al. 2018. The unique GH5 cellulase member in the extreme halotolerant fungus *Aspergillus glaucus* CCHA is an endoglucanase with multiple tolerances to salt, alkali and heat: prospects for straw degradation applications. Extremophiles https://doi.org/10.1007/s00792-018-1028-5.

Liu, H., F.Z.R. de Souza, L. Liu and B.S. Chen. 2018. The use of marine-derived fungi for preparation of enantiomerically pure alcohols. Appl. Microbiol. Biotechnol. 102(3): 1317-1330. doi: 10.1007/s00253-017-8707-5.

Liu, X.D., J.L. Liu, Y. Wei, Y.P. Tian, F.F. Fan, H.Y. Pan, et al. 2011. Isolation, identification and biologic characteristics of an extreme halotolerant *Aspergillus* sp. J Jilin Univ (In Chinese; abstract in English.) 49: 548-552.

Liu, X.D. 2014. Two Stress Tolerance Genes in Halophilic *Aspergillus*: Functional Analysis and Their Application. Ph.D. Thesis, Jilin University, Changchun, China (in Chinese) http://kns.cnki.net/KCMS/detail/detail.aspx?dbcode=CDFD&dbname=CDFDTEMP&filename=1015507655.nh.

Liu, X.D., L. Xie, Y. Wei, X.Y. Zhou, B. Jia, J. Liu, et al. 2014. Abiotic stress resistance, a novel moonlighting function of ribosomal protein RPL44 in the halophilic fungus *Aspergillus glaucus*. Appl. Environ. Microbiol. 80: 4294-4300.

Liu, X.D., Y. Wei, X.Y. Zhou, X. Pei and S.H. Zhang. 2015. *Aspergillus glaucus* aquaglyceroporin gene glpF confers high osmosis tolerance in heterologous organisms. Appl. Environ. Microbiol. 81: 6926-6937.

Liu, Z., R. Weis and A. Gliede. 2004. Enzymes from higher eukaryotes for industrial biocatalysis. Food Technol. Biotechnol. 42: 237-249.

Madhu, K.M., P.S. Beena and M. Chandrasekaran. 2009. Extracellular beta-glucosidase production by a marine *Aspergillus sydowii* BTMFS 55 under solid state fermentation using statistical experimental design. Biotechnol. Bioprocess Eng. 14: 457-466. doi: 10.1007/s12257-008-0116-2.

Mahdy, H.M., H.H. el-Sheikh, M.S. Ahmed and B.M. Refaat. 1996. Physiological and biochemical changes induced by osmolarity in *halotolerant aspergilli*. Acta Microbiol. Pol. 45: 55-65.

Mandeel, Q.A. 2006. Biodiversity of the genus Fusarium in saline soil habitats. J. Basic Microb. 46: 480-494.

Mandujano Gonzalez. V., L. Villa-Tanaca, M.A. Anducho-Reyes and Y. Mercado-Flores. 2016. Secreted fungal aspartic proteases: A review. Rev. Iberoam. Micol. 33: 76-82.

Pasqualetti, M., P. Barghini, V. Giovannini and M. Fenice. 2019. High production of chitinolytic activity in halophilic conditions by a new marine strain of *Clonostachys rosea*. Molecules 24(10): 1880.

Méjanelle, L., J.F. Lòpez, N. Gunde-Cimerman and J.O. Grimalt. 2000. Sterols of melanized fungi from hypersaline environments. Org. Geochem. 31(10): 1031-1040.

Mernitz, G., A. Koch, B. Henrissat and G. Schulz. 1996. Endoglucanase II(EGII) of *Penicillium janthinellum*: cDNA sequence, heterologous expression and promoter analysis. Curr. Genet. 29: 490-495.

Moazeni, M., N. Rashidi, A.R. Shahverdi, F. Noorbakhsh and S. Rezaie. 2011. Extracellular production of silver nanoparticles by using three common species of dermatophytes: *Trichophyton rubrum*, *Trichophyton mentagrophytes* and *Microsporum canis*. Iran. Biomed. J. 16(1): 332-333.

Mohite, P., A.R. Kumar and S. Zinjarde. 2017. Relationship between salt tolerance and nanoparticle synthesis by *Williopsis saturnus* NCIM 3298. World J. Microbiol. Biotechnol. 33: 163. doi: 10.1007/s11274-017-2329-z.

Moubasher, A., S. Abdel–Hafez, M. Bagy and M. Abdel–Satar. 1990. Halophilic and halotolerant fungi in cultivated desert and salt marsh soils from Egypt. Acta Mycologica. 26: 65-81.

Moubasher, A.H., M.A. Abdel-Sater and Z. Soliman. 2018. Diversity of yeasts and filamentous fungi in mud from hypersaline and freshwater bodies in Egypt. Czech Mycol. 70: 1-32.

Mourato, A., M. Gadanho, A.R. Lino and R. Tenreiro. 2011. Biosynthesis of crystalline silver and gold nanoparticles by extremophilic yeasts. Bioinorg. Chem. Appl. 2011: 546074. doi:10.1155/2011/546074.

Mukherjee, P., A. Ahmad, D. Mandal, S. Senapati, S.R. Sainkar, M.I. Khan, et al. 2001. Bioreduction of $AuCl_4$-ions by the fungus *Verticillium* sp. and surface trapping of the gold nanoparticles formed. Angew. Chem. Int. Ed. 40: 3585-3588.

Mukherjee, P., M. Roy, B.P. Mandal, G.K. Dey, P.K. Mukherjee, J. Ghatak, et al. 2008. Green synthesis of highly stabilized nanocrystalline silver particles by a nonpathogenic and agriculturally important fungus *T. asperellum*. Nanotechnology 19: 103-110.

Mulder, J.L., M.A. Ghannoum, L. Khamis and K.A. Elteen. 1989. Growth and lipid composition of some dematiaceous hyphomycete fungi grown at different salinities. Microbiology 135: 3393-3404.

Mulder, J.L. and H. El-Hendawy. 1999. Microfungi under stress in Kuwait's coastal saline depressions. Kuwait J. Sci. Eng. 26: 157-172.

Nakagiri, A. 2012. Culture collections and maintenance of marine fungi. pp. 501-508. *In*: E.B.G. Jones, K. Pang. (eds.). Marine Fungi and Fungal-Like Organisms. De Gruyter, Boston, USA.

Namasivayam, S.K.R., S. Ganesh and S. Avimanyu. 2011. Evaluation of anti-bacterial activity of silver nanoparticles synthesized from Candida glabrata and *Fusarium oxysporum*. Int. J. Med. Microbiol. Res. 1: 130-136.

Narayanan, K.B. and N. Sakthivel. 2010. Biological synthesis of metal nanoparticles by microbes. Adv. Colloid Interface Sci. 156: 1-13.

Nayak, S.S., V. Gonsalves and S.W. Nazareth. 2012. Isolation and salt tolerance of halophilic fungi from mangroves and solar salterns in Goa-India. Indian J. Mar. Sci. 41: 164-172.

Nazareth, S., V. Gonsalves and S. Nayak. 2012. A first record of obligate halophilic aspergilli from the Dead Sea. Indian J. Microbiol. 52: 22-27.

Nazareth, S. and V. Gonsalves. 2014. *Aspergillus penicillioides*: A true halophile existing in hypersaline and polyhaline econiches. Ann. Microbiol. 64: 397-402.

Nouri, H., C.S. Borujeni, R. Nirola, A. Hassanli, S. Beecham, S. Alaghmand, et al. 2017. Application of green remediation on soil salinity treatment: A review on halophytoremediation. Process Saf. Environ. Prot. 107: 94-107.

Oad, F.C., M.A. Samo, A. Soomro, D.L. Oad, N.L. Oad and A.G. Siyal. 2002. Amelioration of salt affected soils. Pakistan J. Appl. Sci. 2: 1-9.

Ohno, T., M. Inoue and Y. Ogihara. 2001. Cytotoxic activity of gallic acid against liver metastasis of mastocytoma cells P-815. Anticancer Res. 21(6A): 3875-3880.

Oren, A. 2002. Halophilic microorganisms and their environments. pp. 207-231. *In*: J. Seckbach. (ed.). Cellular Origin and Life in Extreme Habitats and Astrobiology. Kluwer Academic Publishers, Dordrecht, Boston, London.

Pachiadaki, M.G., V. Rédou, D.J. Beaudoin, G. Burgaud and V.P. Edgcomb. 2016. Fungal and prokaryotic activities in the marine subsurface biosphere at Peru margin and canterbury basin inferred from RNA-based analyses and microscopy. Front Microbiol. 7: 846. doi: 10.3389/fmicb.2016.00846.

Pang, K., R.K.K. Chow, C. Chan and L.L.P. Vrijmoed. 2011. Diversity and physiology of marine lignicolous fungi in arctic waters: A preliminary account. Polar Res. 30: 1-5. doi: 10.3402/polar.v30i0.5859.

Pasqualetti, M., P. Barghini, V. Giovannini and M. Fenice. 2019. High production of chitinolytic activity in halophilic conditions by a new marine strain of *Clonostachys rosea*. Molecules 24: 1880. https://doi.org/10.3390/molecules24101880.

Passarini, M.R.Z., M.V.N. Rodrigues, M. Da Silva and L.D. Sette. 2011. Marine-derived filamentous fungi and their potential application for polycyclic aromatic hydrocarbon bioremediation. Mar. Pollut. Bull. 62: 364-370. doi: 10.1016/j.marpolbul.2010.10.003.

Pavani, K.V., N.S. Kumar and B.B. Sangameswaran. 2012. Synthesis of lead nanoparticles by *Aspergillus* species. Pol. J. Microbiol. 61: 61-63.

Peng, X.P., Y. Wang, P.P. Liu, K. Hong, H. Chen, X. Yin, et al. 2011a. Aromatic compounds from the halotolerant fungal strain of *Wallemia sebi* PXP-89 in a hypersaline medium. Arch. Pharm. Res. 34: 907-912.

Peng, X.P., Y. Wang, K. Sun, P.P. Liu, X. Yin and W.M. Zhu. 2011b. Cerebrosides and 2-pyridone alkaloids from the halotolerant fungus *Penicillium chrysogenum* grown in a hypersaline medium. J. Nat. Prod. 74: 1298-1302.

Piñar, G., D. Dalnodar, C. Voitl, H. Reschreiter and K. Sterflinger. 2016. Biodeterioration risk threatens the 3100 year old staircase of hallstatt (Austria): Possible involvement of halophilic microorganisms. PLoS One 11(2): e0148279. doi: 10.1371/journal.pone.0148279.

Prasad, R., V. Kumar and K.S. Prasad. 2014. Nanotechnology in sustainable agriculture: Present concerns and future aspects. Afr. J. Biotechnol. 13(6): 705-713.

Prasad, R., R. Pandey and I. Barman. 2016. Engineering tailored nanoparticles with microbes: Quo vadis. WIREs Nanomed. Nanobiotechnol. 8: 316-330. doi: 10.1002/wnan.1363.

Qin, Y., X. Wei, X. Song and Y. Qu. 2008. Engineering endoglucanase II from *Trichoderma reesei* to improve the catalytic efficiency at a higher pH optimum. J. Biotechnol. 135: 190-195.

Raheman, F., S. Deshmukh, A. Ingle, A. Gade and M. Rai. 2011. Silver nanoparticles: Novel antimicrobial agent synthesized from an endophytic fungus *Pestalotia* sp. isolated from leaves of *Syzygium cumini* (L). Nano Biomed. Eng. 3: 174-178.

Rai, M., C. Ribeiro, L. Mattoso and N. Duran. 2015a. Nanotechnologies in food and agriculture. Springer-Verlag, Germany, doi: 10.1007/978-3-319-14024-7.

Rämä, T., J. Norden, M.L. Davey, G.H. Mathiassen, J.W. Spatafora and H. Kauserud. 2014. Fungi ahoy! Diversity on marine wooden substrata in the high North. Fungal Ecol. 8: 46-58. doi: 10.1016/j.funeco.2013.12.002.

Riddin, T.L., M. Gericke and C.G. Whiteley. 2006. Analysis of the inter- and extracellular formation of platinum nanoparticles by *Fusarium oxysporum* f. sp. lycopersici using response surface methodology. Nanotechnology 17: 3482-3489.

Rietz, D.N. and R.J. Haynes. 2003. Effects of irrigation-induced salinity and sodicity on soil microbial activity. Soil Biol. Biochem. 35: 845-854.

Sahin, U., S. Eroğlum and F. Sahin. 2011. Microbial application with gypsum increases the saturated hydraulic conductivity of saline-sodic soils. Appl. Soil Ecol. 48: 247-250.

Sanae, F., Y. Miyaichi and H. Hayashi. 2003. Endothelium-dependent contraction of rat thoracic aorta induced by gallic acid. 17(2): 187.

Santos, S.X., C.C. Carvalho, M.R. Bonfa, R. Silva and E. Gomes. 2004. Screening for pectinolytic activity of wood-rotting Basidiomycetes and characterization of the enzymes. Folia Microbiol. 49: 46-52.

Sarkar, S., A. Pramanik, A. Mitra and J. Mukherjee. 2010. Bioprocessing data for the production of marine enzymes. Mar. Drugs 8: 1323-1372. doi: 10.3390/md8041323.

Sastry, M., A. Ahmad, M.I. Khan and R. Kumar. 2003. Biosynthesis of metal nanoparticles using fungi and actinomycetes. Curr. Sci. 85: 162-170.

Satyamurthy, P., P. Jain, R.H. Balasubramanya and N. Vigneshwaran. 2011. Preparation and characterization of cellulose nanowhiskers from cotton fibres by controlled microbial hydrolysis. Carbohydr. Polym. 83(1): 122-129.

Satyamurthy, P. and N. Vigneshwaran. 2013. A novel process for synthesis of spherical nanocellulose by controlled hydrolysis of microcrystalline cellulose using anaerobic microbial consortium. Enzyme Microb. Technol. 52(1): 20-25.

Scervino, J.M., M.P. Mesa, I.D. Mónica, M. Recchi, N.S. Moreno and A. Godeas. 2010. Soil fungal isolates produce different organic acid patterns involved in phosphate salts solubilization. Biol. Fertil. Soils 46: 755-763.

Schubert, K., J.Z. Groenewald, U. Braun, J. Dijksterhuis, M. Starink, C.F. Hill, et al. 2007. Biodiversity in the Cladosporium herbarumcomplex (Davidiellaceae, Capnodiales), with standardization of methods for Cladosporium taxonomy and diagnostics. Stud. Mycol. 58: 105-156.

Schuster, E., N. Dunn–Coleman, J.C. Frisvad and P.W. van Dijck. 2002. On the safety of *Aspergillus niger*: A review. Appl. Microbiol. Biotechnol. 59: 426-435.

Seidl, V., L. Song, E. Lindquist, S. Gruber, A. Koptchinskiy, S. Zeilinger, et al. 2009. Transcriptomic response of the mycoparasitic fungus *Trichoderma atroviride* to the presence of a fungal prey. BMC Genomics 10: 567.

Sheikhloo, Z. and M. Salouti. 2011. Intracellular biosynthesis of gold nanoparticles by the fungus *Penicillium chrysogenum*. Int. J. Nanosci. Nanotechnol. 7: 102-105.

Shelar, G.B. and A.M. Chavan. 2014. Fungus-mediated biosynthesis of silver nanoparticles and its antibacterial activity. Arch. App. Sci. Res. 6: 111-114.

Shi, Y. and J. Zhu. 2016. Saline-alkali soil improvement fertilizer and preparation method and use method thereof. Chinese Patent, CN 105237293 A (in Chinese) http://patentool.wanfangdata.com.cn/Patent/Details?id=CN201510597529.0.

Sukumaran, R.K., R.R. Singhania, G.M. Mathew and A. Panday. 2009. Cellulase production using biomass feed stock and its application in lignocellulose saccharification for bio-ethanol production. Renew. Energy 34: 421-424. http://dx.doi.org/10.1016/j.renene.2008.05.008.

Swe, A., R. Jeewon, S.B. Pointing and K.D. Hyde. 2009. Diversity and abundance of nematode-trapping fungi from decaying litter in terrestrial, freshwater and mangrove habitats. Biodivers. Conserv. 18(6): 1695-1714.

Szabo, I., G. Johansson and G. Pettersson. 1996. Optimized cellulase production by *Phanerochaete chrysosporium*: Control of catabolite repression by fed-batch cultivation. J. Biotechnol. 48: 221-230. http://dx.doi.org/10.1016/0168-1656(96)01512-X.

Tam, P.C.F. 1995. Heavy-metal tolerance by ectomycorrhizal fungi and metal amelioration by *Pisolithus tinctorius*. Mycorrhiza 5: 181-187.

Tamura, M., H. Kawasaki and J. Sugiyama. 1999. Identity of the xerophilic species *Aspergillus penicillioides*: Integrated analysis of the genotypic and phenotypic. J. Gen. Appl. Microbiol. 45: 29-37.

Taneja, K., S. Gupta and R.C. Kuhad. 2002. Properties and application of a partially purified alkaline xylanase from an alkalophilic fungus *Aspergillus nidulans* KK-99. Bioresour. Technol. 85: 39-42.

Tiquia-Arashiro, S.M. and D.F. Rodrigues. 2016. Extremophiles: Applications in Biotechnology. Springer briefs in Microbiology: Extremophilic Microorganisms. Springer International Publishing. ISBN 978-3-319-45214-2, ISBN 978-3-319-45215-9 (eBook) doi: 10.1007/978-3-319-45215-9.

Tomme, P., R.A.J. Warren and N.R. Gilkes. 1995. Cellulose hydrolysis by bacteria and fungi. Adv. Microb. Physiol. 37: 1-81.

Torres, M., Y. Dessaux and I. Llamas. 2019. Saline environments as a source of potential quorum sensing disruptors to control bacterial infections: A review. Mar. Drugs 17(3). pii: E191. doi: 10.3390/md17030191.

Trincone, A. 2010. Potential biocatalysts originating from sea environments. J. Mol. Catal. B. Enzym. 66(3-4): 241-256. https://doi.org/10. 1016/j.molcatb.2010.06.004.

Trincone, A. 2011. Marine biocatalysts: enzymatic features and applications. Mar. Drugs 9: 478-499. doi: 10.3390/md9040478.

Turan, M., N. Ataoglu and F. Sahin. 2006. Evaluation of the capacity of phosphate solubilizing bacteria and fungi on different forms of phosphorus in liquid culture. J. Sustain. Agric. 28: 99-108.

Vahabi, K., G.A. Mansoori and S. Karimi. 2011. Biosynthesis of silver nanoparticles by the fungus Trichoderma reesei. Insciences J. 1: 65-79.

Vala, A.K., S. Shah and R. Patel. 2014. Biogenesis of silver nanoparticles by marine derived fungus *Aspergillus flavus* from Bhavnagar coast, gulf of Khambat, India. J. Mar. Biol. Oceanogr. 3(1): 1-3.

Verma, V.C., R.N. Kharwar and A.C. Gange. 2010. Biosynthesis of antimicrobial silver nanoparticles by endophytic fungus *Aspergillus clavatus*. Biomedicine 5: 33-40.

Vrijmoed, L.L.P. 2000. Isolation and culture of higher filamentous fungi. pp. 1-20. *In*: K.D. Hyde, S.B. Pointing. (eds.). Marine Mycology: A Practical Approach. Fungal Diversity Press, Hong Kong.

Wang, L.L., X.Y. Sun, S.Y. Li, T. Zhang, W. Zhang and P.H. Zhai. 2014. Application of organic amendments to a coastal saline soil in north China: Effects on soil physical and chemical properties and tree growth. PLoS One 9(2): e89185. doi: 10.1371/journal.pone.0089185.

Wang, Q.B., C. Halbrendt and S.R. Johnson. 1996. Grain production and environmental management in China's fertilizer economy. J. Environ. Manage. 47(3): 283-296.

Wang, T., X. Liu, Q. Yu, X. Zhang, Y. Qu, P. Gao, et al. 2005. Directed evolution for engineering pH profile of endoglucanase III from *Trichoderma reesei*. Biomol. Eng. 22: 89-94.

Wei, Y., X.D. Liu, B. Jia, S.H. Zhang, J.L. Liu and W. Gao. 2013. Alkaline-tolerant and halophilic *Aspergillus* strain and application thereof in environmental management. Chinese Patent, CN 103436450 A (in Chinese) http://patentool.wanfangdata.com.cn/Patent/Details?id=CN201310162347.1.

Wei, Y. and S.H. Zhang. 2018. Abiostress resistance and cellulose degradation abilities of haloalkaliphilic fungi: applications for Saline-alkaline remediation. Extremophiles 22: 155-164. 10.1007/s00792-017-0986-3.

Xiao, Z.J., M.H. Zong and W.Y. Lou. 2009. Highly enantioselective reduction of 4-(trimethylsilyl)-3-butyn-2-one to enantiopure (R)-4-(trimethylsilyl)-3-butyn-2-ol using a novel strain Acetobacter sp. CCTCC M209061. Bioresour. Technol. 100(23): 5560-5565. https://doi.org/10.1016/j.biortech.2009.06.006.

Xing, G.X. and Z.X. Zhu. 2000. An assessment of N loss from agricultural fields to the environment in China. Nutr. Cycl. Agroecosystems 57(1): 67-73.

Xu, Z.H., X.P. Peng, Y. Wang and W.M. Zhu. 2011. (22E, 24R)-3β,5α,9α-Trihydroxyergosta-7,22-dien-6-one monohydrate. Acta Cryst. E67: o1141-o1142.

Xue, D.S., H.Y. Chen, D.Q. Lin, Y.X. Guan and S.J. Yao. 2012. Optimization of a natural medium for cellulase by a marine *Aspergillus niger* using response surface methodology. Appl. Biochem. Biotechnol. 167(7): 1963-1972. doi: 10.1007/s12010-012-9734-4.

Yan, J., W.N. Song and E. Nevo. 2005. A MAPK gene from Dead Sea fungus confers stress tolerance to lithium salt and freezing-thawing: Prospects for saline agriculture. Proc. Natl. Acad. Sci. USA 102: 18992-18997.

Yu, H.Y. and X. Li. 2015. Alkali-stable cellulose from a halophilic isolate, *Gracibacillus* sp. SK1 and its application in lignocellulosic saccharification for ethanol. Biomass Bioenerg. 81: 19-25. http://dx.doi.org/10.1016/j.biombioe.2015.05.020

Zajc, J., S. Džeroski, D. Kocev, A. Oren, S. Sonjak, R. Tkavc, et al. 2014a. Chaophilic or chaotolerant fungi: A new category of extremophiles? Front. Microbiol. 5: 708-1-708-5. doi:10.3389/fmicb.2014.00708.

Zajc, J., T. Kogej, E.A. Galinski, J. Ramos and N. Gunde-Cimerman. 2014b. Osmoadaptation strategy of the most halophilic fungus, *Wallemia ichthyophaga*, growing optimally at salinities above 15% NaCl. Appl. Environ. Microbiol. 80(1): 247-256. doi: 10.1128/AEM.02702-13.

Zalar, P., G.S. De Hoog, H.J. Schroers, J.M. Frank and N. Gunde-Cimerman. 2005a. Taxonomy and phylogeny of the xerophilic genus Wallemia (Wallemiomycetes and Wallemiales, cl. et ord. nov.). Antonie Van Leeuwenhoek 87: 311-328.

Zalar, P., M.A. Kocuvan, A. Plemenitas and N. Gunde–Cimerman. 2005b. Halophilic black yeasts colonize wood immersed in hypersaline water. Bot. Mar. 48: 323-326.

Zalar, P., G.S. De Hoog, H.J. Schroers, P.W. Crous, J.Z. Groenewald and N. Gunde-Cimerman. 2007. Phylogeny and ecology of the ubiquitous saprobe *Cladosporium sphaerospermum*, with descriptions of seven new species from hypersaline environments. Stud. Mycol. 58: 157-183.

Zalar, P., J.C. Frisvad, N. Gunde-Cimerman, J. Varga and R.A. Samson. 2008. Four new species of Emericella from the Mediterranean region of Europe. Mycologia 100: 779-795.

Zhang, X., Y. Liu, K. Yan and H. Wu. 2007. Decolorization of an anthraquinone-type dye by a bilirubin oxidase-producing nonligninolytic fungus Myrothecium sp. IMER1. J. Biosci. Bioeng. 104: 104-114.

Zhang, X.Y., Y. Zhang, X.Y. Xu and S.H. Qi. 2013. Diverse deep-sea fungi from the South China Sea and their antimicrobial activity. Curr. Microbiol. 67(5): 525-530. doi: 10.1007/s00284-013-0394-6.

Zhou, N., Y. Zhang, F. Liu and L. Cai. 2016. Gymnoascus species from several special environments, China. Mycologia 108: 179-191.

Zhou, X.Y. 2016. Cloning and abiotic functional analysis of salt-tolerant genes in Halophilic *Aspergillus glaucus*. Dissertation, Jilin University(in Chinese) http://kns.cnki.net/KCMS/detail/detail.aspx?filename=1016091320.nh&dbname=CMFDTEMP.

3

Effect of Stimulating Agents on Improvement of Bioremediation Processes with Special Focus on PAHs

Dushyant R. Dudhagara[1,*], Anjana K. Vala[2] and Bharti P. Dave[3]

INTRODUCTION

Polycyclic aromatic hydrocarbons (PAHs) are a class of organic compounds that contain two or more fused aromatic benzene rings that are arranged in different angularities such as linear, cluster and angular. They are persistent and recalcitrant in the environment due to hydrophobic, low water solubility and high electrochemical stability (Lamichhane et al. 2016). Hence, they are highly toxic, carcinogenic, teratogenic, and have mutagenic properties. Therefore, they pose severe threat to human health as well as the ecosystem (Dudhagara et al. 2016a). Thus, United State Environment Protection Agency (USEPA) has listed 16 PAHs as priority pollutants due to their adverse impact on human health and marine ecosystem. PAHs are discharged into the environment through various routes like natural sources and anthropogenic sources.

Natural sources of PAH include burning of forest, volcanic eruptions, decaying organic matter, etc. The degree of PAHs in the atmosphere depends on environmental conditions like wind, temperature, and humidity. The anthropogenic sources of PAHs including industrial and man-made activities are incomplete combustion of organic materials such as coal processing, burning of crude oil, vehicular emission, agricultural production, fossil fuels, gas manufacturing, garbage and wood (Table 3.1) (Ravindra et al. 2008).

An attempt has already been made to reduce the effect of PAH by different physical and chemical methods in different contaminated sites. However, selected techniques have significant limitations such as high cost, less effectiveness and lack of public acceptance. Therefore, natural attenuation processes, which reduce contamination without human interference, show great promise for the cost-effective removal of a wide variety of organic contaminants. Many innovative methods are required to increase the bioavailability of these pollutants to achieve an enhanced

[1] Department of Life Sciences, Bhakta Kavi Narsinh Mehta University, Junagadh-362263, India.
[2] Department of Life Sciences, Maharaja Krishnakumarsinhji Bhavnagar University, Bhavnagar-364001, India.
[3] School of Sciences, Indrashil University, Gandhinagar.
* Corresponding author: dushyant.373@gmail.com

bioremediation performance. Hence, bioremediation is a more effective approach than traditional methods.

TABLE 3.1 Various industrial activities associated with environmental contamination.

S. No.	Activities	Predominant PAHs	Reference
1	Gasoline spill contamination site/Petroleum product spilled site	Naphthalene, Phenanthrene, Fluoranthene, Pyrene and Benz[a]anthracene	Kao et al. 2015
2	Industrial waste water outlet, Alexandria Eastern Harbour	Chrysene, Pyrene, Benzo[a]pyrene, Perylene, Indeno[1,2,3-cd]pyrene, Dibenz[ah]anthracene	Barakat et al. 2011
3	Agricultural, industrial and waste water discharge	Phenanthrene, Anthracene, Fluoranthene, Pyrene, Chrysene, Benzo[a]pyrene, Indeno[1,2,3-cd]pyrene	Masood et al. 2016
4	Crude oil drilling sites	Acenaphthene, Anthracene, Fluoranthene, Pyrene, Chrysene, Benzo[b] fluoranthene, Benzo[a]pyrene, Indeno[1,2,3-cd]pyrene	Sarma et al. 2016
5	Ship breaking activities	Naphthalene, Acenaphthene, Anthracene, Fluoranthene, Pyrene, Chrysene	Dudhagara et al. 2016a
6	Oil- and coal-contaminated	Anthracene, Fluoranthene, Pyrene, Chrysene, Benzo[b]fluoranthene, Benzo[a]pyrene, Indeno[1,2,3-cd]pyrene	Gosai et al. 2017

The purpose of this paper is to review the current approaches for the bioremediation of PAHs contaminated site and their limitations as well as future prospects for developing innovative approaches to enhance the efficacy of PAH remediation at contaminated sites. This chapter also aims to discuss the interactions of PAHs with soils and bacteria, evaluate at the effectiveness of remediation and to address the issue of augmentation in contaminated soils with chemical surfactants, composted materials, and inorganic nutrients. Additionally, other important issues such as solubility of PAHs and bioavailability will also be discussed. By addressing these issues, this chapter aims to critically assess whether composting can be considered as a viable alternative technology for use in bioremediation strategies.

CHEMICAL CHARACTERISTICS OF PAHs

The most distinct characteristic of PAHs is their aromaticity, and the extended π-electron system resulting in chemical stability, which is why they tend to retain the π-electron conjugated ring systems in reactions. Hydrophobicity is another important property of PAHs. Because of their low solubility in an aqueous environment,

TABLE 3.2 Physical properties of 16 PAHs designated by US EPA as priority pollutants (Bojes and Pope 2007).

S. No.	PAH	No. of Rings	M_r	Melting Point (°C)	Boiling Point (°C)	Water Solubility (mg L^{-1})	Vapor Pressure (Pa)	K_{ow} Value	Structure
1	Naphthalene	2	128.17	80.6	218	31	10.4	3.37	
2	Acenaphthene	3	154.21	95	279	3.47	3.0×10^{-1}	3.92	
3	Acenaphthylene	3	152.20	93.5-94.5	265	3.93	8.93×10^{-1}	4.07	
4	Fluorene	3	166.22	116	295	0.190	8.0×10^{-2}	4.18	
5	Anthracene	3	178.23	217.5	340	0.0434	1.0×10^{-3}	4.54	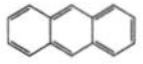
6	Phenanthrene	3	178.23	99.5	340	1.18	2.0×10^{-2}	4.57	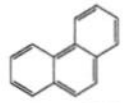
7	Fluoranthene	4	202.26	110.8	375	0.265	1.23×10^{-3}	5.22	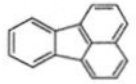
8	Pyrene	4	202.26	156	404	0.013	6.0×10^{-4}	5.18	
9	Benz[a]anthracene	4	228.29	159.8	437.6	0.014	2.8×10^{-5}	5.91	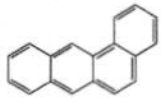
10	Chrysene	4	228.29	255.8	448	0.0018	5.70×10^{-7}	5.86	
11	Benzo[k]fluoranthene	5	252.31	215.7	480	0.00055	7.0×10^{-7}	6.04	
12	Dibenz[a,h]anthracene	5	278.35	265	524	0.0005	1.33×10^{-8}	7.16	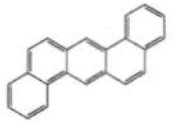
13	Benzo[a]pyrene	5	252.31	176.5	495	0.0038	1.40×10^{-8}	6.25	
14	Indeno[1,2,3-cd]pyrene	6	276.34	162.5	536	0.0620	1.0×10^{-10}	6.58	
15	Benzo[b]fluoranthene	6	252.31	167	357	0.0012	6.67×10^{-5}	6.57	
16	Benzo[g,h,i]perylene	6	276.34	278.3	500	0.00026	1.39×10^{-8}	7.10	

PAHs tend to associate with particles and eventually sink into soil and sediments (Xiao et al. 2003). PAHs are categorized as low molecular weight (LMW) and high molecular weight (HMW) based on their fused benzene ring moieties, carbon and hydrogen atoms. The LMW PAHs are comprised of two to three benzene rings structure and HMW PAHs are comprised of four or more benzene ring structure (Dudhagara et al. 2016a). HMW PAHs resist biodegradation as they are strongly absorbed in particulate matter and possess high thermostability. HMW PAHs, due to their solid state, high molecular weight and higher hydrophobicity (as expressed by high log P value, i.e. between 3 and 5), are more toxic to cells than LMW PAHs (Cerniglia 1992).

PAHs are normally hydrophobic or lipophilic, thus they are miscible in solvents. Furthermore, the aqueous solubility of PAHs decreases when molecular weight increases. Moreover, PAHs show other properties like light sensitivity, heat resistance, conductivity, and corrosion resistance (Kim et al. 2013). PAHs have unique UV spectrum. Consequently, each PAH has specific light intensity. Most PAHs are fluorescent, emitting characteristic wavelengths of light when they are excited. The various physicochemical properties of 16 USEPA (Expand) listed PAHs are shown in Table 3.2.

SOURCE OF PAHs

PAHs are very common and stable form of hydrocarbons having a low hydrogen-to-carbon proportion and typically found in complex form rather than single compounds. PAHs are commonly derived from anthropogenic activities related to pyrolysis and incomplete combustion of organic matter. Some PAHs are also used in the preparation of medicine, synthesis of dyes, paints, pesticides, etc. (Ravindra et al. 2008). The source of PAHs is classified into five categories such as domestic, mobile, industrial, agricultural, and natural sources. The important contribution of domestic sources in the emission of PAHs is burning and pyrolysis gas, garbage, wood, or other organic substances like tobacco. A major source of PAHs is industrial source such as petrochemical, coke and aluminum production, in addition to the manufacturing of cement, rubber, etc. PAHs are frequently used as intermediaries in pharmaceuticals, lubricating materials, agricultural products, photographic products, and other chemical industries (Rengarajan et al. 2015).

Agricultural source includes moorland burning, straws, woods, and stalks. PAHs would be produced from the open burning of biomass. All of these activities involve the burning of organic materials under sub-optimum combustion conditions (Berset et al. 1999, Vane et al. 2013). Hence, agricultural source of PAHs may be a significant contributor at various levels. The natural sources include non-anthropogenic activities such as volcanic eruption, burning of forest, woodland, etc. The amount of PAH production depends on atmospheric conditions, such as wind, temperature, humidity, and fuel characteristics and type, such as moisture content, green wood, and seasonal wood (Ravindra et al. 2008).

MICROBIAL PAH DEGRADATION

Microorganisms can develop various strategies to activate PAH molecules, the initial steps in degradation pathways are the presence and absence of molecular oxygen (Fig. 3.1). The molecular mechanism of PAH degradation has been reviewed by many researchers (Kimes et al. 2014, McGenity 2014, Duran and Cravo-Laureau 2016, Aydin et al. 2017). Under aerobic condition, biodegradation of PAHs requires the presence of molecular oxygen as it is essential for initiating the catabolic reactions. The first activation steps introduce direct incorporation of molecular oxygen into aromatic ring and activation of mono- or di-oxygenases. Monooxygenases are classified on the basis of the presence of cofactor such as the flavin-dependent cytochrome P450 monooxygenase, which requires NADP as coenzymes. P450 monooxygenase is a heme-containing oxygenase that is present in both prokaryotic and eukaryotic organisms. The intermediate product arene oxide is subsequently transformed into either trans dihydrodiol by hydrolase or phenol by non-enzymatic reaction (Cerniglia 1992, Haritash and Kaushik 2009).

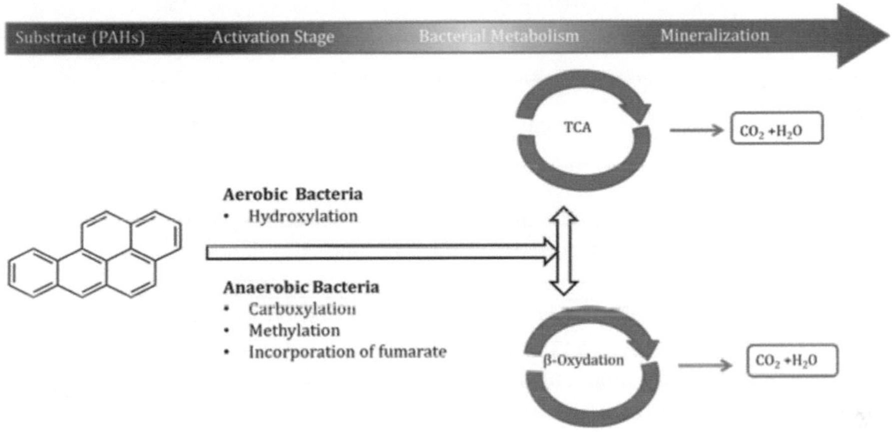

FIGURE 3.1 Catabolic pathway of PAHs operated under aerobic and anaerobic conditions.

The key enzymes of dioxygenase family are aromatic ring-hydroxylating dioxygenases (ARHDs) belonging to the Rieske-type non-heme iron oxygenase. Dioxygenase family primarily oxidize aromatic compound, and hence has greater application in environmental bioremediation of contaminated site. In the first step, dioxygenase catalyzes the aromatic rings by incorporating molecular oxygen to produce *cis* dihydrodiol as primary intermediate through multicomponent enzyme system. Dihydrodiol further converts to catechol or protocatechuate by using NAD as a cofactor for cleavage of the ring. This dihydrodiol is further cleaved by intradiaol or extradiol ring fission, through *ortho*- or *meta*-cleavage pathway, which produces aliphatic products that enter the central metabolism via the tricarboxylic acid cycle (Fig. 3.1 and 3.2) (Cerniglia 1992, Karigar and Rao 2011, Duran and Cravo-Laureau 2016).

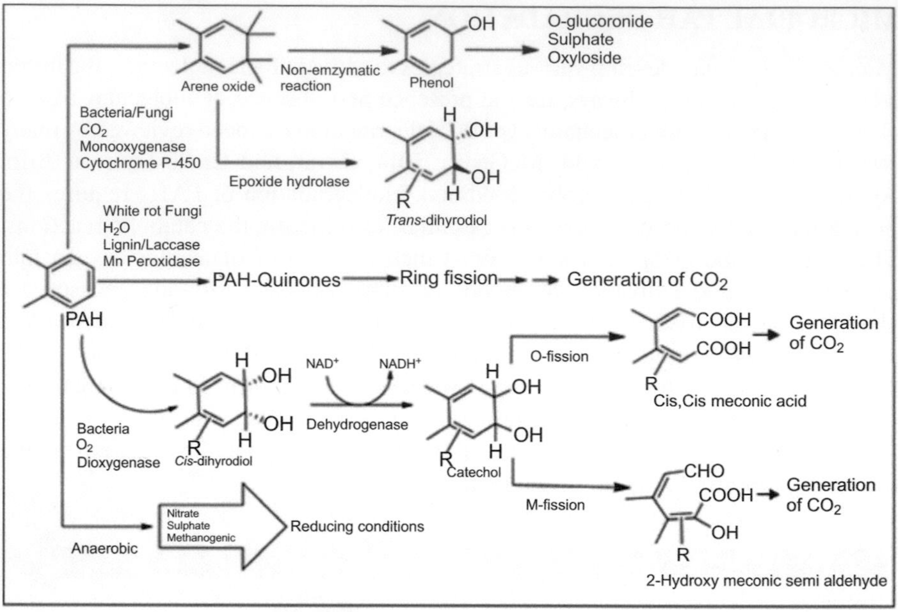

FIGURE 3.2 Microbial metabolisms of PAHs by various routes (Haritash and Kaushik 2009).

Anaerobic degradation mechanism involves direct caboxylation or methylation followed by the addition of fumarate (Fig. 3.1 and 3.2). The addition of carbon dioxide or bicarbonate to an organic acceptor is necessary for the biological reaction. Under anaerobic and reducing conditions, biodegradation of hydrocarbons can be categorized in three steps. The preliminary high concentration of PAHs is partially degraded under nitrate and sulfate-reducing conditions to form LMW organic compound as an intermediate metabolite. These organic compounds generated during anaerobic microbial degradation of PAHs act as ligands for complex insoluble Fe(III) oxides. After that Fe(III) is readily available for iron reducing bacteria, which escalate biodegradation process of PAHs (Schmitt et al. 1996, Haritash and Kaushik 2009).

Interaction between Soil and PAHs

The properties of soil influence the interaction between PAHs and individual soil components. However, soil properties determine the activity of microorganisms, which are responsible for degradation of PAHs. Soil properties such as organic matter, porosity, pH, hydrophobicity, cation exchange capacity, and concentration of clay minerals play a crucial role in bioavailability of PAHs in soil matrices (Oleszczuk and Baran 2006, Dudhagara et al. 2016b). Molecules attached to exchange sites can readily become bioavailable by desorption from these sites. An increase in the number of exchange sites helps the retention of molecules in a bioavailable form. The molecules are more strongly absorbed into the solid surface which is not bioavailable (Fig. 3.3).

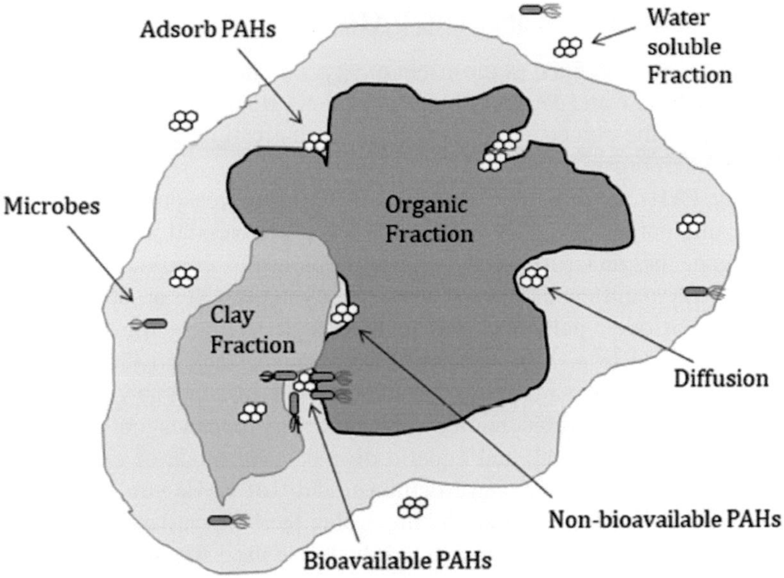

FIGURE 3.3 Potential interactions between soil and PAHs in environmental conditions.

As suggested by Chikere et al. (2016), the interaction between hydrocarbon and soil minerals is very significant when the organic content is >0.1%. Therefore, the organic matter may interact with PAHs by various mechanisms, which include physical and chemical adsorption, solubilization, and partitioning hydrolysis. Physical characteristics of organic matter interact with PAHs including molecular size, molecular configuration, and polarity of the compound. Subsequently, chemical characteristics including dissolved organic matter, fulvic acid, humic acid and humus can affect the mobility of PAHs. The chemical partition of organic matter is generally described by K_{ow} (octanol-water partition coefficient). The K_{ow} values measure the solubility in both aqueous and organic phase (octanol). In addition to that, K_{ow} values >1 indicate that compound is lipophilic or hydrophobic and K_{ow} values <1 indicate that the compound is hydrophilic. The values of K_{ow} are commonly represented in algorithmic form.

The values of K_{ow} may significantly impact the biological degradation of an organic compound; therefore, it plays a vital role in understanding the transport mechanisms and distribution of PAHs in the environment. Thus, the measurement of K_{ow} values is of critical significance for evaluating the fate and potential exposure to chemicals in the environment, and accordingly, estimation of the process of environmental risk assessment (Lu et al. 2008).

Aliphatic hydrocarbons can strongly partition into organic matter and diffuse into the structure of organic matter (Atlas 2005). Thus, hydrocarbon may be sequestered in the soil through sorption to organic matter and minerals or diffuse into the structure of soil (Fig. 3.3).

Interaction between Microbes and PAHs

Bioavailability can be defined as the effect of physico-chemical and microbiological parameters on the rate and degree of biodegradation (Mueller et al. 1996, Alegbeleye et al. 2017). It is the percentage of contaminant that can be readily accessed and degraded by microorganisms.

Generally, PAHs are difficult to degrade due to several reasons including (i) high toxicity to microorganisms, (ii) availability of a preferential source of nutrient to microorganisms, and (iii) unfavorable environmental conditions such as pH, temperature, micronutrient as trace metals, etc. (Alegbeleye et al. 2017). PAHs are more stable than other pollutants due to their high stereochemical stability and carbon double bond (Fig. 3.3). Thus, PAHs are more stable in the environment, and it is difficult for bioavailability, due to some physical factors such as the size of the sediment, and organic matter. Bioavailability depends on the biological diversity including terrestrial, and aquatic diversity and mode of exposure to the contamination. Various studies have demonstrated that PAHs can be biologically degraded under controlled condition by increasing their bioavailability to microbial metabolism (Wick et al. 2011). Various methods are used for the enhancement of bioavailability such as bioaugmentation, biostimulation, biosparging, composting, land farming, use of surfactant, and other solubilizer (Vidali et al. 2001). Several studies have revealed that various cationic (CTAB), anionic (SDS) and non-ionic (Triton X-100, Tergitol NP-10) surfactants could be used for enhancing the mobility of PAHs to the microorganisms. Various surfactants can increase uptake rate of PAHs. The influence of various surfactants was found especially due to the increased bioavailability of PAHs, representing the oxidation of PAHs by the extracellular enzymatic mechanism (Kotterman et al. 1998).

STRATEGIES FOR IMPROVEMENT OF BIOAVAILABILITY

Surfactants

A possible way to enhance the availability of PAHs is use of surface-active-agents (surfactants) as mobilizing agents. Surfactant approaches have been widely used for biodegradation of hydrophobic organic compound by addition of a chemical surfactant, bio-surfactant. The chemical surfactant is categorized in three groups: ionic, cationic and non-ionic surfactants. As compared to other surfactants, non-ionic surfactant like Triton X-100 has performed better for removal of high molecular weight (HMW) PAHs viz. anthracene, fluoranthene, and pyrene, effectively (Rodrigues et al. 2013, Dave et al. 2014). The properties of PAHs molecule tend to concentrate at surface and interface and decrease the level of surface tension and interface tension. The effect of surfactant can be explained by three main mechanisms: (i) increase in the contact area, caused by a reduction in interface tension between aqueous phases and non-aqueous phase, (ii) rise in the solubility of PAHs, which is caused by the presence of critical micelle concentration (CMC) which contains hydrophobic moieties, and (iii) facilitating transport of pollutant from solid phase to aqueous phase, caused by many aspects such as lowering the surface tension of soil particles, which generally enhances the possibilities for interaction of the surfactant with solid particles (Volkering et al. 1995) (Fig. 3.4).

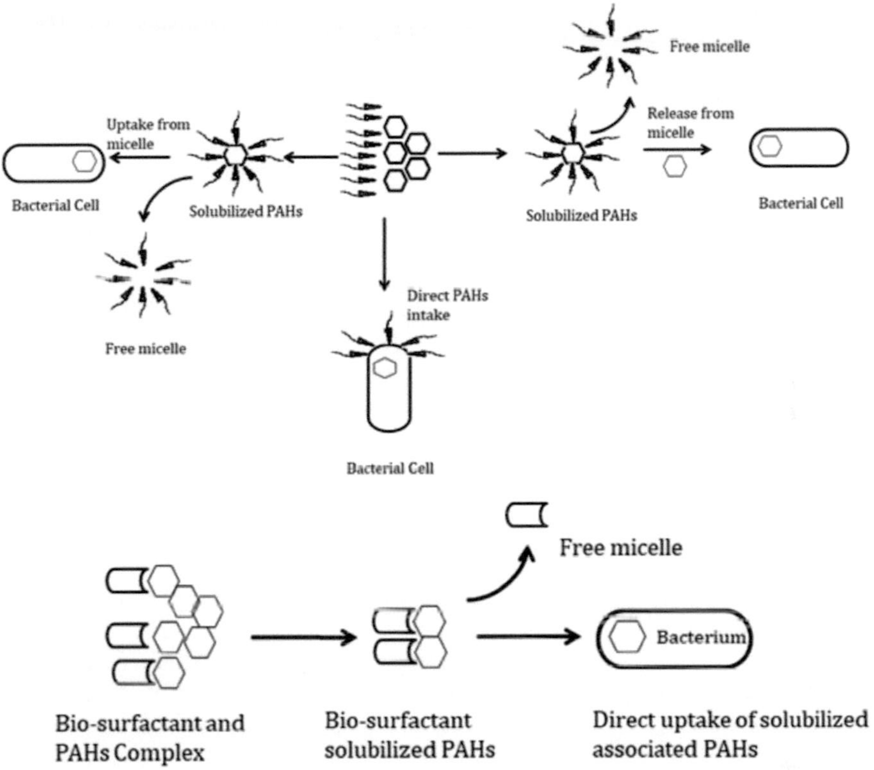

FIGURE 3.4 PAHs' bioavailability enhancement using surfactant shown as (A) mechanisms of direct uptake, uptake from micelle and release from micelle uptake (B) direct uptake from biosurfactant associated PAHs.

The surfactant concentration over and above is indicated by critical micelle concentration (CMC). CMC reduces the surface tension of surfactant and, subsequently, starts initialization of surfactant micelle formation. Various studies have reported the use of chemical surfactant on the removal of Polychlorinated biphenyl (PCB), PAHs, etc. Many investigators used the surfactant for bioremediation of PAHs contaminated soil listed in Table 3.3. Treatment of soil by various chemical surfactants as Triton X-100, Tween-80, Brij-35, Tergitol NP-10, SDS, and CTAB has resulted in removal of alkylated hydrocarbons, PAHs, and PCB from the contaminated soil (Tiehm 1994, Rodrigues et al. 2013, Zhang et al. 2013). Some investigators suggested the use of surfactant for biodegradation of PAHs, while some reported inhibitory effect of surfactant on biodegradation of PAHs (Bueno-Montes et al. 2011, Rodrigues et al. 2013). Thus, it needs to understand the effect of surfactant on biodegradation of PAHs. Rodrigues et al. (2013) investigated the study of three surfactants, Tween-20 (non-ionic), SDS (anionic), and CTAB (cationic), on biodegradation of fluoranthene and anthracene by *Pseudomonas putida*.

TABLE 3.3 Studies on various biostimulating agents for bioremediation of PAHs.

Enhancer	Surfactant Type	Substrate or PAHs	Organisms	Reference
Surfactant	Tween 80 Triton X-100 Tergitol NP-10 (Nonionic)	Naphthalene, Phenanthrene, Anthracene	*Enterobacter* sp., *Pseudomonas* sp., *Stenotrophomonas* sp.	Bautista et al. 2009
	Tween 20 (Nonionic) SDS (Anionic) CTAB (Cationic)	Fluoranthene Anthracene	*Pseudomonas putida*	Rodrigues et al. 2013
	Hexadecyl trimethyl ammonium bromide (CTAB) (Cationic)	Phenanthrene	Mixed culture	Chang et al. 2000
	Tween 80 (Nonionic)	Pyrene	*Mycobacterium frederiksbergense*	Sarma and Pakshirajan 2011
Compost	NPK fertilizer	Crude oil		
	Yard waste type organic kitchen waste	Oil contaminants	Soil microorganisms	Wallisch et al. 2014
	Green waste, and Sawdust	Phenathrene, Benzo[a]pyrene	Bacteria	Teng et al. 2010
	Manure compost	Total petroleum hydrocarbons	Bacterial consortium	Rocchetti et al. 2011
Nutrients	K_2HPO_4 and $(NH_4)_2SO_4$	Total petroleum hydrocarbons	Bacterial consortium	Rocchetti et al. 2011
	Glucose, starch, urea, sodium succinate	Phenathrene, Benzo[a]pyrene	Bacteria	Teng et al. 2010

The presence of non-ionic surfactant as Tween-20 on biodegradation of fluoranthene and anthracene indicated the increase in bioavailability of fluoranthene at a 0.08 mM concentration of Tween-20; it's slightly higher than CMC concentration (0.06 mM). According to Rodrigues et al. (2013), in presence or absence of Tween-20, 70% fluoranthene degradation was achieved, indicating that Tween-20, at a concentration slightly higher than its CMC, had no negative effect on the utilization of fluoranthene by *P. putida* cells. The Tween-20 also had a positive effect on biodegradation of anthracene, which increases the 80% oxygen uptake rate in the presence of chemical surfactant during biodegradation by *P. putida*. A non-ionic surfactant as Triton X-100 showed a significant effect on biodegradation of naphthalene, and anthracene by the consortium such as *Pseudomonas* sp., *Enterobacter* sp. and *Stenotrophomonas* sp. The non-ionic surfactant showed a significant effect on hydrocarbon degradation, which increases the bioavailability and solubility of PAHs (Bautista et al. 2009). Various studies have suggested the positive effect of non-ionic surfactant as Triton X-100, Tween-80, etc. (Kim et al. 2001, Yang et al. 2003, Cheng and Wong 2006). In contrast, some studies reported the inhibitory effect of surfactant on biodegradation of hydrocarbons. The surfactant concentration higher than CMC concentration, which may be due to excess of micelles concentration, resulted in the reduction of bioavailability (Laha and Luthy 1992, Yuan et al. 2000, Mulligan et al. 2001, Bautista et al. 2009).

The effect of an anionic surfactant as SDS on biodegradation of PAHs showed a negative effect on fluoranthene and anthracene degradation in soil matrices. The negative effect of surfactants on the biodegradation of PAHs can depend on the properties of the surfactants and their microorganism characteristics. Many researchers observed that the adhesion of a cell to the substrate is an essential mechanism for the uptake of PAHs; therefore, impact of SDS PAHs biodegradation would be negative. In many cases, the organism can grow in the presence of PAHs but in the presence of higher molecular weight PAHs as fluoranthene, generally organisms are not being able to grow in the presence of SDS. Thus, Rodrigues et al. (2013) suggested the formation of cell aggregation in the presence of anionic surfactant as SDS. Patel et al. (2012) also examined the effect of chemical surfactant like SDS on biodegradation of naphthalene by the bacterial consortium (DV-AL). Makkar and Rockne (2003) suggested that SDS does not show a positive effect on biodegradation of HMW PAHs by addition of a surfactant, which may show a higher toxicity to microorganism on hydrocarbon degradation (Mulligan et al. 2014).

Various researchers studied the effects of cationic surfactant like CTAB on biodegradation of PAHs. CTAB has been widely used as a surfactant in biodegradation of PAHs. CTAB neutralizes the surface of hydrocarbons by the formation of monolayer surrounding the substrate. CTAB is commonly absorbed on the surface and has good adhesive properties for the removal of PAHs.

Compost

Nowadays, bioremediation techniques are being emphasized and improved to remediate the soil polluted with an organic compound. Composting process seems to be useful for the bioremediation of organic compound, specifically PAHs. Thus, the concept of treating the contaminated soils by agriculture compost and organic soil compost or municipal waste compost has been effective for PAHs bioremediation process (Semple et al. 2001, Saraya et al. 2010). Compost materials are mainly based on plant-derived litter material (wood chips, straw) and contains high numbers of inorganic carbon sources, which support the growth of microbes that are capable of degrading hydrocarbons (Antizar-Ladislao et al. 2004). The addition of compost material may stimulate the degradation of pollutants like hydrocarbons. The addition of compost in PAHs polluted soil can facilitate the degradation of PAHs because compost provides essential nutrient and carbon to the contaminated soils. Soil to compost ratio 1:0.5 indicates better results as compared to 1:1. The ratio 1:1 was shown to have a negative effect on biodegradation of PAHs and could inhibit the growth of microorganisms. Thus, the optimum ratio 1:0.5 revealed better results in PAHs biodegradation (Yuan et al. 2009). Therefore, the addition of inorganic supplements resulted in the increment of PAHs degradation rate. In contrast, excessive inorganic supplement inhibited the biodegradation of PAHs and microbial population (Namkoong et al. 2002, Yuan et al. 2009).

Moreover, Saraya et al. (2010) studied the effect of compost on biodegradation of pyrene. Three different types of compost were used for pyrene degradation. Out of these, compost type A showed more significant results due to minimum amendment of compost which supported the biodegradation of PAHs. The minimum mixing ratio of soil to compost (1:0.5) was able to enhance the microbial activity that is capable of degrading HMW PAHs as pyrene by up to 80%. Thus, the mixing ratio suggests a lower degradation as compared to compost type A. Chang et al. (2009) also suggested that the addition of organic compost into the contaminated soil increases the rate of PAHs degradation and excessive supplement could inhibit the degradation of PAHs as well as microorganisms. Hence, the result indicated the improvement of bioremediation process using organic compost techniques. Yuan et al. (2009) results supported the addition of organic supplement as compost in bioremediation of PAHs contaminated soil. The rate of phthalate esters' degradation in compost amended soil resulted in higher degradation (Chang et al. 2009) than reported by Yuan et al. (2002) and Chang et al. (2007). Chang et al. (2009) suggested the feasibility of the remediation of phthalate esters contaminated soil by the addition of organic compost. Abioye et al. (2012) reported that the organic wastes positively enhanced the rate of biodegradation in soil contaminated with burned lubricant oil. Thus, organic waste can be useful for the bioremediation of oil contaminated sites.

Nutrients

Various physical, chemical and environmental factors often limit bioremediation process. Nutrient limitation is often a major problem arising during PAHs bioremediation, especially in soil and sediments. Thus, nutrient requirements of microorganisms are essential for the enhancement of PAHs bioremediation. Addition of nutrients such as nitrogen (N), phosphorus (P) and potassium (K) or oxygen to the contaminated sites for the improvement of the microbial degradation ability is known as biostimulation. Biostimulation refers to an increase of bacterial diversity, present in native contaminated soil by addition of nutrients as NPK. As per literature, the nutrient requirements for microbes are approximately similar to their cell composition. Mainly microorganisms utilize the nutrient for their growth and metabolism of the substrate. As per the literature survey, nutrients have been categorized in two ways, as micronutrient (SCaMn) and macronutrient, (NPK) based on their cellular dried weight. Macronutrients such as nitrogen, phosphorus, potassium, and oxygen are essential for cellular metabolisms in microorganisms and consequently have effect on their growth (Bamforth and Singleton 2005). Micronutrients such as sulfur, zinc, manganese, iron, nickel, cobalt, molybdenum, copper, and chlorine are required in a very low quantity (Brady and Weil 1999, Breedveld and Sparrevik 2000), which is essential for maintenance of ionic or osmotic balance of the cell. Liebeg and Cutright

(1999) reported that phosphorus was the dominant macronutrient in bioremediation of PAH. The optimum values of nutrient are greatly dependent on types of pollution, microorganisms and soil conditions. On the other hand, several researchers reported the negative effect of higher NPK concentrations on biodegradation of hydrocarbons, specifically aromatic ring hydrocarbons (Chaîneau et al. 2005, Chaillan et al. 2006). In most of the cases, the positive impact of nutrient on biodegradation of aromatic hydrocarbons at various extents has been observed (Atlas and Bartha 1992, Morgan and Watkinson 1989, Chaîneau et al. 2000). Generally, addition of nutrients positively influences the microbial growth. Thus, a higher rate of the microbial population is generally reported in NPK fertilized soil (Chaîneau et al. 2000).

Kalantary et al. (2014) reported the highest phenanthrene degradation in the presence of KH_2PO_4 and NH_4NO_3, which directly indicated the role of NP on phenanthrene degradation. Da Silva et al. (2009) have also reported the commercial NPK fertilizer used for removal of petroleum hydrocarbon from marine sediments. Various studies reported the effect of co-substrates on PAH degradation and revealed these additives with sufficient amounts of nitrogen, as peptone and yeast extract to be more effective in enhancing PAH degradation in contaminated sediments (Chen et al. 2008).

FUTURE ROADMAP

Nowadays, trends on innovative bioremediation approach like green biotechnology as nano-remediation, co-culture and/or mixed culture and transgenic approaches are useful for cleanup of contaminated sites. These innovative technologies have successfully remediated the environmental contaminants (Kuppusamy et al. 2016d). However, the PAHs bioremediation still remains unexploited and therefore the primary focus will be on the development of robust, reliable, cost-effective and eco-friendly cleanup technologies.

Nano-remediation

In general, nano-remediation is a newer approach, with great potential for cleanup of contaminated sites, which can also protect the polluted sites from the emerging environment pollutant. The synthesis of magnetic nanoparticles in the presence of Triton X-100 (Mag-PCMA-T complex) can absorb acenaphthene from water. Huang et al. (2016) demonstrated that the complex of nanoparticle provides an effective and suitable approach for remediation of PAHs contaminated sites. Surface coating properties of nanoparticle make them a suitable candidate for the removal of PAHs from an environmental sample. The futuristic approaches involving nano-fertilizers and nano-minerals will prove to be a better alternative to remove the PAHs from contaminated sites. Thus, nano-remediation can integrate with established techniques and improve the remediation efficiency, which denotes the PAHs degradation field scale.

Co-culture Approach

Generally, biodegradation using a pure culture does not represent the actual behavior of bioremediation in natural PAH-contaminated soils because, in nature, bioremediation depends on cooperative metabolic activities of mixed microbial populations. One important advantage of using microbial consortia is that they possess multiple metabolic capacities that increase the efficiency of the biodegradation. The combination of two organisms was more useful for degradation of PAHs than individual species. In natural environmental systems, microbial consortia are more effective as compared to single species in the complex contaminated environment (Ramakrishnan et al. 2011, Ramanan et al. 2016).

Subashchandrabose et al. (2011) reported that the use of cyanobacteria-bacterial consortium is more suitable for the removal of organic pollutants from the environment as compared to fungi-bacteria consortia. Cyanobacteria can synthesize various compounds viz. exopolysaccharide, photon, inorganic nutrients (NPK), proteins, enzyme, excretory products, etc. and generally worked as a growth inducer for associated bacteria which enhanced the degradation rate of organic pollutant present in the environment. Certain microalgae also play a pivotal role in supplying molecular oxygen to other associated species, ultimately enhancing the biodegradation rate of contaminants. Thus, the future studies will be focused on the development of best remedial strategies to remove organic pollutants such as PAHs, PCB, pesticides, etc. very efficiently.

Transgenic Approach

Nowadays genetic engineering techniques have opened a new avenue for the development of new genetically modified microorganisms (GMO). The function of GMO is the use of catalysts through the modification of some enzymatic system, genes degradation pathways, etc. in a single host, which results in the improvement of degradation efficiency (Alegbeleye et al. 2017). The genetically engineered strains perform better results due to higher metabolic activities, predictability, and efficiency of the process and get significantly reduced pollutants. The recombinant organism can degrade a wide range of recalcitrant pollutants rapidly within a short period. By manipulating certain genes, e.g. genes involved in biosurfactant synthesis, bioavailability of pollutant can be enhanced and hence, increase in biodegradation rate of pollutants can be achieved. The gene is also deliberately resistant against environmental stresses such as temperature, pH, and oxygen, which improves the survival rate and catalytic performance (Dua et al. 2002, Kuppusamy et al. 2016c).

However, ecological and environmental concerns and regulatory constraints are major obstacles for testing GMO in the natural ecosystem. These approaches, which are ultimately successful in bioremediation of pollutants, may make a difference in ability to reduce waste, industrial pollution, and enjoy a more sustainable future. This could be the most efficient technology once complete information on bioremediation microbes, their genomes and biochemical mechanisms is revealed. Moreover, all these approaches can be implying the current challenges, future goals, and development of emerging technologies.

CONCLUSIONS

The most common mechanism used in bioremediation experiment is stimulation of the indigenous microbial community by addition of substrate, nutrients, and surfactants as the rate of natural attenuation is generally slow. Thus, the stimulating agent has boosted up the removal of the pollutants. Indigenous soil microbial community can degrade pollutants as PAHs, but their efficiency for degradation of PAHs is limited and requires a long period. Different studies are often based on different concepts of bioavailability. Different methods can be related to a specific availability type, and the selection of a method depends on the purpose of the study. Based on study, the interaction of soil and PAHs in the presence of bacteria requires a Triton X-100 and agricultural compost for accelerating the bioremediation process. Thus, we concluded that the biodegradation of PAHs using a microbial consortium in the presence of surfactant and agricultural compost would be a feasible process for soil bioremediation. Different eco-friendly and modern biological methods will play a significant role in the future developments in this field.

ACKNOWLEDGEMENTS

The Authors are thankful to Science and Engineering Research Board (SERB), Department of Science and Technology, Government of India, New Delhi for providing financial assistance under the national post-doctoral fellowship scheme (grant number PDF/2016/003007).

REFERENCES

Abioye, O.P., P. Agamuthu and A.R. Abdul Aziz. 2012. Biodegradation of used motor oil in soil using organic waste amendments. Biotechnol. Res. Int. 2012.

Alegbeleye, O.O., B.O. Opeolu and V.A. Jackson. 2017. Polycyclic aromatic hydrocarbons: A critical review of environmental occurrence and bioremediation. Environ. Manage. 60(4): 758-783.

Antizar-Ladislao, B., J. Lopez-Real and A. Beck. 2004. Bioremediation of polycyclic aromatic hydrocarbon (PAH)-contaminated waste using composting approaches. Crit. Rev. Environ. Sci. Technol. 34(3): 249-289.

Atlas, R.M. and R. Bartha. 1992. Hydrocarbon biodegradation and oil spill bioremediation. Adv. Appl. Microbiol. Ecol. 12: 287-338.

Atlas, R.M. 2005. Handbook of Media for Environmental Microbiology, Second ed. CRC Press, Taylor and Francis Group, USA.

Aydin, S., H.A. Karaçay, A. Shahi, S. Gökçe, B. Ince and O. Ince. 2017. Aerobic and anaerobic fungal metabolism and Omics insights for increasing polycyclic aromatic hydrocarbons biodegradation. Fungal Biol. Rev. 31(2): 61-72.

Bamforth, S.M. and I. Singleton. 2005. Bioremediation of polycyclic aromatic hydrocarbons: Current knowledge and future directions. J. Chem. Technol. Biotechnol. 80(7): 723-736.

Barakat, A.O., A. Mostafa, T.L. Wade, S.T. Sweet and N.B. El Sayed. 2011. Distribution and characteristics of PAHs in sediments from the Mediterranean coastal environment of Egypt. Mar. Pollut. Bull. 62(9): 1969-1978.

Bautista, L.F., R. Sanz, M.C. Molina, N. González and D. Sánchez. 2009. Effect of different non-ionic surfactants on the biodegradation of PAHs by diverse aerobic bacteria. Int. Biodeterior. Biodegr. 63(7): 913-922.

Berset, J.D., M. Ejem, R. Holzer and P. Lischer. 1999. Comparison of different drying, extraction and detection techniques for the determination of priority polycyclic aromatic hydrocarbons in background contaminated soil samples. Anal. Chim. 383: 263-75.

Brady, N.C. and R.P. Weil. 1999. Soil Organic Matter. pp. 446-490. *In*: Brady, N.C. and R.P. Weil (eds.). The Nature and Properties of Soils. Prentice Hall, Upper Saddle River, New Jersey.

Breedveld, G.D. and M. Sparrevik. 2000. Nutrient-limited biodegradation of PAH in various soil strata at a creosote contaminated site. Biodegradation 11(6): 391-399.

Bueno-Montes, M., D. Springael and J.J. Ortega-Calvo. 2011. Effect of a nonionic surfactant on biodegradation of slowly desorbing PAHs in contaminated soils. Environ. Sci. Technol. 45(7): 3019-3026.

Cerniglia, C.E. 1992. Biodegradation of polycyclic aromatic hydrocarbons. Biodegradation 3: 351-368.

Chaillan, F., C.H. Chaineau, V. Point, A. Saliot and J. Oudot. 2006. Factors inhibiting bioremediation of soil contaminated with weathered oils and drill cuttings. Environ. Pollut. 144(1): 255-265.

Chaîneau, C.H., J.L. Morel and J. Oudot. 2000. Biodegradation of fuel oil hydrocarbons in the rhizosphere of Maize (Zea mays L.). J. Environ. Qual. 29: 569-578.

Chaîneau, C.H., G. Rougeux, C. Ye´pre´mian and J. Oudot. 2005. Effects of nutrient concentration on the biodegradation of crude oil and associated microbial populations in the soil. Soil Biol. Biochem. 37: 1490-1497.

Chang, B.V., T.H. Wang and S.Y. Yuan. 2007. Biodegradation of four phthalate esters in sludge. Chemosphere 69(7): 1116-1123.

Chang, B.V., Y.S. Lu, S.Y. Yuan, T.M. Tsao and M.K. Wang. 2009. Biodegradation of phthalate esters in compost-amended soil. Chemosphere 74(6): 873-877.

Chang, M.C., C.R. Huang and H.Y. Shu. 2000. Effects of surfactants on extraction of phenanthrene in spiked sand. Chemosphere 41(8): 1295-1300.

Chen, J., M.H. Wong, Y.S. Wong and N.F. Tam. 2008. Multi-factors on biodegradation kinetics of polycyclic aromatic hydrocarbons (PAHs) by *Sphingomonas* sp. a bacterial strain isolated from mangrove sediment. Mar. Pollut. Bull. 57(6): 695-702.

Cheng, K.Y. and J.W.C. Wong. 2006. Effect of synthetic surfactants on the solubilization and distribution of PAHs in water/soil-water systems. Environ. Technol. 27: 835-844.

Chikere, C.B., G.C. Okpokwasili and B.O. Chikere. 2011. Monitoring of microbial hydrocarbon remediation in the soil. 3 Biotech. 1(3): 117-138.

Chikere, C.B., A.U. Okoye and G.C. Okpokwasili. 2016. Microbial community profiling of active oleophilic bacteria involved in bioreactor-based crude-oil polluted sediment treatment. Appl. Environ. Microbiol. 4(1): pp. 1-20.

Da Silva, A.C., F.J. de Oliveira, D.S. Bernardes and F.P. de França. 2009. Bioremediation of marine sediments impacted by petroleum. Appl. Microbiol. Biotechnol. 153(1-3): 58-66.

Dave, B.P., C.M. Ghevariya, J.K. Bhatt, D.R. Dudhagara and R.K. Rajpara. 2014. Enhanced biodegradation of total polycyclic aromatic hydrocarbons (TPAHs) by marine halotolerant *Achromobacter xylosoxidans* using Triton X-100 and β-cyclodextrin-A microcosm approach. Mar. Pollut. Bull. 79(1): 123-129.

Dua, M., A. Singh, N. Sethunathan and A. Johri. 2002. Biotechnology and bioremediation: Successes and limitations. Appl. Microbiol. Biotechnol. 59(2): 143-152.

Dudhagara, D.R., R.K. Rajpara, J.K. Bhatt, H.B. Gosai, B.K. Sachaniya and B.P. Dave. 2016a. Distribution, sources and ecological risk assessment of PAHs in historically contaminated surface sediments at Bhavnagar coast, Gujarat, India. Environ. Pollut. 213: 338-346.

Dudhagara, D.R., R.K. Rajpara, J.K. Bhatt, H.B. Gosai and B.P. Dave. 2016b. Bioengineering for polycyclic aromatic hydrocarbon degradation by *Mycobacterium litorale*: Statistical and artificial neural network (ANN) approach. Chemom. Intell. Lab. Syst. 159: 155-163.

Duran, R. and C. Cravo-Laureau. 2016. Role of environmental factors and microorganisms in determining the fate of polycyclic aromatic hydrocarbons in the marine environment. FEMS Microbiol. Rev. 40(6): 814-830.

Gosai, H.B., B.K. Sachaniya, D.R. Dudhagara, R.K. Rajpara and B.P. Dave. 2017. Concentrations, input prediction and probabilistic biological risk assessment of polycyclic aromatic hydrocarbons (PAHs) along Gujarat coastline. Environ. Geochem. Health. 40(2): 653-665.

Haritash, A.K. and C.P. Kaushik. 2009. Biodegradation aspects of polycyclic aromatic hydrocarbons (PAHs): A review. J. Hazard. Mater. 169(1): 1-15.

Huang, Y., A.N. Fulton and A.A. Keller. 2016. Simultaneous removal of PAHs and metal contaminants from water using magnetic nanoparticle adsorbents. Sci. Total Environ. 571: 1029-1036.

Kalantary, R.R., A. Mohseni-Bandpi, A. Esrafili, S. Nasseri, F.R. Ashmagh, S. Jorfi, et al. 2014. Effectiveness of biostimulation through nutrient content on the bioremediation of phenanthrene contaminated soil. J. Environ. Health Sci. 12(1): 143.

Kao, N.H., M.C. Su, J.R. Fan and C.C. Yen. 2015. Investigation of polycyclic aromatic hydrocarbons (PAHs) and cyclic terpenoid biomarkers in the sediments of fishing harbors in Taiwan. Mar. Pollut. Bull. 97(1): 319-332.

Karigar, C.S. and S.S. Rao. 2011. Role of microbial enzymes in the bioremediation of pollutants: A review. Enzyme Res. 2011: 11.

Kim, I.S., J.S. Park and K.W. Kim. 2001. Enhanced biodegradation of polycyclic aromatic hydrocarbons using non-ionic surfactants in soil slurry. J. Appl. Geochem. 16: 1419-1428.

Kim, K.H., S.A. Jahan, E. Kabir and R.J. Brown. 2013. A review of airborne polycyclic aromatic hydrocarbons (PAHs) and their human health effects. Environ. Int. 60: 71-80.

Kimes, N.E., A.V. Callaghan, J.M. Suflita and P.J. Morris. 2014. Microbial transformation of the Deepwater Horizon oil spill—past, present and future perspectives. Front. Microbiol. 5: 603.

Kotterman, M.J., H.J. Rietberg, A. Hage and J.A. Field. 1998. Polycyclic aromatic hydrocarbon oxidation by the white-rot fungus *Bjerkandera* sp. strain BOS55 in the presence of nonionic surfactants. Biotechnol. Bioeng. 57(2): 220-227.

Kuppusamy, S., P. Thavamani, K. Venkateswarlu, Y.B. Lee, R. Naidu and M. Megharaj. 2016a. Remediation approaches for polycyclic aromatic hydrocarbons (PAHs) contaminated soils: Technological constraints, emerging trends and future directions. Chemosphere 168: 944-968.

Kuppusamy, S., P. Thavamani, M. Megharaj, Y.B. Lee and R. Naidu. 2016b. Polyaromatic hydrocarbon (PAH) degradation potential of a new acid tolerant, diazotrophic P-solubilizing and heavy metal resistant bacterium Cupriavidus sp. MTS-7 isolated from long-term mixed contaminated soil. Chemosphere 162: 31-39.

Kuppusamy, S., P. Thavamani, M. Megharaj, Y.B. Lee and R. Naidu. 2016c. Isolation and characterization of polycyclic aromatic hydrocarbons (PAHs) degrading, pH tolerant, N-fixing and P-solubilizing novel bacteria from manufactured gas plant (MGP) site soils. Environ. Technol. Innov. 6: 204-219.

Kuppusamy, S., T. Palanisami, M. Megharaj, K. Venkateswarlu and R. Naidu. 2016d. *In situ* remediation approaches for the management of contaminated sites: A comprehensive overview. In Rev. Environ. Contam. Toxicol. Volume 236 (pp. 1-115). Springer International Publishing.

Laha, S. and R.G. Luthy. 1992. Effects of nonionic surfactants on the solubilization and mineralization of phenanthrene in soil-water systems. Biotechnol. Bioeng. 40: 1367-1380.

Lamichhane, S., K.B. Krishna and R. Sarukkalige. 2016. Polycyclic aromatic hydrocarbons (PAHs) removal by sorption: A review. Chemosphere 148: 336-353.

Liebeg, E.W. and T.J. Cutright. 1999. The investigation of enhanced bioremediation through the addition of macro and micro nutrients in a PAH contaminated soil. Int. Biodeterior. Biodegr. 44(1): 55-64.

Lu, G.N., X.Q. Tao, Z. Dang, X.Y. Yi and C. Yang. 2008. Estimation of n-octanol/water partition coefficients of polycyclic aromatic hydrocarbons by quantum chemical descriptors. Open Chem. J. 6(2): 310-318.

Masood, N., M.P. Zakaria, N. Halimoon, A.Z. Aris, S.M. Magam, N. Kannan, et al. 2016. Anthropogenic waste indicators (AWIs), particularly PAHs and LABs, in Malaysian sediments: Application of aquatic environment for identifying anthropogenic pollution. Mar. Pollut. Bull. 102(1): 160-175.

McGenity, T.J. 2014. Hydrocarbon biodegradation in intertidal wetland sediments. Curr. Opin. Biotechnol. 27: 46-54.

Morgan, P. and R.J. Watkinson. 1989. Hydrocarbon degradation in soils and methods for soil biotreatment. Crit. Rev. Biotechnol. 4: 305-333.

Mueller, J.G., C.E. Cerniglia and P.H. Pritchard. 1996. Bioremediation of environments contaminated by polycyclic aromatic hydrocarbons. Biotechnol. Res. Ser. 6: 125-194.

Mulligan, C.N., R.N. Young and B.F. Gibbs. 2001. Surfactant-enhanced remediation of contaminated soil: A review. Eng. Geol. 60: 371-380.

Mulligan, C.N., S.K. Sharma and A. Mudhoo (eds.). 2014. Biosurfactants: Research Trends and Applications. CRC Press.

Namkoong, W., E.Y. Hwang, J.S. Park and J.Y. Choi. 2002. Bioremediation of diesel-contaminated soil with composting. Environ. Pollut. 119(1): 23-31.

Oleszczuk, P. and S. Baran. 2006. Content of potentially bioavailable polycyclic aromatic hydrocarbons in rhizosphere soil in relation to properties of soils. Chem. Speciat. Bioavailab. 18(1): 39-48.

Oudot, J., F.X. Merlin and P. Pinvidic. 1998. Weathering rates of oil components in a bioremediation experiment in estuarine sediments. Mar. Environ. Res. 45: 113e125.

Patel, V., S. Jain and D. Madamwar. 2012. Naphthalene degradation by bacterial consortium (DV-AL) developed from Alang-Sosiya ship breaking yard, Gujarat, India. Bioresour. Technol. 107: 122-130.

Ramakrishnan, B., M. Megharaj, K. Venkateswarlu, N. Sethunathan and R. Naidu. 2011. Mixtures of environmental pollutants: Effects on microorganisms and their activities in soils. Rev. Environ. Contam. T. Volume 211 (pp. 63-120). Springer New York.

Ramanan, R., B.H. Kim, D.H. Cho, H.M. Oh and H.S. Kim. 2016. Algae-bacteria interactions: Evolution, ecology and emerging applications. Biotechnol. Adv. 34(1): 14-29.

Ravindra, K., R. Sokhi and R. Van Grieken. 2008. Atmospheric polycyclic aromatic hydrocarbons: Source attribution, emission factors and regulation. Atmos. Environ. 42(13): 2895-2921.

Rengarajan, T., P. Rajendran, N. Nandakumar, B. Lokeshkumar, P. Rajendran and I. Nishigaki. 2015. Exposure to polycyclic aromatic hydrocarbons with special focus on cancer. Asian Pac. J. Trop. Biomed. 5(3): 182-189.

Rocchetti, L., F. Beolchini, M. Ciani and A. Dell'Anno. 2011. Improvement of bioremediation performance for the degradation of petroleum hydrocarbons in contaminated sediments. Appl. Environ. Soil. Sci. 2011: 8.

Rodrigues, A., R. Nogueira, L.F. Melo and A.G. Brito. 2013. Effect of low concentrations of synthetic surfactants on polycyclic aromatic hydrocarbons (PAH) biodegradation. Int. Biodeter. Biodegr. 83: 48-55.

Sarma, H., N.F. Islam, P. Borgohain, A. Sarma and M.N.V. Prasad. 2016. Localization of polycyclic aromatic hydrocarbons and heavy metals in surface soil of Asia's oldest oil and gas drilling site in Assam, north-east India: Implications for the bio-economy. Emerg. Contam. 2(3): 119-127.

Sarma, S.J. and K. Pakshirajan. 2011. Surfactant aided biodegradation of pyrene using immobilized cells of Mycobacterium frederiksbergense. Int. Biodeter. Biodegr. 65(1): 73-77.

Sayara, T., M. Sarrà and A. Sánchez. 2010. Optimization and enhancement of soil bioremediation by composting using the experimental design technique. Biodegradation, 21(3): pp. 345-356.

Schmitt, R., H.R. Langguth, W. Püttmann, H.P. Rohns, P. Eckert and J. Schubert. 1996. Biodegradation of aromatic hydrocarbons under anoxic conditions in a shallow sand and gravel aquifer of the Lower Rhine Valley, Germany. Org. Geochem. 25(1-2): 41-50.

Semple, K.T., B.J. Reid and T.R. Fermor. 2001. Impact of composting strategies on the treatment of soils contaminated with organic pollutants. Environ. Pollut. 112(2): 269-283.

Subashchandrabose, S.R., B. Ramakrishnan, M. Megharaj, K. Venkateswarlu and R. Naidu. 2011. Consortia of cyanobacteria/microalgae and bacteria: Biotechnological potential. Biotechnol. Adv. 29(6): 896-907.

Teng, Y., Y. Luo, L. Ping, D. Zou, Z. Li and P. Christie. 2010. Effects of soil amendment with different carbon sources and other factors on the bioremediation of an aged PAH-contaminated soil. Biodegradation. 21(2): 167-178.

Tiehm, A. 1994. Degradation of polycyclic aromatic hydrocarbons in the presence of synthetic surfactants. Appl. Environ. Microbiol. 60(1): 258-263.

Vane, C.H., B.G. Rawlins, A.W. Kim, V. Moss-Hayes, C.P. Kendrick and M.J. Leng. 2013. Sedimentary transport and fate of polycyclic aromatic hydrocarbons (PAH) from managed burning of moorland vegetation on a blanket peat, South Yorkshire, UK. Sci. Total Environ. 449: 81-94.

Vidali, M. 2001. Bioremediation: An overview. Pure Appl. Chem. 73(7): 1163-1172.

Volkering, F., A.M. Breure, J.G. Van Andel and W.H. Rulkens. 1995. Influence of nonionic surfactants on bioavailability and biodegradation of polycyclic aromatic hydrocarbons. Appl. Environ. Microbiol. 61(5): 1699-1705.

Wallisch, S., T. Gril, X. Dong, G. Welzl, C. Bruns, E. Heath, et al. 2014. Effects of different compost amendments on the abundance and composition of alkB harboring bacterial communities in a soil under industrial use contaminated with hydrocarbons. Front. Microbiol. 5.

Wick, A.F., N.W. Haus, B.F. Sukkariyah, K.C. Haering and W.L. Daniels. 2011. Remediation of PAH-contaminated soils and sediments: A literature review. CSES Department, Internal Research Document, 102.

Xiao, Y., X. Tu, J. Wang, M. Zhang, Q. Cheng, W. Zeng, et al. 2003. Purification, molecular characterization and reactivity with aromatic compounds of a laccase from basidiomycete *Trametes* sp. strain AH28-2. Appl. Microbiol. Biotechnol. 60(6): 700-707.

Yang, J.G., X. Liu, T. Long, G. Yu, S. Peng and L. Zheng. 2003. Influence of nonionic surfactant on the solubilization and biodegradation of phenanthrene. Int. J. Environ. Sci. 15: 859-862.

Yuan, S.Y., S.H. Wei and B.V. Chang. 2000. Biodegradation of polycyclic aromatic hydrocarbons by a mixed culture. Chemosphere 41(9): 1463-1468.

Yuan, S.Y., C. Liu, C.S. Liao and B.V. Chang. 2002. Occurrence and microbial degradation of phthalate esters in Taiwan river sediments. Chemosphere 49(10): 1295-1299.

Yuan, S.Y., L.M. Su and B.V. Chang. 2009. Biodegradation of phenanthrene and pyrene in compost-amended soil. J. Environ. Sci. Heal. A. 44(7): 648-653.

Zhang, D., L. Zhu and F. Li. 2013. Influences and mechanisms of surfactants on pyrene biodegradation based on interactions of surfactant with a Klebsiella oxytoca strain. Bioresour. Technol. 142: 454-461.

4

Phycoremediation–A Potential Approach for Heavy Metal Removal

K. Suresh Kumar[1,*], Kamleshwar Singh[1],
Sushma Kumari[1], Pratibha Kushwaha[1] and P.V. Subba Rao[2]

INTRODUCTION

Heavy metals (HMs) are natural elements found in the Earth's crust, having high atomic mass (>23), and high density (at least 5 g cm^{-3}). While some heavy metals perform essential biological functions (e.g. zinc (Zn), copper (Cu), manganese (Mn), nickel (Ni), and cobalt (Co) are micronutrients necessary for plant growth; Cu, selenium (Se), or Zn are essential trace elements, while Co acts as the central atom in the vitamin B_{12} complex), others comprise precious noble elements like gold (Au), silver (Ag), iridium (Ir), rhodium (Rh), or platinum (Pt). However, several HMs (mercury (Hg), cadmium (Cd), arsenic (As), chromium (Cr), thallium (Tl), lead (Pb), etc.) cannot be biologically degraded, thereby either causing environmental pollution or having toxic effects on living organisms (Kumar et al. 2015, Koller and Saleh 2018). HMs of concern can be categorized as toxic metals (Hg, Cr, Pb, Zn, Cu, nickel (Ni), Cd, arsenic (As), Co, tin or stannum (Sn), etc.), precious metals (palladium (Pd), platinum (Pt), Ag, Au, ruthenium (Ru), etc.) and radionuclides (U, Th, Ra, Am, etc.), according to Wang and Chen (2009).

Rise in population and human activities (such as urbanization, industrialization, and unabated agricultural practices) have led to HM enrichment in our natural ecosystems. Indiscriminate disposal of HM-containing industrial and domestic wastes into aquatic bodies threatens all kinds of inhabiting organisms. Anthropogenic sources of HMs include process waste streams from metal plating, mining operations, and semi-conductor manufacturing operations. These HMs, at alarming toxic levels in the environment, ultimately reach the top of the food chain. Having negative impact on both the stability of the aquatic environment (causing increased eutrophication, contamination and being responsible for disappearance of certain species), they also cause damage to humans. According to Natural Health Improvement Center (2018), HMs detrimentally influence behavior by impairing mental and neurological function, influencing neurotransmitter production and utilization, and altering numerous metabolic body processes. Systems in which

[1] Department of Botany, University of Allahabad, Prayagraj 211002, India.
[2] Aquaculture Foundation of India, Madurai, Tamil Nadu, India.
[*] Corresponding author: ksuresh2779@gmail.com

toxic metal elements can induce impairment and dysfunction include the blood and cardiovascular, detoxification pathways (colon, liver, kidneys, skin), endocrine (hormonal), energy production pathways, enzymatic, gastrointestinal, immune, nervous (central and peripheral), reproductive, and urinary (Singh 2005). In addition, they also increase allergic reactions, cause genetic mutation, compete with "good" trace metals for biochemical bond sites, and act as antibiotics, killing both harmful and beneficial bacteria (NHIC 2018). The main damage caused by toxic metals is due to the oxidative free radicals they produce; these free radicals can cause tissue damage throughout the body and cause degenerative diseases.

Arumugam et al. (2018) mentioned that HMs being non-biodegradable can easily accumulate in living organisms or build-up in soil and groundwater. Certain HMs prove to be toxic and carcinogenic, even at minute concentrations. Arumugam et al. (2018) enlisted certain HMs and their toxicity as follows:

- Lead (Pb) is highly toxic to the kidney, nervous system, and reproductive system.
- Mercury (Hg) is a neurotoxin that inhibits the enzymatic activities for normal neurotransmission and causes structural damage.
- Arsenic (As) environmental exposure may cause cancer and non-cancer health effects, leading to the formation of tumours; as is transported from the water bodies to soils and plants, and builds up in many types of food crops (e.g. rice) and aquatic plants, threatening human health.
- Cadmium (Cd) exposure occurs from the manufacture of cement, construction materials, welding alloys, foundries, manufacturing steel and alloys, electroplating industry, lamps, mines, urban waste and industrial waste incineration, coal ash, tanneries, fertilizers, and wood preservatives.
- Nickel (Ni) toxicity causes alterations in plants, e.g. germination, growth of roots, stems, and leaves; it also impacts plant physiological processes, such as photosynthesis, water relations, and mineral nutrition. As nickel has the ability to generate reactive oxygen species that causes oxidative stress, it also influences metabolic reactions in plants. It harms animal liver and also causes spleen injury, lung inflammation, and cardiac toxicity.
- Chromium (Cr) is one of the toxic elements widely used in the industry, particularly in paints and metal platings as corrosion inhibitors, which eventually enter the water bodies via effluents from tanneries, textiles, electroplating, mining, dyeing, printing, photographic, and pharmaceutical industries. The presence of chromium in excess causes genotoxicity and oxidative damage to cells.

Pollution assessment and control requires regular monitoring of HM concentrations in the environment; thereafter, necessary removal or reduction measures need to be undertaken. Especially as the water bodies comprise the maximum sites of contamination, working out a legitimate means to curb the adversities of HMs in the aquatic ecosystems is an absolute requisite in order to lower the possibility of uptake into the food web. Researchers have hence been striving to develop cost effective treatment technologies for removal of HMs.

CONVENTIONAL TECHNIQUES FOR HEAVY METAL REMEDIATION

Conventional methods for removing heavy metals include chemical precipitation, lime coagulation, chemical oxidation/reduction, ion-exchange, reverse-osmosis, membrane separation and solvent extraction (Chisti 2018); besides these, techniques like coagulation/flocculation, electrocoagulation, electro-floatation, and electro-deposition are also used for HM removal. These techniques are often ineffective, expensive, give incomplete metal removal, require voluminous reagent and energy, have limited tolerance to pH change, have moderate or no metal selectivity, need very high or low working levels of metals and also produce toxic sludge or other waste products (Kanamarlapudi et al. 2018). The high cost of activated carbon has motivated scientists to search for new means of low cost adsorption. Several adsorbents such as sawdust, silica and iron oxide (Ajmal et al. 1998), bagasse fly ash (Gupta and Ali 2000), spent activated clay (Weng et al. 2007) and modified goethite (Li et al. 2007) have been used for treatment of metal rich effluents.

Owing to the fact that several biological materials show increased heavy metal binding capacities, some biomasses have been explored for effective HM removal. This innovative technology utilizing the metal-binding capacity of live or dead biological organisms has gained momentum since 1985 due to its simplicity, effectiveness and environmental friendliness (Vieira and Volesky 2000). In fact, by 1990 scientists began recovery of HMs by a process of biological sorption or biosorption recognizing the abundance of biological material, their low cost, reliability, predictability, high efficiency, minimization of chemical or biological sludge, no additional nutrient requirement, regeneration of biosorbent, possibility of metal recovery, higher affinity of the sorbent for the sorbate species and environmental friendliness; metal-sequestering properties of non-viable biomass provide a new approach to remove heavy metals when they occur at low concentrations. The economy of environmental remediation requires that the biomass must come from nature or could even be a waste material; moreover, availability is a major factor to be taken into account. Pine sawdust, sour orange, sugar beet pulp, maize cob, and carrot residue, wheat shell, seaweeds, molds, yeasts, bacteria, microalgae and crab shells comprise few of the biomasses that have been tested for metal sorption with very encouraging results (Khormaei et al. 2007, Vieira and Volesky 2000). While some biomasses bind a wide range of HMs with no specific priority, others are specific for certain types of metals.

Algae are photosynthetic organisms that grow in a range of aquatic habitats, including lakes, ponds, rivers, oceans, and wastewater. They tolerate a wide range of temperatures, salinities, pH and light intensities (Khan et al. 2018). They are broadly classified as Chlorophyta (green algae), Phaeophyta (brown algae) and Rhodophyta (red algae) etc.; however, based on the size they are classified into (i) macroalgae: multicellular, large-size algae, visible with the naked eye, and, (ii) microalgae: microscopic single cells that may be either prokaryotic, like cyanobacteria (Chloroxybacteria), or eukaryotic, like green algae (Chlorophyta).

MICROALGAE

Algae need certain heavy metals for their normal functioning, for example, iron for photosynthesis and chromium for metabolism (Priya et al. 2014). Microalgae sequester, adsorb, or metabolize suitable noxious elements at substantial levels. Recently, with the increasing interest in the use of biomass from microbial sources, researchers have begun exploring the potential of microalgae for heavy metal absorption and remediation. Microalgae have a spectrum of absorption (extracellular) and adsorption (intracellular) mechanisms for coping with heavy metal toxicity. Kaplan (2013) emphasized on the potential utilization of living and dead, prokaryotic and eukaryotic microalgal biomass in removing toxic heavy metals. Microalgae have thus emerged as a potential low-cost alternative to physicochemical treatments (Priya et al. 2014). For example, *Spirulina* sp. has the ability to remover copper (Cu^{+2}; 91%) from municipal wastewater (Salama et al. 2019), while *Chlorella minutissima* could remove 62, 84, 74, and 84% of zinc (Zn^{+2}), manganese (Mn^{+2}), cadmium (Cd^{+2}), and copper (Cu^{+2}) ions, respectively (Yang et al. 2015). Table 4.1 presents the metal remediation potential of several microalgae reported till date.

SEAWEEDS

Marine macroalgae (seaweeds) are plant – like organisms widely distributed in the marine ecosystem. Being ecologically important, they are crucial primary producers in oceanic aquatic food web. They have been commercially used for production of cosmetics, fertilizers, industrial gums and chemicals. They contain polysaccharides, proteins, fatty acids, and are rich in minerals and essential trace elements; they can be directly consumed as food or utilized in various food products. Owing to the versatility of these organisms, and, due to the fact that the natural abundance is unable to cater to the growing demand, seaweed cultivation has been popularized. The global harvest of seaweeds for food and algal products (e.g. agar, alginate, and carrageenan) exceeds 3 million tons annually (Sheng et al. 2004).

PHYCOREMEDIATION USING SEAWEEDS

Over the past years, researches have been directed at identifying readily available, efficient, and economic biomass for effective removal of metals (Sheng 2004). The term phycoremediation comprises the use of algae (micro and macroalgae) for the removal or transformation of pollutants including nutrient and xenobiotics. Seaweeds have been used for remediation of organic and inorganic pollutants present in water and wastewaters from manufacturing industries, factories, landfills, households, textile industries, petrochemical industries, aquaculture, agriculture, etc. (Arumugam et al. 2018). They have emerged as a good candidate for HM remediation due to the various advantages they offer; these include:

- All year round occurrence
- Growth rate faster than higher plants

TABLE 4.1 Metal remediation potential of microalgae.

Metal/ Metal Speciation	Microalgae	pH	Time	Type of Biomass	Uptake (mg g^{-1})	Reference	Country
Hg^{2+}	*Chlamydomonas reinhardtii*	2-7		Wild type	72.2	Mantzorou et al. (2018)	Greece
	Chlamydomonas reinhardtii	6	1	Live	72.2	Zeraatkar et al. (2016)	Iran
	Chlorella vulgaris BCC15	7		Non-living	18	Ahmad et al. (2020)	India
	Cyclotella Cryptica	4		Non-living	11.92	Ahmad et al. (2020)	India
	Phaeodactylum tricornutum	4		Non-living	0.51	Ahmad et al. (2020)	India
	Porphyridium purpureum	4		Non-living	0.51	Ahmad et al. (2020)	India
	Scendesmus subspicatus	4		Non-living	9.2	Ahmad et al. (2020)	India
Cd^{2+}	*Chlorella vulgaris*	4	2	Live	85.3	Zeraatkar et al. (2016)	Iran
	Chlamydomonas reinhardtii	3-7	1	Wild type (Immobilized)	106.6	Mantzorou et al. (2018)	Greece
	Chlamydomonas reinhardtii	7	1		145	Zeraatkar et al. (2016)	Iran
	Chlorella sorokiniana				13.33	Jacob et al. (2018)	Vietnam
	Microsterias denticulata					Jacob et al. (2018)	Vietnam
	Spirulina platensis TISIR 8217	6		Alginate immobilized	70.92	Ahmad et al. (2020)	India
	Spirulina platensis TISIR 8217	4-7		Silica immobilized	36.63	Ahmad et al. (2020)	India
	Chlorella sp.	4.5	2	Immobilized	62.3	Kumar et al. (2018)	India
	Chaetoceros calcitrants	8		Live	1055.27	Ahmad et al. (2020)	India
	Desmodesmus pleiomorphus	4		Live	61.2	Ahmad et al. (2020)	India
	Scendesmus abundans	8		Live	574	Ahmad et al. (2020)	India
	Tetraselmis chuii	8		Live	13.46	Ahmad et al. (2020)	India
	AER Chlorella	3-7		Non-living	7.74	Ahmad et al. (2020)	India
Cr^{3+}	*Chlorella miniata*	4.5	24	Live	41.12	Zeraatkar et al. (2016)	Iran
	Chlorella sorokiniana	4		Live	58.8	Zeraatkar et al. (2016)	Iran
	Chlanydomonas reinhardtii	7		Wild type	7.23 a mol/cell	Mantzorou et al. (2018)	Greece

Contd.

TABLE 4.1 Contd.

Metal/ Metal Speciation	Microalgae	pH	Time	Type of Biomass	Uptake (mg g^{-1})	Reference	Country
	Spirulina sp.				304	Ahmad et al. (2020)	India
Cr^{6+}	*Chlorella vulgaris*	1.5			140	Zeraatkar et al. (2016)	Iran
	Chlanydomonas reinhardtii	1-8	2	Live	18.2	Zeraatkar et al. (2016)	Iran
	Chlanydomonas reinhardtii	2	2	Wild type (Heated)	25.6	Mantzorou et al. (2018)	Greece
	Chlanydomonas reinhardtii	2	2	Wild type (Acid treated)	21.2	Mantzorou et al. (2018)	Greece
	Chlanydomonas reinhardtii	7		Wild type	9.36 a mol/cell	Mantzorou et al. (2018)	Greece
	Dunaliella sp. 1	2	72	Non-living	58.3	Jacob et al. (2016)	Vietnam
	Dunaliella sp. 2	2	72	Non-living	45.5	Zeraatkar et al. (2016)	Iran
	Scendesmus incrassatulus	8.9	24	Live	4.4	Zeraatkar et al. (2016)	Iran
	Spirulina sp.			Live	333	Ahmad et al. (2020)	India
Cu^{2+}	*Chlorella vulgaris*	3.5	0.5	Live	39.19	Ahmad et al. (2020)	India
	Chlorella vulgaris	3.5	0.5		14.48	Zeraatkar et al. (2016)	Iran
	Chlanydomonas reinhardtii	6		Wild type (2137)	0.109	Mantzorou et al. (2018)	Greece
	Chlorella vulgaris	3.5	3		420.67	Zeraatkar et al. (2016)	Iran
	Chlorella vulgaris	3.5	3		714.89	Zeraatkar et al. (2016)	Iran
	Chlorella vulgaris	6		Dry	5.72	Mondal et al. (2019)	India
	Chlorella sp.	7		Immobilized Sodium Alginate	33.4	Ahmad et al. (2020)	India
	Microcystis aeruginosa	4-5		Non-living	2.47	Ahmad et al. (2020)	India
	Phaeodactylum tricornutum	6		Non-living	1.67	Ahmad et al. (2020)	India

Contd.

TABLE 4.1 Contd.

Metal/ Metal Speciation	Microalgae	pH	Time	Type of Biomass	Uptake (mg g^{-1})	Reference	Country
Fe^{3+}	*Chlorella vulgaris*	2		Non-living	24.52	Ahmad et al. (2020)	India
	Microcystis sp.	9.2		Non-living	0.03	Ahmad et al. (2020)	India
Zn^{2+}	*Scendesmus obliquus* (ACO1598)	6-7	24	Live	429.6	Zeraatkar et al. (2016)	Iran
	Chlanydomonas reinhardtii	n.g		Wild type (boiled)	n.g	Mantzorou et al. (2018)	Greece
	Arthrospira platensis	5-5.5		Non-living	33.21	Ahmad et al. (2020)	India
	Desmodesmus pleiomorphus	5		Non-living	360.2	Ahmad et al. (2020)	India
	Hydrodictyon reticulatum	5		Non-living	3.7	Ahmad et al. (2020)	India
	Phaeodactylum tricornutum	6		Non-living	14.52	Ahmad et al. (2020)	India
	Pithophora odeogonia	5		Non-living	8.98	Ahmad et al. (2020)	India
U^{4+}	*Chlorella Vulgaris*	4.4			26.6	Ahmad et al. (2020)	India
U^{6+}	*Chlorella Vulgaris*	4.4	0.08		14.3	Zeraatkar et al. (2016)	Iran
	Chlorella Vulgaris	4.4	96		26.6	Zeraatkar et al. (2016)	Iran
	Chlorella Vulgaris	4.4	96		27	Zeraatkar et al. (2016)	Iran
Pb^{2+}	*Chlamydomonas reinhardtii*	2-7	1	Wild type	96.3	Mantzorou et al. (2018)	Greece
	Arthrospira platensis	5-5.5		Non-living	102.56	Ahmad et al. (2020)	India
	Cyclotella cryptica	6		Non-living	36.68	Ahmad et al. (2020)	India
	Hydrodictyon reticulatum	5		Non-living	24	Ahmad et al. (2020)	India
Ni^{2+}	*Chlorella miniata*	5	2	Immobilized Ca Alginate	28.6	Mondal et al. (2019)	India
	Chlorella sorokiniana	5	0.33		48.08	Zeraatkar et al. (2016)	Iran
	Chlorella vulgaris	5.5	0.5	Live	23.47	Zeraatkar et al. (2016)	Iran
	Arthrospira (Spirulina) platensis	5-5.5		Non-Living	20.78	Mondal et al. (2019)	India

- Inexpensive
- High efficiency in metal removal
- Faster uptake capacity – time saving
- Ecofriendliness
- User friendliness – easy handling
- Ability to be recycled/reused
- High selectivity
- No toxic waste generation
- No synthesis required – Energy saving
- Suitable for both batch and continuous systems
- Live biomass needs minimum nutrients and environmental conditions, and dead biomass does not require specific nutrients or oxygen
- Suitable for aerobic and anaerobic systems
- Could be immobilized; however (unlike microbes), this is not a requisite
- Effective water at high and low metal concentrations
- Simple process for metal recovery and few chemicals needed for desorption
- Effective in case of multi-metal solutions.

There are several reports available enlisting the use of seaweed for HM remediation. Arumugam et al. (2018) recommended the use of seaweed for wastewater treatment as they reduced or removed toxic HMs such as Chromium (Cr), Nickel (Ni), Copper (Cu), Arsenic (As), Cadmium (Cd), Mercury (Hg) and Lead (Pb). For example, *Ulva reticulata* effectively biosorbs copper, cobalt and nickel from aqueous solutions (Vijayaraghavan et al. 2005); it has been successfully used for mercury biosorption too (Zeroual et al. 2003). Moreover, being one of the most commonly occurring macroalgae used for heavy metal biosorption, *Ulva* sp. solves eutrophication problem from the environment. It is also known for nitrogen and ammonia removal in fish aquaculture. Lee et al. (2000) screened 48 species of brown, green and red seaweeds for uptake of hexavalent chromium. Jalali et al. (2002) reported the biosorption of lead by eight species of seaweeds. According to Mehta and Gaur (2005), the ubiquitous seaweed, *Sargassum* sp., is most exploited for heavy metal sorption; the non-living biomass of *Sargassum* species, *Macrocystis pyrifera, Kjellmaniella crassiforia, Undaria pinnatifida* and *Ulva* species are known to effectively remove Cd, Cu, Zn, Cr and Ni. Ortiz–Calderon et al. (2017) evaluated seaweed biosorbent materials, enlisting several brown seaweeds that are highly effective and reliable for HM removal; e.g. *Sargassum* species was efficient in Cu remediation. Several reports on seaweed HM removal and remediation can be seen in Table 4.2 to 4.4. Baghour (2017) enlisted metal sorption by several seaweeds based on type as follows:

Green Seaweed

As: *Codium cuneatum, Maugeotia genuflexa, Rhizoclonium tortuosum, Ulothrix cylindricum*; B: *Caulerpa racemosa*; Ba: *Codium cuneatum*; Cd: *Enteromorpha*

TABLE 4.2 Metal remediation potential of green seaweeds

Metal/ Metal Speciation	Seaweed (Chlorophyta)	pH	Time	Type of Biomass	Metal (mmol g^{-1})	Uptake (mg g^{-1})	Reference	Country
Pb^{2+}	*Ulva fasciata*	5	3 min (65-90%)	Treated		149.3	Nessim et al. (2011)	Egypt
	Ulva fasciata	5	3 min (65-90%)	Dry		80.7	Nessim et al. (2011)	Egypt
	Caulerpa racemosa	5		Dry		40.11	Tamilselvan et al. (2012)	India
	Ulva lactuca	5.0		Dry		68.9	Ibrahim et al. (2016)	Egypt
	Ulva lactuca	5.0		Treated		83.3	Ibrahim et al. (2016)	Egypt
Cu^{2-}	*Cladophora* sp.	5.46		Treated	1.059		Sargin et al. (2016)	Turkey
	Ulva lactuca	5.0		Dry		64.5	Ibrahim et al. (2016)	Egypt
	Ulva lactuca	5.0		Treated		84.7	Ibrahim et al. (2016)	Egypt
Ni^{2+}	*Cladophora* sp.	5.77	60	Treated	0.239		Sargin et al. (2016)	Turkey
Zn^{2+}	*Cladophora* sp.	5.65	60	Treated	0.310		Sargin et al. (2016)	Turkey
	Ulva lactuca	6	60 min	Dry		107.29	Shaaban et al. (2017)	Egypt
	Ulva lactuca	5	30 min (79.53%)	Dry	0.5012		Bulgariu and Bulgariu (2014)	Romania
	Ulva lactuca	5	5 min (88.82%)	Treated	0.6886		Bulgariu and Bulgariu (2014)	Romania
Cd^{2+}	*Ulva fasciata*	5	3 min (65-90%)	Treated		119.1	Nessim et al. (2011)	Egypt
	Ulva fasciata	5	3 min (65-90%)	Dry		43.1	Nessim et al. (2011)	Egypt
	Caulerpa racemosa	5		Dry		10.4	Tamilselvan et al. (2012)	India
	Ulva lactuca	5	30 min (79.88%)	Dry	0.4538		Bulgariu and Bulgariu (2014)	Romania
	Ulva lactuca	5	5 min (88.92%)	Treated	0.6521		Bulgariu and Bulgariu (2014)	Romania
	Ulva lactuca	5.0		Dry		62.5	Ibrahim et al. (2016)	Egypt
	Ulva lactuca	5.0		Treated		84.6	Ibrahim et al. (2016)	Egypt
	Cladophora sp.	5.63	60	Treated	0.240		Sargin et al. (2016)	Turkey
Cr^{3+}	*Caulerpa racemosa*	5		Dry		59.0	Tamilselvan et al. (2012)	India
	Cladophora sp.	4.06		Treated	1.128		Sargin et al. (2016)	Turkey
	Ulva lactuca	5.0		Dry		60.9	Ibrahim et al. (2016)	Egypt
	Ulva lactuca	5.0		Treated		81.9	Ibrahim et al. (2016)	Egypt
Al^{3+}	*Ulva lactuca*	4	60 min	Dry		60.89	Shaaban et al. (2017)	Egypt
Fe^{3+}	*Ulva lactuca*	3	60 min	Dry		62.52	Shaaban et al. (2017)	Egypt
Co^{2+}	*Ulva lactuca*	5	30 min (78.90%)	Dry	0.2406		Bulgariu and Bulgariu (2014)	Romania
	Ulva lactuca	5	5 min (85.19%)	Treated	0.4675		Bulgariu and Bulgariu (2014)	Romania
Cr^{6+}	*Caulerpa racemosa*	5		Dry		65.96	Tamilselvan et al. (2012)	India

TABLE 4.3 Metal remediation potential of brown seaweeds.

Metal/ Metal Speciation	Seaweed (Phaeophyta)	pH	Time	Type of Biomass	Metal Uptake (mmol/g)	Metal Uptake (mg/g)	Reference	Country
Cd^{2+}	*Sargassum muticum*	5	Within 10 min	Dry		38.4	Freitas et al. (2008)	Portugal
	Fucus spiralis	5	Within 10 min	Dry		42.1	Freitas et al. (2008)	Portugal
	Bifurcaria bifurcata	5	Within 10 min	Dry		30.3	Freitas et al. (2008)	Portugal
	Sargassum sp.	5	3 min (65-90%)	Treated		94.3	Nessim et al. (2011)	Egypt
	Sargassum sp.	5	3 min (65-90%)	Dry		79.4	Nessim et al. (2011)	Egypt
	Cystoseira barbata	4	60 min	Treated		116.60	Yalçın et al. (2012)	Turkey
	Sargassum wightii	5		Dry		38.8	Tamilselvan et al. (2012)	India
	Fucus vesiculosus	4		Treated	0.87		Demey et al. (2018)	France
	Pelvetia canaliculata	4.5		Treated	1.25	140	Hackbarth et al. (2015)	Portugal
	Turbinaria conoides	5	60-80 min (90%)	Dry	0.96		Vijayaraghavan et al. (2012)	India
	Lessonia nigrescens	3.7		Dry		109.5	Gutiérrez et al. (2015)	Chile
	Durvillaea antarctica	3.7		Dry		95.3	Gutiérrez et al. (2015)	Chile
	Sargassum oligocystum		8 hr	Treated		153.85	Delshab et al. (2016)	Iran
	Laminaria sp.			Treated		23.16	Son et al. (2018)	Republic of Korea
	Sargassum fusiforme			Treated		19.40	Son et al. (2018)	Republic of Korea
	Saccharina japonica	10.5-11.3		Treated		60.7	Poo et al. (2018)	Republic of Korea
	Sargassum fusiforme	12.14		Treated		37.2	Poo et al. (2018)	Republic of Korea
Pb^{2+}	*Laminaria hyperborea*	5	1 min	Dry	0.21-0.35	50.3	Freitas et al. (2008)	Portugal
	Sargassum muticum	5	Within 10 min	Dry		38.2	Freitas et al. (2008)	Portugal
	Fucus spiralis	5	Within 10 min	Dry		43.5	Freitas et al. (2008)	Portugal
	Bifurcaria bifurcata	5	Within 10 min	Dry		52.7	Freitas et al. (2008)	Portugal
	Sargassum sp.	5	3 min (65-90%)	Treated		185.2	Nessim et al. (2011)	Egypt
	Sargassum sp.	5	3 min (65-90%)	Dry		119.1	Nessim et al. (2011)	Egypt

Contd.

TABLE 4.3 Contd.

Metal/ Metal Speciation	Seaweed (Phaeophyta)	pH	Time	Type of Biomass	Metal Uptake (mmol/g)	Metal Uptake (mg/g)	Reference	Country
	Sargassum wightii	5		Dry		170.74	Tamilselvan et al. (2012)	India
	Cystoseira barbata	4	60 min	Treated		239.82	Yalçın et al. (2012)	Turkey
	Pelvetia canaliculata	4		Treated	1.25	259	Hackbarth et al. (2015)	Portugal
	Fucus vesiculosus	4		Treated	1.09		Demey et al. (2018)	France
Cr^{6-}	*Sargassum wightii*	5		Dry		65.96	Tamilselvan et al. (2012)	India
	Nizamuddina zanardinii	1.0	90 min	Dry		38.68	Koutahzadeh et al. (2013)	Sweden
	Stoechospermum marginatum	1.0	70 min	Dry		46.78	Koutahzadeh et al. (2013)	Sweden
	Cystoseira indica	1.0	150 min	Dry		44.02	Koutahzadeh et al. (2013)	Sweden
	Dictyota cervicornis	1.0	150 min	Dry		34.92	Koutahzadeh et al. (2013)	Sweden
	Padina australi,	1.0	150 min	Dry		31.36	Koutahzadeh et al. (2013)	Sweden
	Sargassum glaucescens	1.0	150 min	Dry		118.33	Koutahzadeh et al. (2013)	Sweden
Cr^{3+}	*Sargassum wightii*	5		Dry		79.60	Tamilselvan et al. (2012)	India
Zn^{2+}	*Laminaria hyperborea*	5.5	Within 10 min	Dry		19.2	Freitas et al. (2008)	Portugal
	Sargassum muticum	5	Within 10 min	Dry		34.1	Freitas et al. (2008)	Portugal
	Fucus spialis	5	Within 10 min	Dry		34.3	Freitas et al. (2008)	Portugal
	Bifurcaria bifurcate	5	Within 10 min	Dry		30.3	Freitas et al. (2008)	Portugal
	Sargassum ilicifolium	5	60 min	Dry	1.39		Tabaraki and Nateghi (2014)	Iran
	Colpomenia sinuosa	6	60 min	Dry		111.57	Shaaban et al. (2017)	Egypt
	Fucus vesiculosus	4		Treated	1.07		Demey et al. (2018)	France
	Laminaria sp.			Treated		22.22	Son et al. (2018)	Republic of Korea
	Sargassum fusiforme			Treated		19.13	Son et al. (2018)	Republic of Korea
	Saccharina japonica	10.5-11.3		Treated		84.3	Poo et al. (2018)	Republic of Korea
	Sargassum fusiforme	12.14		Treated		43.0	Poo et al. (2018)	Republic of Korea

Contd.

TABLE 4.3 Contd.

Metal/ Metal Speciation	Seaweed (Phaeophyta)	pH	Time	Type of Biomass	Metal Uptake (mmol/g)	Metal Uptake (mg/g)	Reference	Country
Cu	*Sargassum vulgare*	4.5		Dry	0.93		Davis et al. (2000)	Canada
	Sargassum filipendula	4.5		Dry	0.89		Davis et al. (2000)	Canada
	Sargassum fluitans	4.5		Dry	0.80		Davis et al. (2000)	Canada
Cu^{2+}	*Sargassum ilicifolium*	5	60 min	Dry	1.27		Tabaraki and Nateghi (2014)	Iran
	Cystoseira indica	5		Treated	1.24		Keshtkar et al. (2015)	Iran
	Sargassum oligocystum		8 hr	Treated		45.25	Delshab et al. (2016)	Iran
	Padina sanctae crucis	6	80 min (96.68%)	Treated		13.996	Foroutan et al. (2018)	Iran
	Fucus vesiculosus	4		Treated	1.44		Demey et al. (2018)	France
	Laminaria sp.			Treated		55.86	Son et al. (2018)	Iran
	Sargassum fusiforme			Treated		47.75	Son et al. (2018)	Republic of Korea
	Saccharina japonica	10.5-11.3		Treated		98.6	Poo et al. (2018)	Republic of Korea
	Sargassum fusiforme	12.14		Treated		94.1	Poo et al. (2018)	Republic of Korea
	Sargassum filipendula	5.0		Treated	0.53		Júnior et al. (2019)	Brazil
Ni^{2+}	*Cystoseira barbata*	4	60 min	Treated		58.08	Yalçın et al. (2012)	Turkey
	Sargassum ilicifolium	5	60 min	Dry	1.36		Tabaraki and Nateghi (2014)	Iran
	Cystoseira indica	5		Treated	0.53		Keshtkar et al. (2015)	Iran
	Fucus vesiculosus	4		Treated	1.03		Demey et al. (2018)	France
Ag	*Sargassum filipendula*	5.0		Treated	0.49		Júnior et al. (2019)	Brazil
Co	*Padinasanctaecrucis*	6	80 min (94.56%)	Treated		13.73	Foroutan et al. (2018)	Iran
Al^{3+}	*Turbinaria conoides*	4	60 min (90%)	Dry	2.37		Vijayaraghavan et al. (2012)	India
	Colpomenia sinuosa	4	60 min	Dry		63.78	Shaaban et al. (2017)	Egypt
Fe^{3+}	*Colpomenia sinuosa*	3	60 min	Dry		65.95	Shaaban et al. (2017)	Egypt
Hg^{2+}	*Sargassum oligocystum*		8 hr	Treated		60.25	Delshab et al. (2016)	Iran
U^{6+}	*Cystoseira indica*	5		Treated	1.85		Keshtkar et al. (2015)	Iran

TABLE 4.4 Metal remediation potential of red seaweeds.

Metal/ Metal Speciation	Seaweed (Rhodophyta)	pH	Time	Type of Biomass	Metal Uptake (mmol g^{-1})	Metal Uptake (mg g^{-1})	Reference	Country
Pb^{2+}	*Jania rubens*	5	60 min	Dry		30.6	Ibrahim (2011)	Egypt
	Pterocladia capillacea	5	60 min	Dry		34.1	Ibrahim (2011)	Egypt
	Corallina mediterranea	5	60 min	Dry		64.3	Ibrahim (2011)	Egypt
	Galaxaura oblongata	5	60 min	Dry		88.6	Ibrahim (2011)	Egypt
	Chondracanthus chamissoi	4		Dry	1.37		Yipmantin et al. (2011)	France
	Porphyra leucosticta	8	120 min	Dry		31.45	Ye et al. (2015)	China
	Kappaphycus sp.	5	120 min	Dry		22.17	Rahman and Sathasivam (2015)	Malaysia
	Kappaphycus alvarezii	4.5	45 min (90%)	Dry	0.51		Praveen and Vijayaraghavan (2015)	India
	Eucheuma denticulatum	5	30 min (75%)	Dry		81.97	Rahman and Sathasivam (2016)	Malaysia
Cu^{2+}	*Kappaphycus* sp.	5	120 min	Dry		19.49	Rahman and Sathasivam (2015)	Malaysia
	Kappaphycus alvarezii	4.5	45 min (90%)	Dry	0.47		Praveen and Vijayaraghavan (2015)	India
	Eucheuma denticulatum	5	30 min (75%)	Dry		66.23	Rahman and Sathasivam (2016)	Malaysia
	Kappaphycus alvarezii	7	70 min	Dry		27.02-1000	Pandya et al. (2017)	India
Fe^{2+}	*Kappaphycus* sp.	5	120 min	Dry		16.92	Rahman and Sathasivam (2015)	Malaysia
	Eucheuma denticulatum	5	30 min (75%)	Dry		51.02	Rahman and Sathasivam (2016)	Malaysia
Zn^{2+}	*Kappaphycus* sp.	5	120 min	Dry		16.23	Rahman and Sathasivam (2015)	Malaysia
	Eucheuma denticulatum	5	30 min (75%)	Dry		43.48	Rahman and Sathasivam (2016)	Malaysia
	Gelidium latifolium	6	60 min	Dry		101.29	Shaaban et al. (2017)	Egypt
Ni^{2+}	*Kappaphycus alvarezii*	4.5	90% 45 min	Dry	0.38		Praveen and Vijayaraghavan (2015)	India
	Kappaphycus alvarezii	7	70 min	Dry		10.30-90.9	Pandya et al. (2017)	India

Contd.

TABLE 4.4 Contd.

Metal/ Metal Speciation	Seaweed (Rhodophyta)	pH	Time	Type of Biomass	Metal Uptake (mmol g^{-1})	Metal Uptake (mg g^{-1})	Reference	Country
Cd^{2+}	*Jania rubens*	5	60 min	Dry		30.5	Ibrahim (2011)	Egypt
	Pterocladia capillacea	5	60 min	Dry		33.5	Ibrahim (2011)	Egypt
	Corallina mediterranea	5	60 min	Dry		64.1	Ibrahim (2011)	Egypt
	Galaxaura oblongata	5	60 min	Dry		85.5	Ibrahim (2011)	Egypt
	Chondracanthus chamissoi	4		Dry	0.76		Yipmantin et al. (2011)	France
	Porphyra leucosticta	8	120 min	Dry		36.63	Ye et al. (2015)	China
	Kappaphycus alvarezii	4.5	45 min (90%)	Dry	0.48		Praveen and Vijayaraghavan (2015)	India
Cr^{6+}	*Kappaphycus alvarezii*	7	70 min	Dry		55.55-500	Pandya et al. (2017)	India
Al^{3+}	*Gelidium latifolium*	4	60 min	Dry		56.79	Shaaban et al. (2017)	Egypt
Fe^{3+}	*Gelidium latifolium*	3	60 min	Dry		61.09	Shaaban et al. (2017)	Egypt
Cr^{3+}	*Jania rubens*	5	60 min	Dry		28.5	Ibrahim (2011)	Egypt
	Pterocladia capillacea	5	60 min	Dry		34.7	Ibrahim (2011)	Egypt
	Corallina mediterranea	5	60 min	Dry		70.3	Ibrahim (2011)	Egypt
	Galaxaura oblongata	5	60 min	Dry		105.2	Ibrahim (2011)	Egypt
Co^{2+}	*Jania rubens*	5	60 min	Dry		32.6	Ibrahim (2011)	Egypt
	Pterocladia capillacea	5	60 min	Dry		52.6	Ibrahim (2011)	Egypt
	Corallina mediterranea	5	60 min	Dry		76.2	Ibrahim (2011)	Egypt
	Galaxaura oblongata	5	60 min	Dry		74.2	Ibrahim (2011)	Egypt

sp., *Cladophora fasicularis, Codium tomentosum*; Co: *Enteromorpha intestinalis, Ulva lactuca*; Cr: *Enteromorpha* sp., *Ulva* sp.; Cu: *Codium tomentosum, Enteromorpha* sp., *Rhizoclonium tortuosum, Ulva lactuca, Ulva* sp.; Fe: *Codium cuneatum, Enteromorpha* sp., *Ulva lactuca*; Hg: *Enteromorpha* sp., *Ulva lactuca*; Mn: *Ulva lactuca*; Ni: *Ulva lactuca, Enteromorpha intestinalis*; Pb: *Cladophora fasicularis, Enteromorpha* sp., *Rhizoclonium tortuosum*; Sr: *Codium cuneatum*; Zn: *Enteromorpha* sp., *Rhizoclonium tortuosum, Ulva lactuca, Ulva reticulata.*

Brown Seaweed

As: *Fucus serratus*; Au: *Ascophyllum nodosum, Chondrus crispus, Palmaria palmata, Palmaria tevera, Rhodymenia palmata, Sargassum natans*; Ba: *Padina durvillaei, Sargassum sinicola*; Cd: *Ascophyllum nodosum, Cystoseira* sp., *Fucus vesiculosus, Padina gymnospora, Sargassum natans, Turbinaria conoides*; Co: *Ascophyllum nodosum*; Cr: *Cystoseira* sp., *Dictyota bartayresiana, Fucus vesiculosus, Sargassum* sp., *Turbinaria conoides*; Cu: *Fucus serratus, Padina pavonica, Rhizoclonium tortuosum, Sargassum boveanum, Sargassum filipendula, Sargassum fluitans, Turbinaria conoides*; Fe: *Fucus vesiculosus, Cystoseira* sp., *Padina durvillaei, Sargassum fluitans, Sargassum sinicola*; Hg: *Cystoseira* sp., *Fucus vesiculosus*; Mn: *Padina gymnospora*; Ni: *Ascophyllum nodosum, Fucus vesiculosus, Padina gymnospora, Sargassum fluitans, Sargassum natans*; Pb: *Ascophyllum nodosum, Cystoseira* sp., *Fucus vesiculosus, Sargassum natans, Sargassum vulgare, Turbinaria conoides*; Sb: *Turbinaria conoides, Sargassum* sp.; Sr: *Padina durvillaei, Sargassum sinicola*; Zn: *Cystoseira* sp., *Fucus vesiculosus, Laminaria japonica, Padina pavonica, Sargassum angustifolium, Sargassum fluitans, Sargassum latifolium.*

Red Seaweed

Cd: *Gelidium floridanum, Gracillaria compressa, Kappaphycus alvarezii, Porphyra* sp., *Pterocladia capillacea*; Co: *Gracillaria compressa, Jania rubens, Kappaphycus alvarezii, Pachymeniopsis* sp., *Polysiphonia lanosa*; Cu: *Gracillaria compressa, Porphyra* sp., *Solieria chordalis*; Fe: *Gracilaria pachidermatica, Laurencia papilosa, Pterocladia capillacea,* Hg: *Gracilaria gracilis, Porphyra* sp.; Ni: *Gracillaria compressa, Gracillaria verrucosa*; Pb: *Gracilaria pachidermatica, Gelidium floridanum, Porphyra* sp., Se: *Gracilaria edulis*; Sr: *Laurencia papilosa*; Zn: *Pterocladia capillacea.*

Phaeophytes seem to be the most exploited when it comes to HM remediation, followed by Rhodophytes and Chlorophytes. Among the brown seaweeds, *Sargassum, Ascophyllum, Ecklonia* and *Laminaria* are reported to have good HM remediation efficiency. The optimum pH and time for removal of various metals vary with species. An overview of the metal removal by various seaweeds has been provided in Tables 4.1, 4.2 and 4.3. The brown seaweeds demonstrate good lead, zinc, cadmium and copper biosoprtion with varying pH ranges (~3.5 to 6) and time (10 min to 6 hours). Although the green and red seaweeds species are not as much evaluated compared to their brown counterparts, they are reported to uptake cadmium; as evidenced in the tables, optimal pH (4 to 6) and time (10-40 minutes) for heavy metal uptake in this case too varied with species. *Gracilaria* sp. is most extensively used for metal uptake studies amongst Rhodophytes. The Chlorophytic seaweeds too have good lead and cadmium uptake efficiency. Essentially, one cannot generalize the efficacy of any seaweed for its HM removal, as there are several factors that determine HM remediation efficiency. These include: concentration of metal and biomass in a solution, pH, temperature, cations, anions, metabolic stage of the organism, adsorbent dose, species of algae, form (live or dead), type of biomass (pre-treated or native) and contact time.

SIGNIFICANCE OF FUNCTIONAL GROUPS IN METAL UPTAKE

Metal accumulation capacity of algal biomass is either comparable or sometimes higher than chemical sorbents (Mehta and Gaur 2005). Metal holding properties have been attributed to several functional groups in various organisms, for example, acetamido groups of chitin, structural polysaccharides of fungi, amino and phosphate groups in nucleic acids, amino, amido, sulfhydryl, and carboxyl groups in proteins, hydroxyls in polysaccharides, and mainly carboxyls and sulfates in the polysaccharides of marine algae that belong to divisions Phaeophyta, Rhodophyta, and Chlorophyta (Baghour 2017). According to Baghour (2017), algal cell wall contains chitin, polysaccharides, proteins and lipids that have different groups such as phenolic, hydroxyl and carboxyl, which can form complexes with HMs. The presence of these functional groups makes it an ideal biosorbent for heavy metal removal. Since metal ions in water are generally in the cationic form, they are adsorbed onto the negatively charged algal cell wall consisting of functional groups [hydroxyl (–OH), phosphoryl ($-PO_3O_2$), amino ($-NH_2$), carboxyl (–COOH) and sulphydryl (–SH) confer to the cell surface, and they are associated with peptidoglycan, teichouronic acid, teichoic acids, polysaccharides and proteins] (Mehta and Gaur 2005). Each functional group has a specific pKa (dissociation constant) and it dissociates into corresponding anion and proton at a specific pH.

Among different cell wall constituents, polysaccharides and proteins have most of the metal binding sites (Kuyucak and Volesky 1989a). The composition of the cell wall is especially important in an accumulation or biosorption process (ability of biological material to accumulate heavy metals through metabolic processes or by physico-chemical uptake). The cell wall containing polysaccharides, proteins and lipids have a number of functional groups (i.e. hydroxyl, carboxyl, amino, ester, sulphydryl, carboxyl terminal, carboxyl internal), which play a key role in the biosorption of cations from aqueous solution; during biosorption, the protons and/ or light metal cations, which are naturally bound to functional groups located on the surface of biomass, are exchanged with metal cations present in aqueous solution. Phosphoryl groups are mainly associated with lipopolysaccharides, lipids and peptidoglycans of cell wall. Amino groups are associated with membrane proteins and peptide component of peptidoglycan. Accumulation is attributed to adsorption of metal onto the cell surface (wall, membrane or external polysaccharides) and binding to cytoplasmic ligands, phytochelatins and metallothioneins, and other intracellular molecules. However, as the distribution and abundance of cell wall components vary with each algal group, the number and kind of functional group also varies (Dixit and Singh 2015). According to Raize et al. (2004), oxygen atom from carboxyl group of alginicacid, sulfur atom from thiol and sulfonate groups of amino acid, sulphate, polysaccharides, fucoidan, and nitrogen atom from amine and amide groups of amino acid, peptidoglycan, act as main binding sites which are present in cell wall matrix of seaweeds.

Phaeophytes are considered most effective and promising substrates for the HM adsorption, for e.g. cobalt biosorption by *Halimeda opuntia*. *Ascophyllum nodosum* performs HM sorption at high pH (Mehta and Gaur 2005). Brown algal cell wall predominately contains cellulose (fibrous skeleton), and alginate and fucoidan (amorphous matrix and extracellular mucilage), with smaller amounts of the sulphated polysaccharide fucoidan; these cell wall constituents (alginate and fucoidan contain carboxyl and sulfate functional groups) are responsible for their HM efficiency. Mehta and Gaur (2005) reported that brown seaweeds contain greatest number of acidic functionalities (both total and weak) on the seaweed surface. According to them, the main mechanisms of heavy metals biosorption by brown seaweeds involve key functional groups such as carboxylic groups (the most abundant acidic functional group in them). Moreover, dried brown algal biomass constitutes highest percentage of titrable sites (>70%) on their major biochemical constituent alginate polymer, which directly correlates to its HM efficiency (Mehta and Gaur 2005). Alginate (D–mannuronic acid and L–guluronic acid units), comprising 40% of the dry matter, exists both within the cell wall and in the intercellular substance; their part in the cell wall may be as high as is mainly involved in metal accumulation by brown algae due to its polyguluronic acid content which shows a high specificity for divalent metal ions such as Pb^{+2}, Cu^{+2}, Cd^{+2}, Zn^{+2}, and Ca^{+2} (Mehta and Gaur 2005). Baghour (2017) reported that biosorption properties of a few algae are accredited to their cell-wall polysaccharides like alginate and fucoidan, which have a high affinity for divalent cations; further, uptake of trivalent cations by few brown algae are attributed to sulfated polysaccharide fucoidan (comprising sulfate esters with L–fucose). Sulfonic acid groups typically play a secondary role, except when metal binding takes place at low pH. Hydroxyl groups are also present in all polysaccharides but they are less abundant and only become negatively charged at pH > 10, thereby also playing a secondary role in metal binding at low pH 9 (Baghour 2017). According to Davis et al. (2003), the carboxylate and hydroxyl present in the biopolymers of biomass cell walls are mainly believed to be responsible for the sequestration of metal ion; the majority of metals of interest (i.e. Cd^{+2}, Co^{+2}, Cu^{+2}, Fe^{+2}, Ni^{+2}, Pb^{+2}) display maximal or near maximal sequestration at pHs near the apparent dissociation constant of carboxylic acids observed in brown algal biomass (pK 0 near 5). The carboxyl and amine groups present in brown algae also play a vital role in metal uptake. Carboxyl groups are the dominant species in the heavy metal sorption by *Sargassum* biomass, especially nickel; however, groups containing nitrogen and sulfur, such as amino/amido and sulfonate/thiol, are also involved in the adsorption of lead and cadmium

(Raize et al. 2004). Kuyucak and Volesky (1989b) demonstrated the binding of cobalt on carbonyl groups of non living *Sargassum natans*; they have also reported Au uptake by *Sargassum natans* (Volesky and Kuyucak 1998). Thiol group plays a vital role in sorption of metals like cadmium at acidic pH (<2) (Sheng et al. 2004). Kratochvil and Volesky (1998) estimated other functional groups, such as the strongly acidic sulfate groups ($R–SO_3^-$) present in cell wall polymers (fucoidan, carrageenans), to contribute 10% of the overall metal binding sites of brown seaweeds. Essentially, both the Pheaophytes and Rhodophytes contain the largest amount of amorphous embedding matrix polysaccharides that make them excellent HM binders.

Green seaweeds could contain water-soluble ulvans, water insoluble cellulose, as well as alkali-soluble linear xyloglucan and a glucuronan; the carboxylic and sulphate groups of these polysaccharides sequester HMs (Mehta and Gaur 2005, Baghour 2017). Chlorophyta comprises of xylans and mannans. The cell wall of green algae contains heteropolysaccharides, which offer carboxyl and sulfate groups for sequestration of heavy metal ions. The protein content of cell wall of green algae is approximately 13%; this protein contains carboxylic (aspartic and glutamic acid) and amino groups (lysine and arginine) (Mehta and Gaur 2005). Particularly, the extracted polysaccharides (12% DW) of *Ulva* sp. is reported to contain 16% sulfate and 15-19% uronic acid. Protein content of green algal cell wall generally ranges from 10 to 70%; according to Mehta and Gaur (2005), aspartic and glutamic acid comprise 12% of the protein content, which corresponds to 0.15 meq g^{-1} of carboxylic groups per dry weight, while lysine and arginine make 13% of the protein (~ 0.08 meq g^{-1} of amino groups).

In addition to xylans and cellulose, rhodophytes contain a number of sulphated galactan (e.g. agar, carrageenan and porphyran; Baghour 2017). The level of sulfation in the carrageenan is known to influence the HM metal binding capacity of seaweed. Moreover, the biosorption of Cu(II) by the dried biomass of two red seaweeds, *Palmaria palmata* and *Polysiphonialanosa*, are attributed to carboxyl and sulphonate functionalities; further, the amino and hydroxyl groups too take part in Cu(II) binding in *P. lanosa* (Baghour 2017). While the presence of uronic acid in *Palmaria decipiens* has been pointed out in the neutral xylan and xylogalactan, Murphy (2007) emphasized that the uronic acid, in all the three algal groups, has been reported to assist cadmium binding. Red algae contain cellulose in the cell wall; nonetheless, their biosorption capacities are attributed mainly to the presence of sulfated polysaccharides made of galactans (He and Chen 2014).

CATEGORIZATION OF METAL UPTAKE

The chief mechanisms for heavy metal binding can be categorized as follows:

Ion Exchange

Mehta and Gaur (2005) reported ion exchange to be the most important mechanism for the biosorption of metal ions by algal biomass. Murphy (2007) indicated ion-exchange to involve a number of physical (electrostatic or London–van der Waals forces) and chemical binding (ionic and covalent) to be involved in HM binding by seaweeds. Essentially, sorption is the binding of a metal to a free site as opposed to a site that was previously occupied by another cation (Murphy 2007). Sorption is a property of both living and dead organisms (and their components) and has been heralded as an encouraging technique for pollutant removal from solution, and/or pollutant recovery, for a number of years; this is due to its efficiency, simplicity, analogous operation to conventional ion exchange technology, and availability of biomass (Gadd 2009). The term biosorption refers to adsorption of metal ions on the dead biomass, and includes metal ion binding on extracellular as well as intracellular ligands. Algae may almost completely remove metal ions from solutions having low metal concentrations (Mehta and Gaur 2005). In fact, seaweed are considered as natural ion-exchange material that primarily contain weak acidic and basic group; according to the ion exchange acid-base equilibrium theory at pH range 2.5 to 5, the binding of heavy metal cations is determined mainly by the state of dissociation of the weak acidic groups (Murphy 2007). Untreated algal biomass contains alkali and alkaline earth metal ions like K^+ and Na^+.

Seaweed biomass generally contains light metal ions such as K^+, Na^+, Ca^{+2}, and Mg^{+2}, which are originally bound to the acid functional groups of the algae and are acquired from seawater (Abdi and Kazemi 2015). When these natural seaweeds are placed in a HM bearing solution, a pH increase along with release of light metal ions is evidenced, which explains the ion exchange mechanism (light metals balanced the uptake of protons and heavy metals; Murphy 2007). During the ion exchange process, calcium, magnesium and hydrogen cations in the algal cell wall matrix are replaced by the heavy metals (Raize et al. 2004). Raize et al. (2004) suggested cadmium cation sequestration mechanism by *Sargassum* sp. to occur due to chelation while nickel cation sequestration mechanism to occur due to ion exchange. Schiewer and Volesky (2000) also explained ion-exchange biosorption mechanism. According to Lobban and Harrison (1994), Ca, Sr and Mg concentrations in brown seaweeds are largely the result of ion exchange between seawater and alginate in the cell walls by a process of cation adsorption called Donnan Exchange System. Reports (Kuyucak and Volesky 1989a, b) suggest that cadmium ions replace the calcium and magnesium ions bound to the cell wall matrix of *Ulva lactuca* and *A. nodosum* to form stronger cross-linking. This occurs due to stronger electrostatic and coordinative bonding between the cadmium and the negatively charged chemical groups in the cell wall polymers (mainly carboxyl groups in the alginic aid monomers). Percival and McDowell (1967) too reported earlier that a gel based on sodium alginate after cross-linking with different divalent metals changed its volume, depending on the metal being used; they suggested that a reduction in gel volume followed the order Pb > Cu > Cd > Ba > Ca, where lead formed the gels of greatest density and calcium gave gels of low density. According to them, concomitant release of Ca^{+2} during Ni^{+2} uptake by *Ecklonia maxima* occurs due to ion exchange. Kuyucak and Volesky (1989b) reported that cobalt biosorption by non-living biomass of the brown marine macroalgae *Ascophyllum nodosum* was predominantly an ion-exchange process, where the carboxyl groups of the cell wall alginate play an important role in cobalt binding. Seaweeds can either be 1) treated to obtain protonated biomass (using a strong acid (HCl)), whereby the proton displaces the light metal ions from the binding sites (i.e. carboxylic, sulfonic, and others), or 2) can be treated using aqueous solution of a given ion at high concentration so that the majority of sites are occupied (e.g. calcium or potassium). In case of protonated seaweed biomass, an ion-exchange occurs between protons and the heavy metal ions at the binding sites (Kuyucak and Volesky 1989a, c). The precise binding mechanism could be either physical (i.e. electrostatic or London–van der Waals forces) or chemical binding (i.e. ionic and covalent). Sorption would therefore refer to binding of a metal cation to a free site as opposed to one that was previously occupied by another cation. This is distinct from adsorption that, strictly speaking, defines binding in terms of a physical rather than chemical surface phenomenon. In the case of biosorption of heavy metals by brown algal biomass, the mechanisms can be viewed, in principle, as being extracellular, or occurring discretely at the cell wall. Intracellular sorption would normally imply bioaccumulation by a viable organism.

Rajfur (2013) explained that the main mechanism of heavy-metal cation sorption in algae biomass is the ion-exchange between the solution and the thallus; their study showed that the algal cell wall not only contained organic compounds, but also ions such as Na^+, K^+, Ca^{+2} and Mg^{+2} ions. They reported that during cobalt biosorption by *Ascophyllum nodosum*, there was release of Na^+, K^+, Ca^{+2} and Mg^{+2} ions from the alga cells to the solution, while during sorption of Cr^{3+} and Pb^{+2} by the alga, an increase of calcium contamination in the solution was observed (Rajfur 2013). Williams and Edyvean (1997) observed a concomitant release of Ca^{+2} during biosorption of Ni^{+2} by the brown alga *Ecklonia maxima*. Nevertheless, the ion exchange capacity is highly variable among different algal species.

Complexation

Some researchers also believe that complexation or coordination are vital in metal sorption by algae, for example *Sargassum vulgaris* binds cadmium by chelation, while it binds lead by combination of ion exchange and chelation (Raize et al. 2004). Generally, a complex is formed by the donation of electrons from a complexing ligand to a metal (Murphy 2007); as the ligand is an electron pair donor, it is described as a Lewis base while the electron acceptor is the metal described as a Lewis acid. Overall, the complexation mechanism is described as a Lewis acid–Lewis base neutralization mechanism. Here, a metal ion in an aqueous solution is surrounded by a shell of water molecules; each of the water molecules is bound to the metal by donating electrons originating from the lone pairs of the oxygen (Murphy 2007). Essentially, a combination of cations (e.g. Zn^{+2}) with molecules or anions containing free electrons pairs (bases) is called coordination or complex formation. The heavy metal cation that binds is often designated as the central atom; it is distinguished from the anion or molecule ligand with which it forms a coordination compound.

According to Abdi and Kazemi (2015), coordination, or complex formation, may be either electrostatic (i.e. Coulombic) or covalent in character. The heavy metal cation that is bound is frequently designated as the central atom, and is distinguished from the anions or molecules with which it forms a coordination compound, the ligands. When the ligand is composed of several atoms, the atom responsible for the basic or nucleophilic nature of the ligand is termed the ligand atom. A base containing more than one ligand atom, a multidentate complex, could occupy more than one coordination position in the complex. The complex formation with multidentate ligands is called chelation, and the complexes are known as chelates. However, lead ions exhibit a higher affinity for algal biomass and their mechanism includes a combination of ion exchange, chelation and reduction reactions accompanied by metallic lead precipitation on the cell wall matrix (Abdi and Kazemi 2015).

Denticity

Ligands are characterized by the number of metal binding groups that they contain; unidentate ligands attach through one coordinating group and tend to result in formation of ionic water soluble complexes, while multi-dentate ligands contain more than one group capable of binding to the metal centre. In case of poly dentate ligands, a reagent that attaches to a single metal centre in this way is termed a chelating agent and the complex is called chelates (where the ring closure with a metal ion occurs). Generally, most chelating ligands contain nitrogen, oxygen and sulfur donors. The number of ligand atoms surrounding the central atom is called the coordination number; this number may vary but usually most metal cations engage in 2, 4, 6 and 8 coordination, especially in the seaweed context (Murphy 2007).

Complex Formation with Biomaterials

Murphy (2007) also describes the process of complexation or coordination of a central heavy metal to a mulit dentate ligand (the surface of an algae). The algal surface behaves as a heterogenous ligand and a complicated system and is treated as a polyfunctional macromolecule. Different metal ions bind differently to biomaterials having varying functional groups depending on the ionic properties such as electronegativity, ionization potential, ionic radius and redox potential of the metals. According to Allen and Brown (1995), the more electronegative metal ions are strongly attracted to the surface; therefore, the sorption of Cu^{+2} is smaller than that for Pb^{+2} due to its lower electronegativity (Allen and Brown 1995). The sorption capacity also depends on the ionic radii of the metal ions; molecules with smaller ionic radii are quickly sorbed on to a fixed surface area of macroalgae; for e.g. smaller ionic radius of Zn^{+2} (0.95A) is responsible for its higher sorption (Cho et al. 1994).

The varying mannuronic acid (M): guluronic acid (G) ratios influence metal binding in alginate containing algae. For example, the presence of high G content (due to its zigzag structure) increases the affinity of alginate for divalent ions. While the alginate rich in G residues have higher divalent metal ion selectivity, the regions rich in M residues are predominately monodentate and therefore weaker (Haug 1961, Murphy 2007).

Hard Soft Acid Base Theory

Any complexation reaction involves a metal and the complexing agent; both of these contribute to the formation of a complex. For a strong bond to form between these two, it is necessary that the characteristics of both of these match. Metals are therefore categorized as being hard or soft acids and ligands are categorized as being hard or soft bases (Howard 1998). According to the hard soft acid base theory, in a competitive situation, hard acids form complexes with hard bases and soft acids form complexes with soft bases. It means that metals that tend to bond covalently preferentially form complexes with ligands that tend to bond covalently; similarly metals that bind electrostatically prefer to form complexes with ligands that bind electrostatically. Intermediate acids will bind with harder or softer bases depending on their oxidation states. Based on Murphy (2007), hard metals include K^+, Li^+, Na^+, Cr^{+3}, Ca^{+2}, Mg^{+2}, Cr^{+6}, while soft metals include Ag^+, Au^+, Hg^+, Hg^{+2}, Cu^+, Cd^{+2}; further intermediate metals include Fe^+, Co^+, Ni^+, Cu^+, Zn^+ and Pb^{+2}. Moreover, the hard ligands include COO^-, H_2O, OH^-, RNH_2, SO_4^{-2}, while the soft ligands include RSH, SCN^- and CN^-. Thus, the Cr(III) and Cr(VI) would involve binding to groups such as carboxyl and ether groups, while Cu(II) would bind to sulfur and nitrogen containing groups.

In a general context, phycoremediation is a cost-effective and environmentally sound technology for the treatment of polluted streams and wastewaters contaminated with metals. According to Olguín and Sánchez–Galván (2012), the most commonly used parameter to assess the metal uptake of biomass is (q) expressed as mg metal g DW^{-1}. Moreover, the bioconcentration factor (BCF) is another widely used factor for evaluation of metal uptake capacity of macrophytes. However, both parameters, the metal uptake (q) and the BCF, cannot be applied to differentiate between the ability of live plants or photosynthetic microorganisms to adsorb the metal onto their surface through passive mechanisms or to accumulate the contaminant at intracellular level through metabolically active mechanisms (Olguín and Sánchez–Galván 2012).

Mamboya (2007) explains that in the marine environment, heavy metals may occur as dissolved free metal ions or as complex ions, chelated with certain inorganic ligands such as Cl^-, OH^-, CO^{3-}, and NO^{3-}; HMs can also form complexes with organic ligands such as fulvic acid, amines, humic acids, and proteins. They can be present in various particulate forms: as colloids or aggregates, bound into particles, precipitated as metal coatings on particles, incorporated into organic matter such as algae, and held in crystalline detrital particles (Mamboya 2007). Essentially, the physical and chemical forms of heavy metals in the marine environment are governed by environmental variables such as salinity, temperature, pH, redox potential, organic and particulate matter, biological activities, and metal properties (Lobban and Harrison 1994).

METAL AND SEAWEED: THE MECHANISMS

At times, algae outperform other biosorbents in removing metals from solutions; seaweed biomass serves as biotraps for the removal of heavy metals from industrial effluents (Mehta and Gaur 2005). The uptake of metals by seaweed can be categorized as follows:

Biosorption

Biosorption is a physicochemical metabolically-independent process based on a variety of mechanisms including absorption, adsorption, ion exchange, surface complexation and precipitation; it can be accomplished using dead biomass, fragments of cells and tissues, and passive uptake by live cells as a metabolically-independent process which could involve surface complexation onto cell walls or other outer layers (Abbas et al. 2014, Fomina and Gadd 2014). According to Ahalya et al. (2003), during non-metabolism dependent biosorption, metal uptake is by physico-chemical interaction between the metal and the functional groups present on the cell surface. This is based on physical adsorption, ion exchange and chemical sorption, which is not dependent on the cell metabolism. The cell wall of the biomass (which is mainly composed of polysaccharides and proteins) has abundant metal binding groups such as carboxyl, sulphate, phosphate and amino groups that facilitate metal binding; this non-metabolism dependent process is relatively rapid and can be reversible (Ahalya et al. 2003).

Although several algal species are known, only a few of them have been investigated for their metal sorption ability and subsequent utilization in aquatic pollution. The metal sorption ability of seaweeds varies with species and environmental conditions. Due to their high alginate content, brown algae such as *Ascophyllum* and *Sargassum* have been reported to sorb more metal than other algae (Mehta and Gaur 2005). According to Kumar and Kaladharan (2006), *Sargassum wightii* showed maximum metal uptake (18% to 29% of DW) at pH 4-5; the metal adsorption was fast with 70-80% taking place within 30 min. The affinity of various algal species for binding of various metal ions demonstrates different hierarchies. In general, metal ions with greater electronegativity and smaller ionic radii are preferentially sorbed by algal biomass (Mehta and Gaur 2005); Pb is sorbed maximally compared to other metals in a majority of algal species.

A commercial product, AlgaSORB, consisting of gel encapsulated algal cell wall is reported to have a remarkable affinity for Hg, Pb, Cd, Cr, Cu, Zn, Ni, Ag and Au (Darnall 1989). According to Ortiz–Calderon et al. (2017), commercial biosorbents like Metagener and RAHCO Bio-Beads could be used for HM removal from wastewaters (mainly electroplating and mining industry); further, BV–SORBEX™ (Montreal, Canada) sorbents based on metal-biomass binding could also be used. These include different-sized powders and granules made of different types of biomass, including seaweeds.

Earlier, most biosorption studies were initially oriented towards removal of HM cations, but recently, interest in binding of anions to biomass has been a growing area of research. Anthropogenic wastes from mining, electroplating, power generation, are often loaded with anionic complexes such as chromate, vandate, selenate and gold cyanide, which can be recovered by precipitation, ion exchange or activated carbon sorption. Nevertheless, anionic metal complexes effectively bind to biomass containing abundant amino groups such as chitin containing biosorbents. For example, *Ecklonia* sp. and *U. lactuca* have been used for removal of Cr(VI) (Murphy 2007).

Accumulation

Transport of the metal across the cell membrane yields intracellular accumulation, which is dependent on the cell's metabolism (Ahalya et al. 2003). This means that this kind of biosorption may take place only with viable cells. It is often associated with an active defense system of the organism, which reacts in the presence of toxic metal. In other words, bioaccumulation is a function of living organisms dependent on a variety of physical, chemical and biological mechanisms including both intra- and extracellular processes where passive uptake plays only a limited and not very well-defined role (Fomina and Gadd 2014). Seaweeds accumulate high concentrations of metals depending on their concentration in the external environment. The concentration factor for heavy metals varies greatly in different algal species, but it increases as the metal concentration in the water decreases (Mehta and Gaur 2005). The capacity of seaweed to accumulate metals depends on (i) the bioavailability of metals in the surrounding water and (ii) the uptake capacity of the seaweed. According to Murphy (2007), the process of uptake takes place in two ways, i.e. an initial rapid (passive) uptake followed by a much slower (active) uptake. During the passive uptake, metal ions are adsorbed onto the cell surface within a relatively shorter time (few seconds or minutes); this process is metabolism independent. However, the active uptake is metabolism-dependent, causing the transport of metal ions across the cell membrane into the cytoplasm. At times, the transport of metal ions might occur through passive diffusion because of metal-induced increase in permeability of the cell membrane. Specifically, the first is a surface reaction in which metals are absorbed by algal surfaces through electrostatic attraction to negatives sites; this step is independent of factors influencing metabolism (temperature, light, pH or age of the plant), but it is influenced by the relative abundance of elements in the surrounding water. Zn uptake occurs by this process. However, the second is a slower active uptake wherein the metal ions are transported across the cell membrane into the cytoplasm. This form of uptake is more dependent upon metabolic processes (it seems to be the relevant one for Cu, Mn, Se and Ni), and it is subject to variations due to changes in temperature, light, or age of the plant (Besada et al. 2009). Moreover, the intrinsic and extrinsic parameters that influence the accumulation of metals by algae include: cellular activity, exposure time, chelating species, and environmental factors (pH, salinity, organic matter, and temperature).

While adsorption contributes to more (>80%), accumulation too has a significant role in the total metal accumulation by algal cells. Nevertheless, structural differences in species influence metal uptake capacity of seaweeds; several seaweeds accumulate high levels of trace metals from the environment (sometimes, even larger quantities than those found in water samples which they inhabit), making them a good tool for monitoring heavy metals (Mehta and Gaur 2005). According to Lobban and Harrison (1994), metal accumulation in some algae is more dependent upon metabolic processes and is subject to changes in temperature, light, pH, nitrogen availability or age of the organism; for e.g. Zn uptake by *Ascophyllum nodosum* and Co uptake by *Laurencia corallopsis* occur by active transport.

FACTORS INFLUENCING METAL BIOSORPTION AND ACCUMULATION

The capacity of seaweeds to accumulate or sorb HMs depends on a variety of factors. The factors involving the metal include: the form of metal (including its solubility), concentration of HM, and whether the metal is in a mixture or individual form; further features of the sorbent material influencing metal sorption and uptake include: type of biomass (genus and species), form of biomass (metabolic stage, whether live, dead or pretreated), and the concentration of biomass. Environmental or ambient factors such as pH, temperature and presence of competing ions also need to be considered. A brief understanding, regarding factors affecting HM accumulation/ sorption by seaweeds and limitations, would enhance their utility as bioindicators or sorbents for HM removal in the environment and industry.

Functional Groups

The seaweed cell surface provides an array of ligands with various functional groups that enable binding of metals. The number of sites on the surface, their accessibility and chemical state, and the affinity between the site and metal, are factors that play pivotal role in HM binding, for e.g. cadmium binds to polyuronic acids. Uronic acids are present in all the three division of macroalgae; in addition, they contain protein with carboxyl and amino groups that also help HM binding. Murphy (2007) described that alginic acid is the most abundant cell wall polysaccharide offering carboxyl (–COOH) groups for metal binding in Phaeophyta, while fucoidan is the second abundant group that offers sulphonic acid; the $R{-}OSO_3^-$ represents 10% of metal binding site in *Sargassum* (Murphy 2007). Further, the cell wall matrix of Chlorophytes comprises sulfated heteropolysaccharides offering carboxyl and sulphonate groups for metal binding (for e.g. *Ulva* sp. has 16% sulfate and 15-16% uronic acids). Most Rhodophytes contain galactans with varying sulphation, pyruvation and methylation of hydroxyl groups and anhydride bridge. Sulphonate groups of carrageenan and sulfated galactans of agar offer anionic sites for binding polyvalent cations (Murphy 2007). The biochemistry of the seaweed influences the type and quantity of HM sorbed or accumulated.

Initial Metal Ion Concentration

In marine environment, the concentration of heavy metals is largely governed by the biological, chemical, and physical characteristics of the surrounding seawater. Mehta and Gaur (2005) stated that the sorption and removal of heavy metals by seaweed largely depends on the initial concentration of metals in the aqueous medium. Generally, the metal sorption initially increases with increase in metal concentration in the solution, and then on it reaches saturation after a certain concentration of metal. However, unlike sorption the removal of a metal generally decreases with its increasing concentration in the solution. According to Abbas et al. (2014), the initial concentration provides an important driving force to overcome all mass transfer resistance of metal between the aqueous and solid phases. They stated that the optimum percentage of metal removal can be taken at low initial metal concentration; at a given concentration of biomass, the metal uptake increases with increase in initial concentration.

pH

Metals in seawater could exist in either particulate or dissolved form; moreover, their form too could be determined by the properties of seawater and factors such as pH, salinity, redox potential, ionic strength, alkalinity, persistent organic and particulate organic matter, and biological activity (Mamboya 2007). In any natural or laboratory (batch or a continuous) system, pH could determine the type and quantity of HM binding; for e.g. at times there could be selective sorption of specific metals at a distinct pH.

According to Mehta and Gaur (2005), since a majority of the metal binding groups of algae are acidic (e.g. carboxyl), their availability is pH dependent; these groups generate a negatively charged surface at acidic pH, and electrostatic interactions between cationic species and the cell surface is responsible for metal biosorption. Nevertheless, at extreme pH (alkaline and acidic), a decreased sorption is expected. High concentration of H^+ ions at low pH decrease metal sorption by competitively excluding them from binding to ligands on the cell surface. Sorption of Cr(VI), Cd and Cs on *Padina* sp., *Sargassum* sp. and *Padina australis* occurs at pH 2 (Mehta and Gaur 2005). Yu and Kaewsarn (1999) reported only slight Cu sorption by *Durvillaea potatorum* at pH below 2, but it increased with a rise in pH. Hashim and Chu (2004) observed that at pH 3, the maximum adsorption capacity of algae (qm) ranged as follows: red < green < brown. According to Mehta and Gaur (2005), at pH 5.0, generally cell surfaces possess a net negative charge on the surface that favors the binding of Pb(II) to surface ligands. At pH below isoelectric point, cells have a positive charge; this inhibits the approach of positively charged Pb(II) ions. Chromium sorption is extremely pH dependent; higher pH favours Cr(III) removal while lower pH values facilitate Cr(IV) removal (Bellú et al. 2008). As Cr(IV) is anionic, it does not bind at high pH when the overall algal surface charge is negative; it binds to the algal surface at low pH when algal surface is positively charged.

Murphy (2007) stated that the dependence of metal uptake on pH is related to both surface functional groups of the biomass and the metal chemistry in a solution. Both the sulfonate and carboxyl groups are acidic and thus the optimum solution pH for maximum binding would be related to the pKa of these surface groups, they state that at low pH, these groups are protonated and therefore less available for binding of metals, but as the pH increases dissociation occurs on these groups resulting in increased negative charge, which leads to increased bisorption of metal cations on the surface, which explains increased binding of HMs with increasing pH. Murphy (2007) precisely mentioned that many divalent ions such as Cd^{+2}, Cu^{+2} and Pb^{+2} display maximum binding at pH value near the apparent pKa value of the seaweed carboxyl groups (pKa approximately 5), indicating the importance of these groups. Nevertheless, the same cannot be implied for the affinity towards all HMs.

In fact, hydroxyl groups in biomasses become negatively charged at high pH, thereby contributing to HM removal at high pH.

On the whole, Mehta and Gaur (2005) reported that seaweeds particularly have considerable potential to adsorb anionic species of certain metals; for instance, algal cells can also adsorb Cr(VI) with considerable ease at low pH values (<2). Removal of Cr(VI) by algae is carried out by the reduction of this anionic form to the cationic Cr(III) form under strong acidic conditions. Park et al. (2004) observed that thermally treated *Ecklonia* biomass reduces Cr(VI) to Cr(III) at low pH, while the native biomass efficiently could be used for Cr(III) removal at higher pH.

Even Abbas et al. (2014) reported that in general the heavy metal uptake for most of the biomass types declines significantly when pH of the metal solutions is decreased from pH 6.0 to 2.5. However, at pH below 2, there is minimum or negligible removal of metal ions from solutions. The metal uptake increases when pH increases from 3.0 to 5.0. Furthermore, the optimum value of pH is crucial in order to get the highest metal sorption, and this capacity will decrease with further increase in pH value. Abbas et al. (2014) mentioned that a competition between cations and protons for binding sites could result in reduced biosorption of metals like Cu, Cd, Ni, Co and Zn at low pH values.

Bilal et al. (2018) particularly reviewed seaweed, suggesting pH to be the most important factor in sorption studies. According to them, increase in pH increases the biosorption of metal ions; however, too great an increase in pH could cause precipitation, which should be avoided. Though the optimum pH varies for different biosorption systems, they suggested that an increase in pH up to 5 caused an increase in biosorption capacity (98%), but a further increase in pH led to reduced capacity. Essentially, the pH of the medium controls the protonation and deprotonation of functional groups, which affects the biosorption capacity. In other words, at low pH, carboxylic groups, being acidic, exist in a protonated state due to the presence of excess H^+ and H_3O^+. Therefore, repulsive forces of these protonated groups with positively charged heavy metal ions are responsible for the lower biosorption capacity at low pH. However, with an increase in pH, functional groups such as amine, carboxyl, and hydroxyl groups are exposed by deprotonation, which increases electrostatic attraction with heavy metal ions due to a negative charge. A greater increase in pH could lead to the formation of hydroxide anionic complexes and precipitation which could probably be the reason for the low biosorption capacity (Bilal et al. 2018).

Metal Speciation

According to Mehta and Gaur (2005), the availability of metal ions for binding onto algae depends on chemical speciation, which in turn is dependent on pH of the solution. Metal ions in a natural water sample could be bound onto biological materials and different kinds of solid surfaces. According to them, the free metal ion concentration is regulated by complex interactions between metal ions, ligands and major ions. It is also controlled by pH of the solution, and the extent of hydrolysis and stability constant (β) of metal-ligand complexes. It is usually assumed that the free, hydrated cupric ion is the physico-chemical form of copper most toxic to aquatic organisms, and that complexes of copper are much less toxic. Nevertheless, Florence et al. (1984) studied copper, lead, cadmium and zinc, stating that in most natural waters, a large fraction of each of these metals are bound in highly stable complexes by unidentified organic ligands; in addition, many of the water samples contain lipid-soluble metal.

Mehta and Gaur (2005) particularly stated that the free metal ion concentration is regulated by complex interactions between metal ions, ligands and major ions. It is also regulated by pH of the solution, and the extent of hydrolysis and stability constant (β) of metal-ligand complexes. They reported that the concentrations of free ions of various metals like Pb(II), Cu(II), Hg(II), Al(III) and Fe(III) in freshwater remain very low in the pH range 7.5-9.0 due to complexation with inorganic ligands. However, decrease in pH of the solution generally increases the concentration of free metal ions. Especially, for efficient removal of heavy metals by an algal biosorbent, the ratio of free metal ion to total metal concentration should remain high. The ratio of free metal ion to total metal concentration in a solution can be determined by the free ligand concentration and stability constant (β). Therefore, the ionic speciation of metals in solution is important since the metal uptake often depends on the sorption system pH; moreover, higher pH values usually result in higher metal cation uptakes due to lowered metal solubility. According to Mehta and Gaur (2005), the availability of metal ions for binding onto algae depends on chemical speciation, which in turn is determined by pH of the solution. Metals in wastewaters occur in a variety of chemical forms, namely, free aquo ions, complexed with inorganic and organic ligands, and adsorbed on particulate phases. Wastewaters may have inorganic ligands like HCO^{-3}, CO_3^{2-}, Cl^-, SO_4^{2-}, HS^-, and organic compounds, such as acetic acid, oxalic acid, amino acids, etc. for metal binding, in natural water.

Davies et al. (2000) evaluated *Sargassum* reporting that for the cadmium nitrate system, Cd^{+2} is the dominant species present at pH lower than approximately pH=5.0 and this depends on the total molarity of the salt in solution. At higher concentrations, however, the Cd^{+2} predominance is shifted to lower pH values. Nevertheless, it remains the dominant species up to a maximum of pH 5.5. According to them, at pH values greater than 5.0, hydroxides of cadmium become more dominant and it is known that small quantities of biomass polysaccharide are leached into solution which may further impede kinetics of precipitation. Davies et al. (2000) also mentioned that the copper sulfate system includes two copper species dominant at lower pH values and occur, respectively, at successive pH maxima. Selvan et al. (2012) investigated *Sargassum wightii* (brown) and *Caulerpa racemosa* (green) for biomass concentration and pH–dependent biosorption; they observed varying results in case of Cr(VI) and Cr(III). Murphy et al. (2008) too observed that metal sorption was highly pH dependent with maximum Cr(III) and Cr(VI) sorption occurring at pH 4.5 and pH 2, respectively. Extended equilibrium times were required for Cr(VI) binding over Cr(III) binding (180 and 120 min, respectively), thus indicating possible disparities in binding mechanism between chromium oxidation states. The red seaweed *P. palmata* revealed the highest removal efficiency for both Cr(III) and Cr(VI) at low initial concentrations.

Biomass

Parameters such as environment, biological competition, nutrient level, growth rate, illumination, etc. too greatly influence metal sorption by algae. In fact, when we consider metal binding in case of live seaweed, variations in growth conditions that possibly bring about changes in composition of the cell surface are crucial, as they could in turn influence metal binding capacity and characteristics of the seaweed. In a general review, Abbas et al. (2014) mentioned that the nature of the biomass or derived product is vital in metal sorption studies; the form of biomass (freely–suspended cells, immobilized preparations, living biofilms, etc.) as well as treatments (chemical (alkali) or physical (boiling, drying, autoclaving and mechanical disruption)) affect binding properties. They reported that growth and age of the biomass, and nutrition influence changes in cell size, cell wall composition, extracellular product formation, etc. thereby affecting biosorption. Further, the surface area to volume ratio is another factor that should be considered according to Abbas et al. (2014); it is important for individual cells or particles as well as immobilized biofilms. Moreover, the biomass concentration should also be considered in case of biosorption efficiency with a reduction in sorption per unit weight occurring with increasing biomass concentration.

Biomass concentration influences the quantity of metal uptake from a solution. Abbas et al. (2014) explained that at a given equilibrium concentration, the biomass adsorbs more metal ions at low cell densities than at high densities. Thus, electrostatic interaction between the cells plays an important role in metal uptake. At lower biomass concentration, the specific uptake of metals is increased because an increase in biosorbent concentration leads to interference between the bindings sites; however, high biomass concentration restricts the access of metal ions to the binding sites. In other words, an increase in biomass concentration reduces metal sorption per gram of biomass, although this is generally attributed to a shift in the sorption equilibrium. Metal uptake (Mq) generally increases with the metal concentration until it reaches a plateau value (Mqmax) which corresponds to the binding capacity of the biomass.

According to Schiewer and Volesky (2000), biosorption by seaweeds is not usually non-specific physical sorption or microprecipitation, but rather binding to a limited number of binding sites, their number determines the capacity. Hamdy (2000) evaluated *Sargassum asperifolium, Cystoseiratrinode, Turbinaria decurrens* (brown algae), and *Laurencia obtusa* (red algae) and reported a decreased sorption of Cr, Co, Ni, Cu and Cd by four different algae with increasing biomass concentration. A probable explanation for such a relationship between biomass concentration and sorption may be limited availability of metal, increased electrostatic interactions, interference between binding sites and reduced mixing at higher biomass concentrations.

Nevertheless, growth rate is an influential factor when it comes to biomass concentration and using live seaweed for metal remediation; Rice (1984) observed

Cd and Rb levels decreasing and Mn levels increasing as the specific growth rate increases. This probably indicates the metabolic regulation of these metals or possibly the presence of a "dilution factor" as a result of an increase in the heavy metal-to-biomass ratio (Mamboya 2007).

Temperature

Temperature influences the metabolic rate of organisms, and hence also their heavy metal uptake (Mamboya 2007). However, contrasting results have been obtained regarding the effect of temperature on sorption of heavy metals by algae. Metal uptake by live cells is considerably affected by variations in temperature owing to dependence on metabolism. Altered metal uptake by live organisms with change in temperature regime has been emphasized by Mehta and Gaur (2005); in fact, they cite reports showing maximum uptake occurring at specific optimal temperature optima. Even in case of seaweed biomass, several reports suggest the role of optimum temperature for heavy metal binding. For example, Rahman and Sathasivam (2015) reported that the solution temperature plays a vital role on the metal ions' biosorption; the biosorption increased with the increase of solution temperature, i.e. rate of Zn^{+2}, Cu^{+2}, Pb^{+2}, and Fe^{+2} biosorption by the dried biomass of *Kappaphycus* sp. rapidly reached a maximum of 61.74, 89.12, 86.68, and 40.06%, respectively, at 50°C. This phenomenon indicated that the biosorption process of the metal ions onto the biomass was endothermic. They summarized that at higher temperature the activation of the biosorbent surfaces is enlarged, facilitating more active sites for biosorption of the metal ions. Moreover, an easy mobility and enhanced accessibility of metal ions from the bulk solution to the biomass active sites could also be the possible reason for the maximum biosorption of metal ions at higher temperatures. Even Cossich et al. (2002) found that in case of *Sargassum* sp. at pH 3.0, an increase of temperature from 30°C to 40°C produced a definite increase in biosorption capacity, which was not observed when the temperature increased from 20°C to 30°C. In their study at pH 4.0 and at high equilibrium concentrations, biosorption capacity increased significantly with temperature, which was not the case at low equilibrium concentrations. Therefore, they suggest that the fact that chromium uptake is strongly affected by pH and increases with temperature is in agreement with the ion–exchange hypothesis. Tsezos and Volesky (1981), Kuyucak and Volesky (1989c), and Aksu and Kutsal (1991) reported a slight increase in cation sorption by powdered seaweed biomass with increase in temperature (4 to 55°C). Increased metal sorption with increase in temperature suggests that metal biosorption by algae is an endothermic process. Contrarily, some studies indicate exothermic nature of metal sorption by algae (Aksu 2001, Benguell and Benaissa 2002). Increased biosorption of heavy metals with increasing temperature could be ascribed to bond rupture that perhaps enhanced the number of active sites involved in metal sorption or higher affinity of sites for metal (Mehta and Gaur 2005). As per the study of Park and Gamberoni (1997), at higher temperatures the ions can be adsorbed more actively on adsorption sites. Cruz et al. (2004) observed that there was slight decrease in Cd^{+2} sorption by *Sargassum* biomass with an increase in ambient temperature. Interestingly, certain reports show that there is no effect of temperature on metal sorption (for e.g. Norris and Kelly 1979, Zhao et al. 1994). Zumdahl (1992) too mentioned that seasonal variation in temperature does not affect heavy metal accumulation.

Bilal et al. (2018), in a detailed review on macroalgae, suggested that temperature is an important parameter influencing the sorption process; any change in temperature alters thermodynamic parameters, resulting in variation in sorption capacity. In endothermic sorption processes, an increase in temperature leads to an increase in biosorption, while in an exothermic sorption processes, an increase in temperature decreases biosorption.

Presence of Ions

Metal sorption by an algal biosorbent is also affected by the presence of ions (both cations and anions). The presence of other ions (apart from the metal ion being studied) significantly affects metal sorption by seaweeds (Mehta and Gaur 2005). In reality, industrial wastewaters could contain a variety of impurities that could affect metal biosorption. Most of the industrial effluents have high concentrations of metal ions like Al; even though Al is not a major environmental problem, its presence in solution interferes with sorption of many other metals (Lee et al. 2004). Wastewaters that need to be treated could comprise impurities that contain light ions; reduced metal uptake in the presence of light metals has been reported and attributed to competition for cellular binding sites, or precipitation or complexation by carbonates, bicarbonates or hydroxides of Ca and Mg (Rai et al. 1981).

Presence of salts (sodium chloride), surfactants and some chelating agents also influence metal removal (Cho et al. 1994). Mehta and Gaur (2005) suggested the presence of sodium ion to decrease Ni, Co and Cs accumulation by some algae and cyanobacteria, while a suppression of sorption of Cr(VI) and Va occurs in the presence of Na and Cl. High concentrations of monovalent cations Na^+ and K^+ increase the ionic strength, which in turn decreases metal biosorption capacity of the biomass (Greene et al. 1987, Ramelow et al. 1992). The inhibitory effect of Na is more pronounced with weakly bound metals such as Zn or Ni. It is important to note that Na^+ and K^+, being monovalent cations, do not compete directly with covalent binding of heavy metals by the biosorbent (Dixit and Singh 2015). In general, most of the reports suggest inhibitory effect of light metal ions on sorption of heavy metals by biosorbent; however, some show no effect, for e.g. Cs sorption capacity of *Padina australis* did not change in the presence of Na or K ions (Jalali-Rad et al. 2004). Johanssen et al. (2015) reported that selenate uptake by *Gracilaria* biomass was severely impacted by the presence of low concentrations of sulfate in the wastewaters.

Multi-metal Systems

On the other hand, a multi-metal system comprising several metals leads to interactive effects on physiological and biochemical processes, for example growth, metal uptake, etc., of various organisms. Mehta and Gaur (2005) suggested that there could be competitive interactions amongst metals for binding onto sorption sites. Cd and Fe interfere with sorption of each other in case of *Sargassum* sp., in a binary metal solution (Figueira et al. 1997). Similarly, Lee et al. (2004) reported that 1 mM concentrations of Pb and Al at pH 4.5 mutually reduced Al and Pb uptake by *Laminaria japonica* to 22.3 and 83%, respectively. Lee et al. (2004) found that Al decreases sorption of Pb, Cu, Cd and Zn by 78, 80, 86 and 89%, respectively, in *Laminaria japonica*. However, Al caused no significant decrease in Cr(III) sorption. According to Lee et al. (2004), as Al is trivalent, it did not compete with divalent metal ions for binding on surface sites. Essentially, at times the adsorption capacity of a single metal in case of a particular seaweed varies when an additional metal is added to the solution; for e.g. Apiratikul et al. (2004) reported that the overall Cd and Cu adsorption capacity of *Caulerpa lentifollifera* was higher than adsorption of Cd alone, but was lower than adsorption of Cu in single component system. This indicated that binding sites for Cd are not the same as that for Cu. According to them, *Caulerpa lentifollifera* had common binding sites for Cu and Pb; moreover, Cu and Pb showed competitive inhibition for sorption of each other in the binary system. The competitive interaction between two metals depended on the concentration of algal biomass in the solution. Mehta and Gaur (2005) suggested that although metal sorption ability of the biomass decreases in multi-metal system, multi-metal solutions can be remediated to the same level as single metal solutions at high biomass concentrations. When more than one metal is present in a system, the evaluation of biosorption results, interpretations and representation becomes much more complicated.

On the other hand, Mehta et al. (2002) demonstrated that acid pretreatment not only increased metal sorption but was also able to alleviate the inhibitory effect of other metal ions on sorption of the metal of interest. Evenmore, Rice and Lapointe (1981) reported that light and nitrogen availability positively affected rates of uptake of Fe, Mn, Zn, Cd, and Rb in *Ulva fasciata*, and it has been demonstrated that uptake of Cd in *Ulva fasciata* increased with increased ambient concentration of nitrate in the growth medium. In environments with high nutrient levels, metal uptake can be inhibited because of complex formation between nutrient and metalions (Mamboya 2007).

Time

The time required for metal removal depends on the valency of the metal as well as the affinity of the biomass for the typical metal, which in turn depends on its chemical composition; for e.g. Cr(VI) binding requires longer exposure times than Cr(III) to ensure that equilibrium has been reached (Murphy et al. 2008). The binding time could range from a period of 10 minutes to 6 hours. Bilal et al. (2018) mentioned that the contact time of biosorbent influences total biosorption. An increase in contact time up to the optimum contact time increases biosorption; however, later it becomes relatively constant. Occupancy of all active sites causes saturation of biomass, leading to an equilibrium state. The optimum time for each biomass varies, for e.g. in case of red algae it could be 60 min, while it could be 300 min for immobilized algal mass.

Rajfur (2013) described that time was a priority in heavy-metal sorption kinetics by algae; the time needed to attain the dynamic equilibrium depends, among others, on the type of algae and the degree of biomass fragmentation. For example, biomass of the marine alga *Ecklonia maxima*, with the thalli of 1.2 mm, attained the equilibrium during sorption of Cu, Pb and Cd after roughly 60 min, while for thalli of the size of 0.075 mm after grinding, the equilibrium stabilised after approximately 10 min. Further, in case of *Caulerpa lentillifera*, equilibrium was reached after 20 min. However, in case of biosorbents (DP95Ca and ER95Ca) based on *Durvillaea potatorum* and *Ecklonia radiata*, 90% sorption of Cu and Pb ions have been reported within 30 min (equilibrium stabilized after approx. 60 min) (c.f. Rajfur 2013). The marine alga *Ulva lactuca* sorbs 80% of Cr^{6+} ions within the first 20 min of the process (equilibrium stabilized after approx. 40 min; El–Sikaily et al. 2004).

Other Factors

Mamboya (2007) explained that salinity too affects heavy metal bioavailability, but he stated that though most studies of salinity describe its effects on heavy metal accumulation by animals, information concerning the effects on macroalgae is scarce. Nevertheless, humic substances in the aquatic environment impact the accumulation of metal ions. Mamboya (2007) reported that the bioavailability and toxicity of HMs are reduced through complex formation with dissolved organic matter (DOM); this reduces the concentration of free ionic metals in the aquatic environment. In fact, DOM could block the accumulation of some heavy metals by blocking the algal surface sites (Mamboya 2007).

LIVE SEAWEED vs. NON-LIVING SEAWEED BIOMASS

Bioremediation is a state of art technique that utilizes biological mechanisms to eradicate hazardous contaminants using microorganisms, plants, algae or their products, to restore polluted environments. One of the promising techniques for the removal of metals is the use of living or non-living organisms and their derivatives. A wide variety of organisms (both living and non-living) have been found to be capable of sequestering trace levels of metal ions from dilute aqueous solutions.

The ability of live seaweed to accumulate metals has been established in many studies; these include growth of seaweed in polluted estuaries where they have been shown to reduce the survival of crustaceans (Neveux et al. 2018). In fact, the ability of seaweeds to accumulate metals within their tissues has led to their widespread use as biomonitors of metal availability in marine ecosystem, for example *Enteromorpha* and/or *Cladophora* have been utilized to measure heavy metal levels in many parts of the world; Chlorophyta and Cyanophyta are Arsenic and Boron hyper–absorbents and hyper–accumulators, reducing these polltants from the water bodies, but Phaeophytes are particularly efficient accumulators of metals due to high levels of sulfated polysaccharides and alginates within their cell walls for which metals show a strong affinity (Chekroun and Baghour 2013). Nielsen et al. (2005) suggested that brown algae such as *Fucus serratus* often dominate the vegetation of heavy metal contaminated habitats. On the other hand, cultivation of seaweed for metal remediation purposes and integrated ocean based aquaculture where fish and invertebrates are co-cultured are also carried out nowadays. In this context, Neveux et al. (2018) cited the example of *Sargassum* cultivated in aquaculture effluents that had higher growth rate and the cultivated biomass had two to three fold higher Cu, Zn, Pb and Cr than the reference location. On the other hand, *Laminaria digita,* co-cultivated in open ocean adjacent to salmon farm, had biomass rich in Cu, Mn and V, which were sourced from the feed used in the fish farm located adjacently. Henriques et al. (2015) studied bioaccumulation and biosorption of mercury by *U. lactuca*, *G. gracilis* and *F. vesiculosus* under environmentally realistic conditions. All seaweeds showed huge bioaccumulation capabilities, reaching up to 209 μg of Hg per gram of macroalgae (d.w.), which corresponds to 99% of Hg removed from the contaminated seawater. According to them, *U. lactuca* was the fastest to accumulate Hg, and led to final levels of Hg in solution equal or lower than the legal value for drinking water. They suggested that the use of living *U. lactuca* instead of its dried biomass was more advantageous, since internal accumulation, together with the organism growth, allowed to obtain lower residual concentrations of Hg in seawater. Henriques et al. (2015) evidenced the tremendous potential of living macroalgae, particularly *U. lactuca,* in the removal of Hg or for improvement of saline waters contaminated with metals. Nevertheless, most studies on metal accumulation by algae have been performed with living organisms for environmental, toxicological and pharmaceutical purposes rather than with an industrial application in mind. The use of live biomass has certain demerits; potential for desorptive metal recovery is limited since metal may be intracellularly bound, and metabolic extracellular products may form complexes with metals to retain them in solution.

Of late, the attention has shifted to non-living algae for metal removal and/or recovery; they do not require a nutrient supply, and could be used for multiple sorption desorption cycles. Moreover, considering the huge availability, use of dead biomass seems to be convenient. Although brown algae have gained prominence as good biosorbents because of their high sorption capacity, red and green algae have also been reported as potential sorbents. Essentially, all of them are readily available in the marine environment, have high sorption capacity, are autotrophic, require minimum nutrients and produce a large biomass compared to other microbial biosorbents.

Romera et al. (2007) reported maximum sorption of Cd (32.3 mg g^{-1}), Pb (63.7 mg g^{-1}), and Zn (21.6 mg g^{-1}), by *Asparagopsis armata*, while *Codium vermilara* could uptake Cd (21.8 mg g^{-1}), Pb (63.3 mg g^{-1}), and Zn (23.8 mg g^{-1}). Ungureanu et al. (2015) saw maximum adsorption capacity of Sb(III) by the algae *Sargassum muticum,* i.e. 5.5 mg g^{-1}, at pH 5. Nigro et al. (2002) studied adsorption of copper, zinc and cadmium using two dried seaweeds *Ecklonia maxima* and *Laminaria pallida* (order Laminariales) and Kelpak waste (also made from *Ecklonia maxima*), a byproduct from the manufacture of the seaweed concentrate Kelpak. They observed that optimum adsorption occurred at pH 3 and pH 7; Kelpak waste had equal or superior adsorption ability compared to dried *Ecklonia maxima* and *Laminaria pallida*, particularly for copper. Optimum adsorption occurred at temperatures of 20°C and 30°C. In their study, heavy metal adsorption trends by the individual seaweed biosorbent remained constant regardless of the species of anion present. Drying (fan air and oven drying at 85°C) of seaweed prior to adsorption cycles resulted in more efficient ion uptake, particularly after additional rehydration; further, ion uptake was the most efficient after 2-4 adsorption cycles. Though there are several reports on the utilization of dry or dead seaweed biomass as biosorbents, it is essential to consider that dead cells cannot be used where biological alteration in valency of a metal is sought. Moreover, degradation of organometallic species is not possible with dead biomass. Another shortcoming associated with dead biomass is that there is no scope for biosorption improvement through mutant isolation.

On the other hand, utilization of freeze-dried biomass is another option; certain reports (Winter et al. 1994, Bengtsson et al. 1995) suggest that freeze-dried biomass has a higher metal sorption potential than the live biomass. Burdin and Bird (1994) evaluated metal sorption by *Gracilaria tikvahiae*, *Gelidium pusillum*, *Agardhiella subulata* and *Chondrus crispus*, reporting greater accumulation of Pb in living thalli of all the four species as compared to lyophilized thalli. They observed greater accumulation of Ni, Cu and Zn by lyophilized than living thalli of *Gracilaria*, *Gelidium* and *Chondrus* and added that there was no difference between lyophilized and living thalli of the selected seaweeds for accumulating Cd. However, Neide et al. (2000) observed no change in metal sorbing potential of live, freeze-dried and oven-dried biomass. Nevertheless, one should also consider the cost incurred in freeze drying seaweeds.

PRETREATED BIOMASS

HM uptake by algal biomass could be improved by physical or chemical treatments that modify the seaweed cell surface structure and provide additional binding sites (Mehta and Gaur 2005). Ortiz–Calderon et al. (2017) described that physical treatments such as heating/boiling, freezing, crushing, and drying usually lead to an enhanced level of metal ion biosorption. These treatments provide more surface area to increase the biosorption capacity and release the cell contents that could probably bind to metal ions. On the other hand, among the chemical treatments used, the most common algal pretreatments include the use of $CaCl_2$, formaldehyde, glutaraldehyde, NaOH, and HCl. Pretreatment with $CaCl_2$ causes calcium binding to alginate that plays an important role in ion exchange, while formaldehyde and glutaraldehyde strengthen the crosslinking between hydroxyl groups and amino groups. NaOH increases the electrostatic interactions of metal ion cations and provides optimum conditions for ion exchange, while HCl replaces light metal ions with a proton and also dissolves polysaccharides of cell wall, or denatures proteins, increasing the binding sites for the biosorption process (Ortiz–Calderon et al. 2017). Zhao et al. (1994) reported various pretreatments (1 M HCl, 1 M HNO_3, 0.1 M NaOH, 1 M NaOH, acetone, and 60°C water at times from 15 to 60 min and temperatures of 25 and 60°C) to increase metal (Pb, Cu, Zn, Cd, Cr, Mn, Ni and Co) binding ability of certain seaweed biomass. According to them, both acidic and basic treatment increased the uptake of Au, Ag, and Hg by four strains of seaweed to nearly 100%. On the other hand, several studies show that $CaCl_2$ pretreatment is most suitable and economic for activation of algal biomass; for example, *Padina* sp. pretreated with 0.2 M $CaCl_2$ is reported to bind 0.5 mmol g^{-1} Cd (98% removal within 35 minutes; Kaewsarn and Yu (2001)). HCl pretreatment of biomass is reported to be effective in enhancing metal sorption from single as well as binary metal solutions (Mehta and Gaur 2005). Enhanced Ni sorption capacity of *Sargassum* sp. from 181.2 to 250 mg g^{-1} following acid pretreatment has been reported by Kalyani et al. (2004). On the other hand, xanthanation of biomass is another novel approach to increase biosorption of Pb by algal biomass; Xanthanation of biomass increased sulfur content of the *Undaria pinnatifida* biomass from 0.15% to 13.7%, and xanthanated biomass sorbed three times more Pb than the native biomass (Kim et al. 1999). However, Kim et al. (1999) demonstrated that acid treatment decreased sulfur content of the xanthanated biomass indicating its instability in acidic condition. Under acidic condition, the xanthate is instable, so HCl or HNO_3 should not be used for desorption of bound metal. Nevertheless, agents like methanol, acetic acid, NaOH and hot water decrease metal sorption (Mehta and Gaur 2005).

REGENERATION AND REUSE OF BIOMASS

Metal sequestration generally involves mechanisms such as ion exchange, chelation, and adsorption by physical forces and ion entrapment; moreover, the sorbent could either be living and uptake would include both passive and active methods, or it could be non-living where only passive uptake would occur (Stirk and van Staden 2002). In any case, regeneration and reuse of a seaweed-based biosorbent in single and multi-metal systems is always preferred. This makes the process feasible for large scale environmental and industrial application. It could be achieved by desorbing the biomass with a suitable eluent or desorbing solution; then, the biomass could be used in multiple sorption–desorption cycles. Lowering the pH of the metal–loaded biomass suspension displaces HM cations by protons from the binding sites (Mehta and Gaur 2005); organic and mineral acids, bases, salts and metal chelators are also used for metal desorbing. The tested desorbing agents include hydrochloric acid, sodium hydroxide, acetic acid, citric acid, mono- and dibasic ammonium phosphate, ammonium sulfate, ammonium chloride, sodium carbonate, calcium chloride and EDTA. Zhou et al. (1998) recovered approximately 99.5% of Cu and Cd from *Laminaria japonica* and *Sargassum kjellmarianum,* respectively, using hydrochloric acid (HCl) and EDTA. In their study, Zhou et al. (1998) observed an increase in Cu concentration in desorbing solutions after elution of *Sagassum kjellmanianum* with 20 ml of HCl (0.11 N) and EDTA (0.1 N). Mehta and Gaur (2005) cited examples wherein $CaCl_2$ (0.1 M in HCl or HNO_3) is best suited for desorbing Cu bound on *Ulva reticulate*, complete desorption of Cs from *Sargassum glauescens* and *Cystoseria indica* is achieved by treating it with 0.5 N NaOH or 1 M KOH, 50% reduction in Cu sorption capacity of *Sargassum bacularia* occurs on desorption wash with HCl, and 74% decrease in Co uptake capacity of the HCl–eluted *Ascophyllum nodosum* occurs during cobalt elution. Stirk and van Staden (2002) reported that dried *Ecklonia maxima* could be used for more than four adsorption–desorption cycles, whereas the sorbent derived from the Kelpak (a seaweed concentrate used in agriculture) could only be used effectively for up to three adsorption–desorption cycles. Bakir et al. (2010) particularly evaluated a seaweed–waste material resulting from the processing of *Ascophyllum nodosum* showing efficient removal of Zn(II), Ni(II) and Al(III), both in single and multi-metal waste streams; the regeneration of the sorbent was accomplished with very little loss in metal removal efficiency (RE) for both single and multi-metal systems. Values of 92, 96 and 94% RE were achieved for Zn(II), Ni(II) and Al(III), respectively, for the 5th sorption cycle in single metal aqueous solutions. Although an absolute comparison of different desorbing agents cannot be made due to varying experimental conditions used by different researchers, it is reasonable to conclude that $CaCl_2$ and mineral acids like HCl are highly efficient in desorption of metal cations from algal biosorbents. Therefore, the efficiency of each seaweed sorbent would vary based on the conditions and applicability; however, the fact that they can be efficiently utilized for repeated cycles is promising.

HEAVY METAL DETECTION AND QUANTIFICATION

Various methodologies ranging from gravimetric or titrimetric measurements, to the use of automated instruments could be employed for metal analysis. Spectrometric methods, based on the absorption or emission of electromagnetic radiation by atomic or molecular species, are generally used as a tool for metal quantification. Further, flame atomic absorption (FAAS), graphite furnace atomic absorption (GFAAS) and inductively coupled plasma optical emission (ICP/OES), or mass spectrometry (ICP/MS), are widely used for heavy metal detection and quantification; while the first two techniques are essentially used for single–element detection, ICP/OES and ICP/MS facilitate multi-metal analysis (Vandecasteele and Block 1997). ICP spectrometric excitation sources were first introduced in the 1960s, and since then they have evolved dramatically (Wendt and Fassel 1965). ICP/OES could be used to quantify all the elements except argon; it determines elements present in trace to high concentrations efficiently. On the other hand, FT–IR (Fourier Tranform–Infra Red) analysis offers excellent information on the nature of the bonds present on the seaweed surface and allows identification of different functional groups on the cell surface which are capable of interacting with metal ions (Raize et al. 2004). Generally, IR spectra of seaweeds are relatively complex, reflecting the complex nature of the biomass surface (Fourest and Volesky 1996); however, FT–IR spectroscopy helps detect vibrational frequency changes in seaweeds caused during metal uptake. Another technique, the scanning electron microscopy (SEM) involves scanning of the surface of a solid sample with a beam of energetic electrons in a roster pattern; here, several signals are produced from the surface, including backscattered, secondary and Auger electrons, X-ray fluorescence photons, and other photons of various energies (Skoog et al. 1992). The two key features of SEM include back scattering and secondary electrons. An image is obtained by measuring the intensity of the secondary electrons generated where the beam intercepts the sample surface. It could visualize the surface morphological changes including shrinking and layering in the seaweed after metal binding. Additionally, an EDX (Energy Dispersive X-ray) instrument can be attached to an SEM to provide supplementary information. As the SEM electron beam strikes the sample surface, X-rays are produced. An X-ray photon impinging on the surface of the EDX detector produces electron hole pairs which are detected as a single pulse by the liquid nitrogen cooled preamplifier. The pulse energy is determined by the X-ray energy, which in turn is determined by the element being examined (Skoog et al. 1992). The EDX analyzer produces a spectrum of the elements present in the targeted areas of the samples allowing detectable elements to be quantified or mapped. Hence, this plays an important tool in metal uptake studies.

On the whole, in case of appropriate utility of seaweed for metal removal from the environment and wastewaters, emphasis should be given to (i) selection of strains with high metal sorption capacity, (ii) adequate understanding of sorption mechanisms, (iii) development of easy handling methods, (iv) development of better models for predicting metal sorption, (v) pretreatment of the algae or genetic manipulation of algae for increased number of surface groups or over expression of metal binding proteins, and (vi) economic feasibility.

CONCLUSIONS

The efficiency of seaweeds to accumulate or sorb heavy metals (trace to high concentrations) makes them a suitable candidate for remediation and recovery in natural aquatic bodies as well as wastewaters. Seaweeds efficiently aid metal uptake from single metal solutions as well as mixtures; they also efficiently remove metals from effluents containing several other ingredients. Nevertheless, each seaweed has a unique sorption and accumulation capacity which needs to be appropriately understood. A thorough understanding of the mechanism of metal binding and uptake would add to the efficiency and commercial viability of this technology for metal removal. Seaweeds, due to their wide availability, abundance of biomass, variety of speices, ecofriendliness, and ease of handling, form the best source for a commercial viable biosorbent technology projected for metal uptake or scavenging. Though several researchers have focused efforts on development of seaweed based biosorbents, this technology still has a long way to go. This review (acknowledging the utility of seaweed in metal binding) aims to spread awareness regarding applicability of seaweeds as a tool for a better and green environment. It offers a state-of-the-art technology involving the use of algae as ecofriendly, cost-effective biosorbents for the removal of heavy metals.

REFERENCES

Abbas, S.H., I. Ismail, T.M. Mostafa and A.H. Sulaymon. 2014. Biosorption of heavy metals: A review. J. Chem. Sci. Technol. 3(4): 74-102.

Abdi, O. and M. Kazemi. 2015. A review study of biosorption of heavy metals and comparison between different biosorbents. J. Mater. Environ. Sci. 6(5): 1386-1399.

Ahalya, N., T.V. Ramachandra and R.D. Kanamadi. 2003. Biosorption of heavy metals. Res. J. Chem. Environ. 7(4): 71-79.

Ahmad, S., A. Pandey, V.V. Pathak, V.V. Tyagi and R. Kothari. 2020. Phycoremediation: Algae as eco-friendly tools for the removal of heavy metals from wastewaters. pp. 53-76. *In*: R. Bharagava, G. Saxena. (eds.). Bioremediation of Industrial Waste for Environmental Safety. Springer, Singapore.

Ajmal, M., A.H. Khan and S. Ahmad. 1998. Role of sawdust in the removal of copper(II) from industrial wastes [J]. Water. Res. 32(10): 3085-3091.

Aksu, Z. and T. Kutsal. 1991. A bioseparation process for removing lead(II) ions from waste water by using *C. vulgaris*. J. Chem. Technol. Biotechnol. 52(1): 109-118.

Aksu, Z. 2001. Biosorption of reactive dyes by dried activated sludge. Equillibrium and kinetic modeling. Biochem. Eng. J. 7: 79-84.

Allen, S.J. and P.A. Brown. 1995. Isotherm analyses for single component and multi-component metal sorption onto lignite. J. Chem. Tech. Biot. 62: 17-24.

Apiratikul, R., T.F. Marhaba, S. Wattanachira and P. Pavasant. 2004. Biosorption of binary mixtures of heavy metals by green macroalga, *Caulerpa lentillifera*. Songklanakarin J. Sci. Technol. 26: 199-207.

Arumugam, N., S. Chelliapan, H. Kamyab, S. Thirugnana, N. Othman and N.S. Nasri. 2018. Treatment of wastewater using seaweed: A review. Int. J. Environ. Res. Public Health 15: 2851-2868.

Baghour, M. 2017. Effect of seaweeds in phyto-remediation. pp. 47-83. *In*: E. Nabti. (ed.). Biotechnology in Agriculture, Industry and Medicine: Biotechnological Applications of Seaweeds. Nova Science Publishers, Inc., New York.

Bakir, A., P. McLoughlin and E. Fitzgerald. 2010. Regeneration and reuse of a seaweed-based biosorbent in single and multi-metal systems. Cln. Soil. Air. Water. 38(3): 257-262.

Bellú, S., S. García, J.C. González, A.M. Atria, L.F. Sala and S. Signorella. 2008. Removal of chromium(VI) and chromium(III) from aqueous solution by grainless stalk of corn. Sep. Sci. Technol. 43(11-12): 3200-3220.

Bengtsson, L., B. Johansson, T.J. Hackett, L. Mchale and A.P. Mchale. 1995. Studies on the biosorption of uranium by *Talaromyces emersonii* CBS 814. 70 biomass. Appl. Microbiol. Biotechnol. 42: 807-811.

Benguell, B. and H. Benaissa. 2002. Cadmium removal from aqueous solution by chitin: Kinetic and equilibrium studies. Water Res. 36: 2463-2474.

Besada, V., J.M. Andrade, F. Schultze and J.J. González. 2009. Heavy metals in edible seaweeds commercialised for human consumption. J. Mar. Syst. 75: 305-313.

Bilal, M., T. Rasheed, J.E. Sosa–Hernández, A. Raza, F. Nabeel and H. Iqbal. 2018. Biosorption: An interplay between marine algae and potentially toxic elements: A review. Mar. drugs. 16(2): 65.

Bulgariu, L. and D. Bulgariu. 2014. Enhancing biosorption characteristics of marine green algae (*Ulva lactuca*) for heavy metals removal by alkaline treatment. J. Bioprocess Biotech. 4: 146.

Burdin, K.S. and K.T. Bird. 1994. Heavy metal accumulation by carrageenan and agar producing algae. Bot. Mar. 37: 467-470.

Chekroun, K.B. and M. Baghour. 2013. The role of algae in phytoremediation of heavy metals: A review. J. Mater. Environ. Sci. 4(6): 873-880.

Chisti, H.T.N. 2018. Heavy metal sequestration from contaminated water: A review. J. Mater. Environ. Sci. 9(8): 2345-2355.

Cho, D.Y., S.T. Lee, S.W. Park and A.S. Chung. 1994. Studies on the biosorption of heavy metals onto *Chlorella vulgaris*. J. Environ. Sci. Health: Environ. Sci. Eng. 29: 389-409.

Cossich, E.S., C.R.G. Tavares and T.M.K. Ravagnani. 2002. Biosorption of chromium(III) by *Sargassum* sp. biomass. Electron. J. Biotechnol. 5(2): 6-7.

Cruz, C.C.V., A.C.A. Da Costa, C.A. Henriques and A.S. Luna. 2004. Kinetic modeling and equilibrium studies during cadmium biosorption by dead *Sargassum* sp. biomass. Bioresource Technol. 91: 249-257.

Darnall, D.W. 1989. AlgaSORB–R A new biotechnology for removing and recovering heavy metals ions from groundwater and industrial wastewater. Hazardous Waste Treatment: Biosystems for Pollution Control. 113-124. Proceeding of 1989 A and WMA/EPA International symposium.

Davis, T.A., B. Volesky and R.H.S.F. Vieira. 2000. *Sargassum* seaweed as biosorbent for heavy metals. Water Res. 34: 4270-4278.

Davis, T.A., B. Volesky and A. Mucci. 2003. A review of the biochemistry of heavy metal biosorption by brown algae. Water. Res. 37: 4311-4330.

Delshab, S., E. Kouhgardi and B. Ramavandi. 2016. Data of heavy metals biosorption onto *Sargassum oligocystum* collected from the northern coast of Persian Gulf. Data Brief 8: 235-241.

Demey, H., T. Vincent and E. Guibal. 2018. A novel algal-based sorbent for heavy metal removal. Chem. Eng. J. 332: 582-595.

Dixit, S. and D.P. Singh. 2015. Pycoremediation: Future prospective of green technology. pp. 9-21. *In*: B. Singh, K. Bauddh, F. Bux. (eds.). Algae and Environment Sustainablity: Developments in Applied Phycology book series (DAPH, Volume 7). Springer Pub., New Delhi, India.

do Nascimento Júnior, W.J., M.G.C. da Silva and M.G.A. Vieira. 2019. Competitive biosorption of Cu^{2+} and Ag^{+} ions on brown macro-algae waste: Kinetic and ion-exchange studies. Environ. Sci. Pollut. Res. 26: 23416-23428.

El-Sikaily, A., A. El Nemr, A. Khaled and O. Abdelwehab. 2007. Removal of toxic chromium from wastewater using green alga *Ulva lactuca* and its activated carbon. J. Hazard. Mater. 148: 216-228.

Figueira, M.M., B. Volesky and V.S.T. Ciminelli. 1997. Assessment of interference in biosorption of heavy metals. Biotechnol. Bioeng. 54: 334-350.

Florence, T.M., B.G. Lumsden and J.J. Fardy. 1984. Algae as indicators of copper speciation. *In*: C.J.M. Kramer, J.C. Duinker. (eds.). Complexation of Trace Metals in Natural Waters. Developments in Biogeochemistry, Vol 1. Springer, Dordrecht. https://doi.org/10.1007/978-94-009-6167-8-37.

Fomina, M. and G.M. Gadd. 2014. Biosorption: Current perspectives on concept, definition and application. Bioresour. Technol. 160: 3-14.

Foroutan, R., H. Esmaeili, M. Abbasi, M. Rezakazemi and M. Mesbah. 2018. Adsorption behaviour of Cu(II) and Co(II) using chemically modified marine algae. Environ. Technol. 39(21): 2792-2800.

Fourest, E. and B. Volesky. 1996. Contribution of sulfonate groups and alginate to heavy metal biosorption by the dry biomass of *Sargassum fluitans*. Environ. Sci. Technol. 30: 277-282.

Freitas, O.M., R.J. Martins, C.M. Delerue–Matos and R.A. Boaventura. 2008. Removal of Cd(II), Zn(II) and Pb(II) from aqueous solutions by brown marine macro algae: Kinetic modelling. J. Hazard. Mater. 153(1-2): 493-501.

Gadd, G.M. 2009. Biosorption: Critical review of scientific rationale, environmental importance and significance for pollution treatment. J. Chem. Technol. Biotechnol. 84(1): 13-28.

Greene, B., R. McPherson and D. Darnall. 1987. Algal sorbents for selective metal ion recovery. Metal Speciation, Separation and Recovery, J.W. Patterson, R. Pasino. Lewis, Chelsea, MI 315-338.

Gupta, V.K. and I. Ali. 2000. Utilization of bagasse fly ash (a sugar industry waste) for the removal of copper and zinc from wastewater. Sep. Purif. Technol. 18: 131-140.

Gutiérrez, C., H.K. Hansen, P. Hernández and C. Pinilla. 2015. Biosorption of cadmium with brown macroalgae. Chemosphere 138: 164-169.

Hackbarth, F.V., F. Girardi, A.A.U. de Souza, J.C. Santos, R.A.R. Boaventura, V.J.P. Vilar, et al. 2015. Ion exchange prediction model for multi-metal systems obtained from single-metal systems using the macroalga *Pelvetia canaliculata* (Phaeophyceae) as a natural cation exchanger. Chem. Eng. J. 260: 694-705.

Hamdy, A. 2000. Biosorption of heavy metals by marine algae. Curr. Microbiol. 41(4): 232-238.

Hashim, M.A. and K.H. Chu. 2004. Biosorption of cadmium by brown, green and red seaweeds. Chem. Eng. J. 97(2-3): 249-255.

Haug, A. 1961. Affinity of some divalent metals to different types of alginates. Acta Chemica Scandinavica. 15(8). p. 1794.

He, J. and J.P. Chen. 2014. A comprehensive review on biosorption of heavy metals by algal biomass: Materials, performances, chemistry and modelling simulation tools. Bioresour. Technol. 160: 67-078.

Henriques, B., L.S. Rocha, C.B. Lopes, P. Figueira, R.J.R. Monteiro, A.C. Duarte, et al. 2015. Study on bioaccumulation and biosorption of mercury by living marine macroalgae: Prospecting for a new remediation biotechnology applied to saline waters. Chem. Eng. J. 281: 759-770.

Howard, A.G. 1998. Metal complexes in solution. pp. 36-51. *In*: Alan G. Howard (ed.). Aquatic Environmental Chemistry. OUP Oxford. ISBN-13: 978-0198502838.

Ibrahim, W.M. 2011. Biosorption of heavy metal ions from aqueous solution by red macroalgae. J. Hazard. Mater. 192: 1827-1835.

Ibrahim, W.M., A.F. Hassan and Y.A. Azab. 2016. Biosorption of toxic heavy metals from aqueous solution by *Ulva lactuca* activated carbon. Egypt. J. Basic Appl. Sci. 3(3): 241-249.

Jacob, J.M., C. Karthik, R.G. Saratale, S.S. Kumar, D. Prabakar, K. Kadirvelu, et al. 2018. Biological approaches to tackle heavy metal pollution: A survey of literature. J. Environ. Manag. 217: 56-70.

Jalali, R., H. Ghafourian, Y. Asef, S.J. Davarpanah and S. Sepehr. 2002. Removal and recovery of lead using nonliving biomass of marine algae. J. Hazard. Mater. 92(3): 253-262.

Jalali-Rad, R., H. Ghalocerian, Y. Asef, S.T. Dalir, M.H. Sahafipour and B.M. Gharanjik. 2004. Biosorption of cesium by native and chemically modified biomass of marine algae: Introduce the new biosorbent for biotechnology application. J. Hazard. Mater. 116: 125-134.

Johansson, C.L., N.A. Paul, R. de Nys and D.A. Roberts. 2015. The complexity of biosorption treatments for oxyanions in a multi-element mine effluent. J. Environ. Manage. 151: 386-392. http://dx.doi.org/10.1016/j.jenvman.2014.11.031.

Kaewsarn, P. and Q. Yu. 2001. Cadmium(II) removal from aqueous solutions by pre-treated biomass of marine alga *Padina* sp. Environ. Pollut. 112: 209-213.

Kalyani, S., P.S. Rao and A. Krishnaiah. 2004. Removal of nickel(II) from aqueous solutions using marine macroalgae as the sorbing biomass. Chemosphere 57: 1225-1229.

Kanamarlapudi, S.L.R.K., V.K. Chintalpudi and S. Muddada. 2018. Application of Biosorption for Removal of Heavy Metals from Wastewater. Biosorption. p. 69. https://www.intechopen.com/books/biosorption/application–of–biosorption–for–removal–of–heavy–metals–from–wastewater on 8th December 2019.

Kaplan, D. 2013. Absorption and adsorption of heavy metals by microalgae. pp. 439-447. *In*: Amos Richmond, Qiang Hu. (eds.). Handbook of Microalgal Culture: Applied Phycology and Biotechnology. John Wiley & Sons, Ltd. by Blackwell Publishing Ltd.

Keshtkar, A.R., M. Mohammadi and M.A. Moosavian. 2015. Equilibrium biosorption of wastewater U(VI), Cu(II) and Ni(II) by the brown alga *Cystoseira indica* in single, binary and ternary metal systems. J. Radioanal. Nucl. Chem. 303: 363-376.

Khan, M.I., J.H. Shin and J.D. Kim. 2018. The promising future of microalgae: Current status, challenges and optimization of a sustainable and renewable industry for biofuels, feed and other products. Microb. Cell Fact. 17: 36. doi:10.1186/s12934-018-0879–x.

Khormaei, M., B. Nasernejad, M. Edrisi and T. Eslamzadeh. 2007. Copper biosorption from aqueous solutions by sour orange residue. J. Hazard. Mater. 149(2): 269-274.

Kim, Y.H., J.Y. Park, Y.J. Yoo and J.W. Kwak. 1999. Removal of lead using xanthated marine brown alga *Undaria pinnatifida*. Proc. Biochem. 34: 647-652.

Koller, M. and H.M. Saleh. 2018. Introductory Chapter: Introducing Heavy Metals. Heavy Metals. 1. https://www.intechopen.com/books/heavy–metals/introductory–chapter–introducing–heavy–metals.

Koutahzadeh, N., E. Daneshvar, M. Kousha, M.S. Sohrabi and A. Bhatnagar. 2013. Biosorption of hexavalent chromium from aqueous solution by six brown macroalgae. Desalin. Water Treat. 51: 6021-6030.

Kratochvil, D. and B. Volesky. 1998. Advances in the biosorption of heavy metals. Trends. Biotechnol. 16(7): 291-300.

Kumar, K.S., H.U. Dahms, E.J. Won, J.S. Lee and K.H. Shin. 2015. Microalgae: A promising tool for heavy metal remediation. Ecotoxicol. Environ. saf. 113: 329-352.

Kumar, M., A. Singh and M. Sikandar. 2018. Study of sorption and desorption of cd (ii) from aqueous solution using isolated green algae *chlorella vulgaris*. Appl. Wat. Sci. 8(225): 1-11.

Kumar, V.V. and P. Kaladharan. 2006. Biosorption of metals from contaminated water using seaweed. Curr. Sci. 90(9): 1263-1267.

Kuyucak, N. and B. Volesky. 1989a. The mechanism of cobalt biosorption. Biotechnol. Bioeng. 33: 219-235.

Kuyucak, N. and B. Volesky. 1989b. Accumulation of cobalt by marine alga. Biotechnol. Bioeng. 33: 809-814.

Kuyucak, N. and B. Volesky. 1989c. The mechanism of cobalt biosorption. Biotechnol. Bioengin. 33: 823-831.

Lee, D.C., C.J. Park, J.E. Yang, H.Y. Jeong and H.I. Rhee. 2000. Screening of hexavalent chromium biosorbent from marine algae. Appl. Microbiol. Biotechnol. 54(3): 445-448.

Lee, H.S., J.H. Suh, B.I. Kim and T. Yoon. 2004. Effect of aluminium in two-metal biosorption by an algal biosorbent. Minerals Eng. 17: 487-493.

Li, W., S. Zhang and X. Shan. 2007. Surface modification of goethite by phosphate for enhancement of Cu and Cd adsorption, Colloids. Surf. A 293: 13-19.

Lobban, C.S. and P.J. Harrinson. 1994. Seaweed Ecology and Physiology. Cambridge University Press, Cambridge, pp. 366.

Mamboya, F.A. 2007. Heavy Metal Contamination and Toxicity. Studies of Macroalgae from the Tanzanian Coast. PhD Thesis. University of Stockholm, Stockholm, p. 48.

Mantzorou, A., E. Navakoudis, K. Paschalidis and F. Ververidis. 2018. Microalgae: A potential tool for remediating aquatic environments from toxic metals. Int. J. Environ. Sci. Technol. 15: 1815-1830.

Mehta, S.K., A. Singh and J.P. Gaur. 2002. Kinetics of adsorption and uptake of Cu^{2+} by *Chlorella vulgaris*: In fluence of pH, temperature, culture age and cations. J. Environ. Sci. Health Part A. 37: 399-414.

Mehta, S.K. and J.P. Gaur. 2005. Use of algae for removing heavy metal ions from wastewater: Progress and prospects. Crit. Rev. Biotechnol. 25(3): 113-152.

Mondal, M., G. Halder, G. Oinam, T. Indrama and O.N. Tiwari. 2019. Bioremediation of organic and inorganic pollutants using microalgae. pp. 223-235. *In*: V.K. Gupta, A. Pandey. (eds.). New and Future Developments in Microbial Biotechnology and Bioengineering. Elsevier, Amsterdam.

Murphy, V. 2007. An investigation into the mechanisms of heavy metal binding by selected seaweed species. Ph.D. Thesis. Waterford Institute of Technology. https://repository.wit.ie/973/1/An_Investigation_into_the_Mechanisms_of_Heavy_Metal_Binding_by_Selected_Seaweed_Species.pdf.

Murphy, V., H. Hughes and P. McLoughlin. 2008. Comparative study of chromium biosorption by red, green and brown seaweed biomass. Chemosphere 70(6): 1128-1134.

National health improvement centre (NHIC). 2018. Detoxification – Heavy Metals. Accessed from https://www.nhicwestmi.com/detoxification–heavy–metals on 7th December 2019.

Neide, E., E.N. Carrilho and T.R. Gilbert. 2000. Assessing metal sorption on the marine alga *Pilayella littoralis*. J. Environ. Monit. 2: 410-415.

Nessim, R.B., A.R. Bassiouny, H.R. Zaki, M.N. Moawad and K.M. Kandeel. 2011. Biosorption of lead and cadmium using marine algae. Chem. Ecol. 27: 579-594.

Neveux, N., J.J. Balton, A. Bruhn, D.A. Roberts and M. Ras. 2018. The bioremediation potential of seaweeds: Recyling nitrogen, phosphorus and other waste products. pp. 217-241. *In*: Stephane La Barre, Stephen S. Bates. (eds.). Blue Biotechnology: Production and Use of Marine Molecules. Wiley VCH, Germany.

Nielsen, H.D., T.R. Burridge, C. Brownlee and M.T. Brown. 2005. Prior exposure to Cu contamination influences the outcome of toxicological testing of Fucus serratus embryos. Mar. Pollut. Bull. 50: 1675-1680.

Nigro, S.A., W.A. Stirk and J. van Staden. 2002. Optimising heavy metal adsorbance by dried seaweeds. S. Afr. J. Bot. 68: 333-341.

Norris, P.R. and D.P. Kelly. 1979. Accumulation of cadmium and cobalt by *Saccharomyces cerevisiae*. J. Gen. Microbiol. 99: 317-324.

Olguín E.J. and G. Sánchez–Galván. 2012. Heavy metal removal in phytofiltration and phycoremediation: The need to differentiate between bioadsorption and bioaccumulation. New Biotechnol. 30: 3-8.

Ortiz-Calderon C., H. Cid Silva and D.B. Vásquez. 2017. Metal Removal by Seaweed Biomass, Biomass Volume Estimation and Valorization for Energy. Jaya Shankar Tumuluru. p. 361. IntechOpen. doi: 10.5772/65682. https://www.intechopen.com/books/biomass–volume–estimation–and–valorization–for–energy/metal–removal–by–seaweed–biomass.

Pandya, K.Y., R.V. Patel, R.T. Jasrai and N. Brahmbhatt. 2017. Biosorption of Cr, Ni and Cu from industrial dye effluents onto *Kappaphycus alvarezii*: Assessment of sorption isotherms and kinetics. Int. J. Eng. Res. Gen. Sci. 5(4): 137-148.

Park, D., Y.S. Yun, H.Y. Cho and J.M. Park. 2004. Chromium biosorption by thermally treated biomass of brown seaweed. *Ecklonia* sp. Ind. Eng. Chem. Res. 43: 8226-8232.

Park, Y.H. and L. Gamberoni. 1997. Cross-frontal exchange of Antarctic intermediate water and Antarctic bottom water in the Crozet basin. Deep-Sea Res. Part 2 Top Stud. Oceanogr. 44: 963-986.

Percival, E. and R.H. McDowell. 1967. Chemistry and Enzymology of Marine Algal Polysaccharides. Academic Press, London. pp. 99-126.

Poo, K.M., E.B. Son, J.S. Chang, X. Ren, Y.J. Choi and K.J. Chaea. 2018. Biochars derived from wasted marine macro-algae (*Saccharina japonica* and *Sargassum fusiforme*) and their potential for heavy metal removal in aqueous solution. J. Environ. Manag. 206: 364-372.

Praveen, R. and K. Vijayaraghavan. 2015. Optimization of Cu(II), Ni(II), Cd(II) and Pb(II) biosorption by red marine alga *Kappaphycus alvarezii*. Desalin. Water Treat. 55: 1816-1824.

Priya, M., N. Gurung, K. Mukherjee and S. Bose. 2014. Microalgae in removal of heavy metal and organic pollutants from soil. pp. 519-537. *In*: Surajit, Das. (ed.). Microbial Biodegradation and Bioremediation. Elsevier ISBN 9780128000212, https://doi.org/10.1016/B978-0-12-800021-2.00023-6.

Rahman, M.S. and K.V. Sathasivam. 2015. Heavy metal adsorption onto *Kappaphycus* sp. from aqueous solutions: The use of error functions for validation of isotherm and kinetics models. BioMed. Res. Int. ID 126298: 1-13.

Rahman, M.S. and K.V. Sathasivam. 2016. Heavy metal biosorption potential of a Malaysian Rhodophyte (*Eucheuma denticulatum*) from aqueous solutions. Int. J. Environ. Sci. Technol. 13(8): 1973-1988.

Rai, L.C., J.P. Gaur and H.D. Kumar. 1981. Phycology and heavy-metal pollution. Biol. Rev. 56(2): 99-151.

Raize, O., Y. Argaman and S. Yannai. 2004. Mechanisms of biosorption of different heavy metals by brown marine macroalgae. Biotechnol. Bioeng. 87(4): 451-458.

Rajfur, M. 2013. Algae – Heavy Metals Biosorbent. Ecol. Chem. Eng. S. 20(1): 23-40.

Ramelow, G.J., D. Fralick and Y. Zhao. 1992. Factors affecting the uptake of aqueous metal ions by dried seaweed biomass. Microbios. 72: 81-93.

Rice, D.L. and B.E. Lapointe. 1981. Experimental and outdoor studies with *Ulva fasciata* Delile 11. trace metal chemistry. J. Exp. Mar. Biol. Ecol. 54: 1-11.

Rice, D.L. 1984. A simple mass transport model for metal uptake by marine macroalgae growing at different rates. J. Exp. Mar. Biol. Ecol. 82: 175-182.

Romera, E., F. González, A. Ballester, M. Blázquez and J. Munoz. 2007. Comparative study of biosorption of heavy metals using different types of algae. Bioresour. Technol. 98: 3344-3353.

Salama, E., H. Roh, S. Dev, M.A. Khan, R.A.I. Abou-Shanab, S.W. Chang, et al. 2019. Algae as a green technology for heavy metals removal from various wastewater. World J. Microbiol. Biotechnol. 35: 75.

Sargın, I., G. Arslan and M. Kaya. 2016. Efficiency of chitosan-algal biomass composite microbeads at heavy metal removal. React. Funct. Polym. 98: 38-47.

Schiewer, S. and B. Volesky. 2000. Biosorption by Marine Algae. Bioremediation 139-169.

Selvan, T., S. Kumar and K. Kannabiran. 2012. Biosorption of Cr(VI), Cr(III), Pb(II) and Cd(II) from aqueous solutions by *Sargassum wightii* and *Caulerpa racemosa* algal biomass. J. Ocean Univ. China. 11: 52-58.

Shaaban A.E.-S.M., R.K. Badawy, H.A. Mansour, M.E. Abdel-Rahman and Y.I.E. Aboulsoud. 2017. Competitive algal biosorption of Al^{3+}, Fe^{3+} and Zn^{2+} and treatment application of some industrial effluents from Borg El-Arab region, Egypt. J. Appl. Phycol. 29: 3221-3234.

Sheng, P.X., Y. Ting, J.P. Chen and L. Hong. 2004. Sorption of lead, copper, cadmium, zinc and nickel by marine algal biomass: Characterization of biosorptive capacity and investigation of mechanisms. J. Colloid. Interface. Sci. 275: 131-141.

Singh, V.P. 2005. Metal toxicity and tolerance in plants and animals. Sarup and Sons. pp. 262-307.

Skoog, D.A., F.J. Holler and T. Nieman. 1998. Principles of Instrumental Analysis (5th Ed.). New York.

Son, E.B., K.M. Poo, J.S. Chang and K.J. Chae. 2018. Heavy metal removal from aqueous solutions using engineered magnetic biochars derived from waste marine macro-algal biomass. Sci. Total Environ. 615: 161-168.

Stirk W.A. and J. van Staden. 2002. Desorption of cadmium and the reuse of brown seaweed derived products as biosorbents. Bot. Mar. Vol. 45: 9-16.

Tabaraki, R. and A. Nateghi. 2014. Multimetal biosorption modeling of Zn^{2+}, Cu^{2+} and Ni^{2+} by *Sargassum ilicifolium*. Ecol. Eng. 71: 197-205.

Tamilselvan, N., K. Saurav and K. Kannabiran. 2012. Biosorption of Cr(VI), Cr(III), Pb(II) and Cd(II) from aqueous solutions by *Sargassum wightii* and algal biomass. J. Ocean Univ. China 11(1): 52-58.

Tsezos, M. and B. Volesky. 1981. Biosorption of uranium and thorium. Biotechnol. Bioeng. 23: 583-604.

Ungureanu, G., S. Santos, R. Boaventura and C. Botelho. 2015. Biosorption of antimony by brown algae *S. muticum* and *A. nodosum*. Environ. Eng. Manag. J. 14: 455-463.

Vandecasteele, C. and C.B. Block. 1997. Introduction (Chapter 1). pp. 1-8. *In*: Vandecasteele, C. and C.B. Block. (eds.). Modern Methods for Trace Element Determination; Wiley, Chichester, UK.

Vieira, R.H.S.F. and B. Volesky. 2000. Biosorption: A solution to pollution. Int. Microbiol. 3: 17-24.

Vijayaraghavan, K., J. Jegan, K. Palanivenu and M. Velan. 2005. Biosorption of copper, cobalt and nickel by marine green alga *Ulva reticulata* in a packed column. Chemosphere 60: 419-426.

Vijayaraghavan, K., S. Gupta and U.M. Joshi. 2012. Comparative assessment of Al(III) and Cd(II) biosorption onto *Turbinaria conoides* in single and binary systems. Water. Air. Soil. Pollut. 223: 2923-2931.

Volesky, B. and N. Kuyucak. 1998. Biosorbent for gold, US Patent No. 4,769,223.

Wang, J. and C. Chen. 2009. Biosorbents for heavy metals removal and their future. Biotechnol. Adv. 27: 195-226.

Wendt, R.H. and V.A. Fassel. 1965. Induction-coupled plasma spectrometric excitation source. Anal. Chem. 37(7): 920-922.

Weng, C.H., C.Z. Tsai, S.H. Chu and Y.C. Sharma. 2007. Adsorption characteristics of copper(II) onto spent activated clay. Sep. Purif. Technol. 54: 187-197.

Williams, C.J. and R.G.J. Edyvean. 1997. Ion exchange and nickel biosorption by seaweed materials. Biotechnol. Prog. 13: 424-428.

Winter, C., M. Winter and P. Pohl. 1994. Cadmium adsorption by non-living biomass of the semi-macroscopic brown alga, *Ectocarpus siliculosus*, grown in actinic mass culture and localisation of the adsorbed Cd by transmission electron microscopy. J. Appl. Phycol. 6: 479-487.

Yalçın, S., S. Sezer and R. Apak. 2012. Characterization and lead(II), cadmium(II), nickel(II) biosorption of dried marine brown macro algae *Cystoseira barbata*. Environ. Sci. Pollut. Res. 19(8): 3118-3125.

Ye, J., H. Xiao, B. Xiao, W. Xu, L. Gao and G. Lin. 2015. Bioremediation of heavy metal contaminated aqueous solution by using red algae *Porphyra leucosticta*. Water Sci. Technol. 72: 1662-1666.

Yipmantin, A., H.J. Maldonado, M. Ly, J.M. Taulemesse and E. Guibal. 2011. Pb(II) and Cd(II) biosorption on *Chondracanthus chamissoi* (a red alga). J. Hazard. Mater. 185: 922-929.

Yu, Q. and P. Kaewsarn. 1999. Fixed-bed study for copper(II) removal from aqueous solutions by marine alga Durvillaea potatorum. Environ. Technol. 20: 1005-1008.

Zeraatkar, A.K., H. Ahmadzadeh, A.F. Talebi, N.R. Moheimani and M.P. McHenry. 2016. Potential use of algae for heavy metal bioremediation, a critical review. J. Environ. Manag. 181: 817-831.

Zeroual, Y., A. Moutaouakkil, F.Z. Dzairi, M. Talbi, P.U. Chung, K. Lee, et al. 2003. Biosorption of mercury from aqueous solution by *Ulva lactuca* biomass. Bioresour. Technol. 90(3): 349-351.

Zhao, Y., Y. Hao and G.J. Ramelow. 1994. Evaluation of treatment techniques for increasing the uptake of metal ions from solution by non-living seaweed algal biomass. Environ. Monit. Assess. 33: 61-70.

Zhou, J.L., P.L. Huang and R.G. Lin. 1998. Lin sorption and desorption of Cu and Cd by macroalgae and microalgae. Environ. Pollut. 101: 67-75.

Zumdahl, S.S. 1992. Chemical Principles. D.C. Health and Company, Canada.

5

Microalgae Immobilization and Use in Bioremediation

Cátia Palito, Telma Encarnação, Poonam Singh, Artur J.M. Valente and Alberto A.C.C. Pais*

INTRODUCTION

Microalgae are a large and heterogeneous group of prokaryotic and eukaryotic photosynthetic microorganisms, found in almost every habitat on Earth. These life forms display a high photosynthetic efficiency (Priyadarshani et al. 2012, Salgueiro et al. 2016). Their photosynthetic mechanism is usually compared to terrestrial plants, but because they are simpler in structure, they convert solar energy to biomass more efficiently (Priyadarshani et al. 2012). It is known that most of the global photosynthesis is carried out by these microorganisms that are part of the largest aquatic trophic chain (Moreno-Garrido 2008). They can be found in both freshwater and marine waters (Priyadarshani et al. 2012). There is a wide variety of microalgae, but this diversity is not yet fully explored (Borowitzka 2013). Its composition varies according to the species and growth conditions, being constituted by proteins (6 to 52%), lipids (7 to 23%) and carbohydrates (5 to 23%) (Narasimhan 2010). In recent years, the use of microalgae in biotechnology has increased both in the food industry and in products with ecological applications. They as well have applications in cosmetics and pharmaceuticals (Kaparapu 2017, Moreno-Garrido 2008, Pulz and Gross 2004, Vandamme et al. 2015). These microorganisms are an important source of chemicals, such as carotenoids, polyunsaturated fatty acids and phycobilins (Borowitzka 2013). However, relatively small size of the microalgae cells could be considered a challenge in their use in biotechnological processes (Kaparapu 2017, Moreno-Garrido 2008, Vandamme et al. 2015). As they have a large potential for obtaining valuable products (including chemicals and fuels) (Borowitzka 2013, Talukder et al. 2014), it is necessary to minimize the costs associated with the harvesting methods and, as such, to develop more effective methods (Talukder et al. 2014).

The methods used for microalgae harvesting include natural sedimentation, filtration, and centrifugation. The latter is associated with a significant consumption of energy and is considered cost-effective only if microalgae are used for a high-value product (Abu Sepian et al. 2017, Salim et al. 2011, Talukder et al. 2014).

CQC, Department of Chemistry, University of Coimbra, 3004-535 Coimbra, Portugal.
* Corresponding Authors: pais@ci.uc.pt

Thus, in order to reduce the costs, the flocculation method was introduced with an objective of recovering the microalgae biomass in wastewater. It consists of the use of a flocculating agent that causes the dispersed cells to aggregate, which facilitates sedimentation (Salim et al. 2011). However, this method also possesses some shortcomings; the flocculants generally used, such as aluminum and ferric chloride, can contaminate, with metal salts, the biomass in the downstream process (Talukder et al. 2014).

Microalgae in natural environments and at low concentrations, for long periods of time are constantly exposed to pollutants. Pollutants include heavy metals, pharmaceuticals, pesticides, among others, and may also be subjected to high concentrations when pollutants are discharged into effluents and coastal waters (Priyadarshani et al. 2012). Due to their ability for removing certain substances from effluents, microalgae are also used in bioremediation processes (both in open systems and closed systems) (El-Sheekh et al. 2016, Gressler et al. 2014). For this purpose, it can be advantageous to immobilize the microalgae cells. The technique of immobilization emerged for environmental applications, aiming at removing undesirable substances from water (González-delgado et al. 2016). In this technique, the cells are restricted from moving freely in the aqueous medium, usually resorting to a solid matrix (Das and Adholeya 2015, Kaparapu et al. 2016, Malik 2002, Mcintosh 1985, Ng et al. 2017). The immobilization aims to keep the cells alive for an extended period of time, allowing to gain biotechnological benefits both at the metabolic level, and in the removal of residual water pollutants (de-Bashan and Bashan 2010). As a consequence, microalgae immobilization has been the subject of an increasing number of studies.

The main objective of this chapter is to describe the different methods of immobilization and the associated advantages and disadvantages. The most frequently used materials for immobilization of cells within polymer matrices is addressed, and important physical considerations are also discussed. Finally, a detailed approach to the main pollutants that are the object of bioremediation processes is presented.

Methods and Materials in Immobilization

One of the many challenges that modern biotechnology faces is the development of methods of immobilization in which cells do not lose their physiological and functional activity (Valuev et al 1994). This activity should be maintained as it is required during the optimization of the processing material used for immobilization. The immobilization of cells, in certain conditions and depending on the species, may have beneficial consequences for the cells. It includes an increasing resistance to toxic environments, and improving the removal or recovery of metals (Eroglu et al. 2015, Paul 2014).

Microalgal Immobilization Methods

Cellular immobilization aroused great interest a few decades ago (Mcintosh 1985), and since then some methods have been described (Das and Adholeya 2015, Malik 2002, Mcintosh 1985) (see also Fig. 5.1):

I. **Adsorption and adherence:** Adsorption and adherence is a basic method, based solely on the physical interaction between the microorganism and a support surface, in which weak, reversible bonds are formed. The forces involved are predominantly electrostatic; they include van der Waals forces, ionic interactions and hydrogen bonding, but hydrophobic interactions are also possible. This type of immobilization allows to protect the cells from harmful agents and preserves their physiological activity when choosing suitable absorbents (Fig. 5.1a). It is a simple and non-costly method, no chemical changes are required, nutrients and chemicals are in contact with cells, and regeneration of the system is possible. The drawbacks include the occurrence of cell leakage from the polymeric matrix, and cell overloading that can lead to a reduced activity.

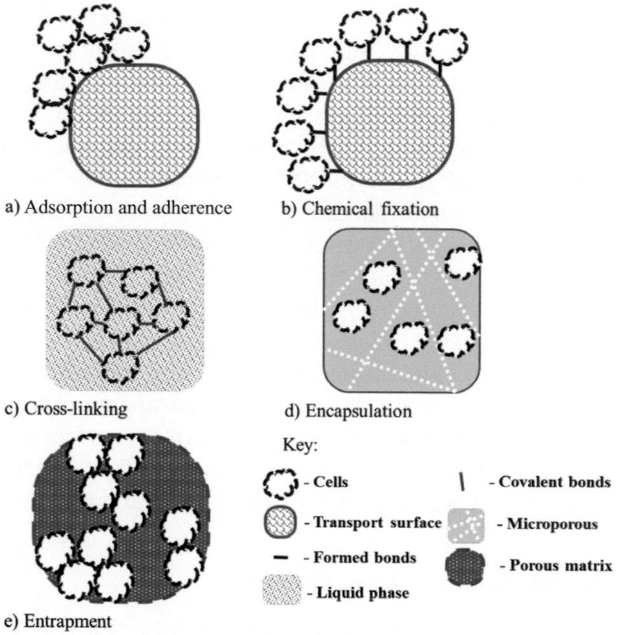

FIGURE 5.1 Illustration of different methods of immobilization.

II. **Chemical fixation:** In chemical fixation, the immobilization of the microorganisms is done by the attachment of reactive groups to the cell surface, thus forming strong bonds to the support surface (Fig. 5.1b).

III. **Cross-linking:** For cross-linking of microorganisms, multifunctional reagents are used to establish the link between bio-macromolecules (covalent bonds). The reagents used are usually cytotoxic and may cause damage to the cells (Fig. 5.1c).

IV. Encapsulation (microencapsulation): The method of encapsulation is based on the fact that the pores of the material enable other substances to diffuse either into or out of the cells, and these can roam free within their compartment. By diffusion through the barrier, the nutrients are supplied and the products removed. However, there are limitations in the diffusion and it can reduce the efficiency of this method (Das and Adholeya 2015, Martins et al. 2013, Willaert and Baron 1996) (Fig. 5.1d).

V. Entrapment: When resorting to entrapment, suspended microorganisms are immobilized in a porous polymer. The fact that the polymer is porous allows the pollutants and other metabolites to diffuse into their membranes. This method is considered irreversible, creating a protective barrier to the immobilized cells, guaranteeing their viability for a longer period of time. There are several synthetic and natural polymers, discussed below, that are used to encapsulate microorganisms (Fig. 5.1e).

The method of cell immobilization by "adsorption" is a common method used for enzyme but not very common for microalgae. The preparation of polyurethane and hydrophilic polyvinyl (PV-50) foam cubes and the adhesion of microalgae cell, used in the removal of nitrates and phosphates, were reported in earlier studies. The cubes transferred to algal suspensions were successfully colonized by adsorption. Then they were subsequently transferred into batch or continuous-flow bioreactors, for the removal of nitrates and phosphates. Although the method was implemented with success, the removal of the nutrients from water was not highly efficient.

More recently, the techniques of immobilization were divided into "passive immobilization" and "active immobilization" (Kaparapu 2017, Martins et al. 2013, Vasilieva et al. 2016), as shown in the diagram of Figure 5.2. Passive immobilization refers to the natural tendency of microalgae to adhere and grow on natural or synthetic surfaces, whereas active immobilization involves methods such as chemical fixation and gel encapsulation (synthetic and natural polymers) (Das and Adholeya 2015, Martins et al. 2013, Moreno-garrido 2013, Vasilieva et al. 2016).

"Chemical fixation" is a method of active immobilization, but it tends to cause damage to the cell surface, effectively reducing cell viability, due to the chemical interactions (Moreno-Garrido 2008). Lastly, "entrapment" has been used for many decades because it is a technique in which cells remain viable, allowing the increase of biomass. The mixing of the cells with the polymers can have various shapes and sizes, although the most commonly used are small beads (Willaert and Baron 1996).

Matrices for Immobilization

Matrices for immobilization include ionic hydrogels, such as alginates and carrageenans, thermogels, such as agar, agarose, and cellulose, and synthetic polymers, such as polyacrylamide, polyvinyl alcohol (PVA) and polyurethanes.

Currently, due to their physico-chemical characteristics, hydrogels are being used in various studies because they possess the mechanical elasticity and the ability to form networks allowing the retention of large amounts of solvent. The polymer in the hydrogel can thus be used to trap different chemical species with environmental benefit (González-delgado et al. 2016).

The following scheme summarizes the synthetic and natural polymers mostly used in the immobilization of microalgae. These polymers must be insoluble, with high mechanical stability and diffusivity, their use must be simple, they must have high capacity of retention of biomass and, finally, a low cost (Leenen et al. 1996). Synthetic polymers are considered to be more stable than natural polymers in wastewater. Although natural polymers possess a higher diffusion rate for nutrients and other substances, they are more vulnerable to biodegradation (Cohen 2001, de-Bashan and Bashan 2010, Eroglu et al. 2015). The polymer used must be, in any case, hydrophilic so that water may diffuse (de-Bashan and Bashan 2010) inside the matrix.

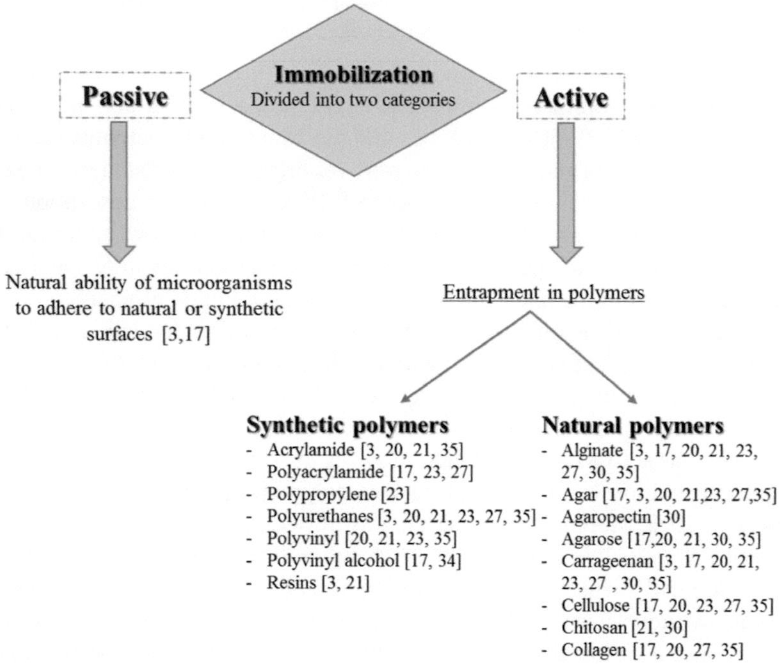

FIGURE 5.2 Scheme with the main materials used to entrap microorganisms - synthetic and natural polymer.

The entrapment in natural matrices is the most common in microalgae immobilization techniques (Moreno-Garrido 2008). Alginate is one of the most widely used polymers for the immobilization of microalgae, being very advantageous due to its low cost, high diffusivity, and ability to preserve cell viability (De-Bashan and Bashan 2008). The protocols described for its use are simple and fast to implement (Wang et al. 2018). Alginate is a natural polymer (De-Bashan and Bashan 2008, Soo et al. 2017, Vasilieva et al. 2016) that can be obtained from marine macroalgae like *Sargassum* spp., *Macrocystispyrifera, Laminariahyperborea, Laminariadigitata, Ascophyllumnodosum, Durvilleaantarctica*, etc. The bacterial sources are *Azotobactervinelandii* and *Pseudomonades* (Ng et al. 2017). The polymeric chains contain two groups, β-D-manuronic acid (M) and α-L-guluronic acid (G). G-G blocks, G-M blocks and M-M blocks can be present in different ratios and in different molecular weights, based on the species of algae or bacteria it was extracted from. The physical and chemical properties of alginate depend strongly upon monomeric composition and arrangements, molecular weight and the divalent cation applied. The biding of divalent cations is an important step in the stability and rigidity of the alginate matrices. The alginate-divalent cation interactions are selective and dependent on the arrangements of the polymeric chain; GG blocks bind Ba^{2+}, Ca^{2+} and Sr^{2+}, MG blocks bind Ca^{2+}, and MM blocks bind Ba^{2+} (Fig. 5.3).

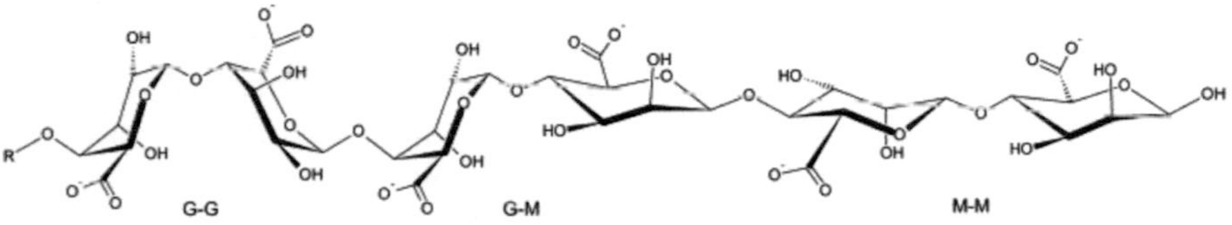

FIGURE 5.3 Chemical structure of alginate.

The incorporation of microalgae into alginate beads aims to immobilize these cells of significant commercial value in a practical way. It can also be considered as a tool for genetic manipulation by immobilizing recombinant bacteria in order to improve the respective stability. Considering as a key objective the ease with which microalgae are collected after treatment (De-Bashan and Bashan 2008), sometimes alginate is associated with other chemical agents. For example, Ca-alginate is widely used in the formation of beads containing microalgae because it does not limit their growth and it does not constitute a toxic material for cells. However, it is not very stable in both seawater and wastewater (Vasilieva et al. 2016). The instability of calcium alginate in seawater is due to the fact that it comes into contact with cation chelating agents. Nonetheless, this stability can be increased by varying the concentration of alginate and that of the crosslinking solution used (usually $CaCl_2$) (Soo et al. 2017). The natural polymer agar is used due to the non-toxic character, possesses a low melting point, and has the ability to form a gel at low concentrations (Vasilieva et al. 2016). It is widely used by microbiologists because it is resistant to degradation by microorganisms (Nussinovitch 2010) and is a thermoreversible gel (Moreno-Garrido 2008). Carrageenan, another natural polymer, is also widely used, although it displays less stability in water, when compared to alginates (Vasilieva et al. 2016). This polymer has two alternative structures: D-galactose 4-sulfate and 3,6-anhydro-D-galactose 2-sulfate, and its strength increases with the concentration of 3,6-anhydro-D-galactose (Leenen et al. 1996). The polymer, in the presence of cationic compounds such as metal ions, amines, amino acid derivatives, and water-miscible organic solvents, precipitates as a gel (Moreno-Garrido 2008).

Synthetic polymers appear to be the most suitable for wastewater treatment because they are more stable under adverse conditions (Cohen 2001), having a better mechanical performance compared to natural polymers. However, they are not as easily biodegradable (Martins et al. 2013).

Polyacrylamide, polyvinyl alcohol, polyethylene glycol, and poly carbamoyl sulfate have been synthesized to serve as base carriers, but in order to emphasize their stability there are other synthetic polymers which have been more exploited, such as polypropylene, polyethylene, polyvinyl chloride, polyurethane and polyacrylonitrile (Martins et al. 2013). Polyurethane is one of those highlighted by promoting the viability of the imprisoned microorganisms and PVA has been widely used due to their biocompatibility.

With the most diversified polymers available, while dealing with industrial projects of immobilization of microalgae, one must take into account certain aspects of the matrix. This includes the material, which must be cheap and available in large quantities; the immobilization method should be simple and maintain the metabolic activity. A high efficiency of the immobilized cells is also desired and the whole process involved should also be simple and easy to handle (Klein and Ziehr 1990). Therefore, the requirements for an efficient microalgae immobilization system include the preservation of integrity and stability (Eroglu et al. 2015, Paul 2014). The transparency (Eroglu et al. 2015, Paul 2014), biocompatibility (Klein and Ziehr 1990), resistance to more aggressive environments (salinity, toxicity, pH) (Eroglu et al. 2015, Paul 2014), allowance of high population density, cost-effectiveness, recovery of cells, and retention of biomass are an advantage too (Eroglu et al. 2015).

Despite the evolution of the methods, there are still some barriers to the efficient use of immobilized microalgae. These rapidly developing microorganisms generate large amounts of dissolved oxygen, which accumulates within their microenvironment. This hampers the respective development because there is less fixation of photosynthetic carbon. Thus, as the cells are together, light penetration and diffusion of CO_2 (Wang et al. 2018) are made more difficult.

Main Uses and Advantages/Disadvantages of Immobilization

The use of microalgae has been increasing since they are used for different purposes, such as water treatment (removal of nutrients, metals and organic pollutants) and biomass production (Eroglu et al. 2015, Kaparapu et al. 2016, Shyam et al. 2012). Its use has also increased at the industrial level, namely pharmaceutical and cosmetic industry, food and aquaculture (Joo et al. 2001, Kaparapu et al. 2016).

The contamination of water with heavy metals and emerging pollutants, such as pharmaceuticals and pesticides, is becoming a concern worldwide since it has consequences on the ecosystem and human health (Xiong et al. 2018, Zeraatkar et al. 2016).

The immobilization of microalgae has shown a high potential, providing solutions for some problems associated to their use. One of the main problems of microalgae production is associated with the harvesting from wastewaters. By immobilization, microalgae occupy less space due to their tightening in a solid matrix, and becomes easier to handle. Another advantage is the increased resistance of cells, from specific species to more aggressive environments, such as high toxicity, salinity and pH (Eroglu et al. 2015, Paul 2014). This was confirmed by studies in which it was observed that cells survived longer when agglomerated with immobilization techniques than in their free form (Das and Adholeya 2015, Paul 2014). This technique could provide advantages at cellular level, such as an increase in the reaction rates due to the increase in cell density and membrane pre-flexibility, resulting in better operational stability (Shyam et al. 2012). As such, immobilization of the cells could increase the biosorption capacity and bioactivity of the biomass (Eroglu et al. 2015). Immobilization is also important because it allows to increase the cost-effectiveness, allowing recovery of the cells in less destructive manners (Eroglu et al. 2015). Microalgae have thus become very promising in the treatment of water and in the production of biomass, also because these microorganisms display a fast growth and the ability to fixate CO_2, capture nitrogen and phosphate and possess a photosynthetic mechanism that assists in these processes (Alzate 2018, Wang et al. 2018).

Using co-immobilization (Kaparapu et al. 2016, Moreno-Garrido 2008), the microalgae are associated with other microorganisms, such as bacteria and yeasts that promotes their growth. The co-immobilization system (bacterial-microalgae) is able to increase nutrient removal from wastewater, compared to microalgae immobilization alone. Note that the bacteria is not able to remove the nutrients from the waters, its function being solely to aid in the development of the microalgae (Moreno-Garrido 2008). The main objective of this technique is to join the biocatalytic activity of the two microorganisms as only one (O'Reilly and Scott 1995). Microalgae are able to synthesize O_2 from the photolysis of water and other organic compounds that are assimilated by bacteria, and bacteria release vitamins, phytohormones and still CO_2, which is beneficial for the growth of microalgae (Eroglu et al. 2015). This will increase the usefulness of microalgae immobilized in bioremediation applications, mainly for the treatment of wastewater (Cruz et al. 2013).

The techniques of immobilization also have their disadvantages. One of the main ones in the use of these photosynthetic microorganisms is associated to situations of insufficient lighting (Cohen 2001). However, several others can be indicated: in the case of adsorption, there may be loss of the immobilized cells and the quantity of biomass is limited (Elakkiya et al. 2016), a limited transfer of oxygen and nutrients can occur (Jia and Yuan 2016, Katchalski-Katzir 1993, Willaert and Baron 1996), some polymeric materials may increase diffusion restriction (Cohen 2001, Martins et al. 2013), immobilization could be associated to additional costs of immobilization matrices (Jia and Yuan 2016, Katchalski-Katzir 1993, Martins et al. 2013), matrices possess usually low mechanical strength (Jia and Yuan 2016, Willaert and Baron 1996), and the immobilization techniques are not easily amenable to scale-up (Jia and Yuan 2016).

What Happens to the Immobilized Microalgae Cells?

It should be noted that the ability of the microorganisms to attach to a surface varies with: (i) culture composition, (ii) age (growth stage). Normally, the best immobilization phase is found for cells that are in exponential phase (Vasilieva et al. 2016).

After immobilization, some characteristics have been reported (see Figure 5.4):

- an increase in pigment content (chlorophyll percentage can increase up to twice as much as free cells) (de-Bashan and Bashan 2010, Moreno-Garrido 2008, Vasilieva et al. 2016);
- the variation in lipid and fatty acid contents (that usually increase) (Vasilieva et al. 2016);
- the possibility of causing stress in the cells, determining a decrease in population. This effect varies with the species being studied (Vasilieva et al. 2016), and in some cases immobilized cells have a higher growth rate (de-Bashan and Bashan 2010, Moreno-Garrido 2008);
- a higher resistance for immobilized cells to changes in pH, temperature or ionic strengths of the medium and also a higher tolerance to toxic substances (Vasilieva et al. 2016)·
- some impact on the metabolic activity of cells (de-Bashan and Bashan 2010, Malik 2002, Vasilieva et al. 2016), with positive consequences in biotechnological applications (Vasilieva et al. 2016);
- variations in the shape and size of cells may occur (Malik 2002, Moreno-Garrido 2008, Vasilieva et al. 2016);
- an effect upon the photosynthetic activity, which decreases because of insufficient lighting (de-Bashan and Bashan 2010, Vasilieva et al. 2016).

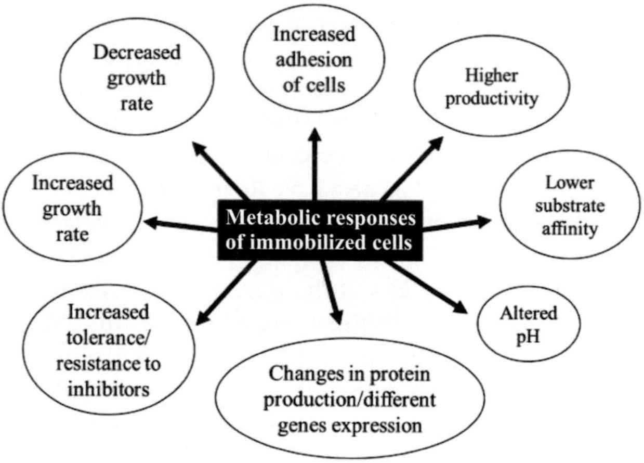

FIGURE 5.4 Schematic representation of some metabolic responses from immobilized cells.

Generally, the effects depend on the species, the matrix and the culture conditions that are used. It should be stressed that cells that are bound in natural environments show a higher growth, while the cells that undergo immobilization processes may have this factor limited. The former increase in growth, in bound cells, may be due to the absorption of nutrients and particles dissolved in the surface (Zur et al. 2016).

Impact of Immobilization

The physical confinement or location of the cells in each space is carried out by resorting them to the techniques of immobilization. This might change cell´s activity or cell´s viability. The goal is to convert the free (homogeneous) cells to heterogeneous conditions. An important aspect in the immobilization processes is the size of the immobilized material, since it is influential in the characteristics of the flow and in the cell's activity. The optimum size defines not only the material, but also the method and the reactor to be used. Smaller particles have a greater surface area per unit of volume which implies an improvement of the rate of diffusion and mass transfer (Mcintosh 1985).

Electrostatic extrusion is considered an immobilization technique for different cell types, used, for its advantages, in medicine. This technique, relative to the others, provides much smaller particles, such as alginate microspheres with 100 or 50 μm in diameter. This technique is reproducible, controllable, with low polydispersity and can easily be used in sterile conditions. Its production is carried out in conditions of little stress, and does not require the use of organic solvents that can interfere in the activity of cells (Obradovic et al. 2006).

The development of electrostatic droplets is based on the electrostatic forces that rupture a liquid filament at the tip of the needle or capillary, forming a charged stream of small droplets. This process depends on several parameters (Obradovic et al. 2006) that include the applied electrostatic potential, needle diameter, distance and geometry of the electrode, flow rate of the polymer solution, superficial tension, density and viscosity (Liu et al. 2009). The hydrogels used for immobilization are mechanically flexible and soft in solution, being able to immobilize microalgae cells and enzymes, maintaining their biological functions. The water-soluble polymers are capable of producing nano and microfibers through electro-spinning, resulting in nano hydrogels and microstructures. Thus, if these hydrogels have mechanical and biocompatible properties, together with the high surface ratios for fiber volume, they can be used as an effective support material for enzymes and also for other microorganisms (Liu et al. 2009), such as microalgae. Beads are the most commonly form used, although there are other options that have good results such as disks, chips, flakes, blocks and thin films (Mcintosh 1985).

As regard to a population, the growth rate in immobilized cell systems is possibly the most studied parameter. Different studies have been published with contradictory results (Martins et al. 2013, Zur et al. 2016). Studies show increased, unchanged and decreased growth rates, compared to free cells. Possibly, the decrease in growth rate can be explained by the fact that there may be limitations in mass transfer and due to modifications of its microenvironment, such as nutrient/oxygen concentration and osmotic pressure (Martins et al. 2013). The immobilized cells, in wastewater, absorb the organic compounds on the surface of the supports and later penetrate through their pores. Consequently, the microorganisms release extracellular enzymes into the pre-hydrolysis of the xenobiotics, then transposing pollutants through the cell membrane for oxidation (Zur et al. 2016).

Physical Considerations

In microalgae immobilization, the studies involving physical interactions are rare. These are more common for other microorganisms, such as bacteria and yeast. Therefore, this is a study field to be explored.

The immobilization procedure of the microalgae cells on the matrix can be interpreted by physical-chemical electrostatic interactions. The structure of the cell wall of microalgae is composed by polysaccharides, proteins, and lipids, which contains functional groups. These include amino, carboxyl, sulphate, phosphoryl and hydroxyl groups. As a result, in their natural medium, the algal surface is negatively charged. The functional groups of cell wall react with the functional groups on polymeric matrices, forming chemical bonding, such as covalent bonding and hydrogen bonding, and Van der Waals interaction, influencing the adhesion between the polymers and cells. A theory that can be used to describe the electrostatic interactions between microalgae cells and matrices for immobilization is the DVLO theory, developed simultaneously by Derjaguin and Landau and Verwey and Overbeek (Verwey 1947, Verwey and Overbeek 1955). It was originally developed to describe the interaction energy between non-biological colloidal spherical particles and surface in a liquid environment, and has since been applied to living cell organisms. The theory has been applied in the interpretation of the mechanisms involved in the harvest of microalgae by bioflocculation, or in the interpretation of the electrostatic double layer forces between the virus particle and the membrane surface. The theory has also been applied in suspended bacteria approach to a solid substratum in an aqueous medium.

The DLVO theory is based on the sum of two terms: an electrostatic repulsive double-layer force and an attractive van der Waals force. The first is calculated through the Poisson-Boltzmann (PB) equation. The PB equation describes the electrical double layer potential in an ionic solution and with a charge density. The van der Waals force involves three contributions; one is the quantum mechanical dispersion interactions, the London term, and can be described in terms of the dipoles and multipoles produced on one atom or molecule polarizing another atom or molecule, by quantum fluctuations; the second contribution, the so-called Keesom term, derives from the dipole-dipole interaction and the third contribution, the Debye term, comes from dipole-induced-dipole interactions. The van der Waals forces occur between polar and non-polar molecules and, compared with the double-layer forces, can be significantly influenced by pH or salt concentration. The adhesion forces of the microalgae cells are affected by variations in the culture medium composition and concentration, since they can strongly influence the surface energy of the immobilization matrix.

IMMOBILIZED MICROALGAE FOR POLLUTANTS REMOVAL

Due to anthropogenic activities, lakes, rivers and streams, seas and oceans of the earth are heavily polluted. Water pollution is a serious environmental threat, mostly caused by pollutants released from industrial and sewage discharges and runoffs from agriculture. This not only affects habitats and entire ecosystems, but also, specifically, human health. Many of the pollutants found in water bodies are considered disruptors to the endocrine system of humans and wildlife; they are called endocrine disrupting chemicals (EDCs) (Encarnação et al. 2019, Snyder et al. 2003). Through the urban cycle of water, pollutants such as bisphenols, organochlorine pesticides, heavy metals and so many others are found in ground, surface and drinking waters (Riva et al. 2018, Snyder et al. 2003). The conventional drinking and wastewater treatments, which include aeration, sedimentation, flocculation, coagulation, flotation, activated sludge, ozonation, chlorination and oxidation processes, do not completely remove most of these pollutants (Snyder et al. 2003). Also, excessive concentration of nitrates and phosphates in the water bodies leads to eutrophication, leading to the inevitable consequences of having microalgal blooms, sometimes with toxic and lethal species, that reduce the oxygen present in the water causing the death of the native flora and fauna.

One potential contribution to solve the water pollution problem consists of the use of biological agents, together with existing methods. The biological treatment processes include bacteria, fungi, protozoa, enzymes and microalgae. The bioremediation using microalgae has several advantages over the conventional methods. The potential solution that concerns all aspects of the environment and the economic valorization of the resulted biomass is among the most important advantages of this treatment.

As mentioned in the previous sections, the use of microalgae immobilization technique has several advantages when compared with free cells. The first report on the use of the microalgae immobilization technique, used for bioremediation, dates back to 1985, and was based on bacteria immobilization techniques, to efficiently remove nitrogen and phosphorous (Laval and Louis 1985). In the mentioned study, the authors compared the results of free cells and in an immobilized matrix, and concluded that the growth and uptake rates were similar in both cases. Note that the majority of the studies focus on the removal of heavy metals, and nitrates and phosphates from wastewater and sewage effluents (Table 5.1).

The removal of contaminants with viable live cells happens by adsorption or incorporation into the cell. Biosorption is a technique that can be used to remove pollutants using live, inactive or non-living biomass. The microalgal cell membrane contains functional groups that are good binders to heavy metal ions. The carboxyl, hydroxyl, carbonyl and phosphodiester functional groups present in polysaccharides, proteins and lipids form stable complexes with heavy metals. The removal of contaminants using non-living cells' system has the advantage over free cells of being metabolism independent, and no growth media are required nor controlled growth parameters, such as pH and temperature and, as such, promotes a lower operating cost.

The adsorption efficiency and the biosorption equilibrium of a pollutant can be described by several well-known models, including the Langmuir and Freundlich adsorption isotherms (Ali and Gupta 2007, Atkins 1978, Rengaraj and Moon 2002, Selvi et al. 2001). The biosorption process is affected by various environmental conditions, namely, pH, temperature, salinity and the nature of the pollutant and the contact time. The bioremediation capacity is dependent on those parameters and on the species involved.

The most commonly used natural polymer for immobilization of biological agents is alginate. Alginate has been used, as the matrix of microalgae immobilization, for the removal of different heavy metals such as mercury, cadmium, lead and copper. The co-immobilization of microalgae and other microorganisms in alginate beads is also possible. The immobilization of the bacteria *Azospirillumbrasilense* and the microalgae *Chlorella vulgaris* in the alginate beads was demonstrated in studies for the removal of ammonium and phosphorus ions from synthetic wastewater (Luz et al. 2002). The co-immobilization could be applied during tertiary wastewater treatment. The immobilization of microalgae in alginate has been often used in studies of removal of nitrates and phosphates from wastewater (Table 5.1). However, for wastewater treatment purposes, alginate matrices present degradation by microorganisms found in wastewater and, as a consequence, the strength of the beads and the bioremediation efficiency are affected (Cruz et al. 2013). Also, the presence of several unknown chemicals in the wastewater can affect the stability of the beads. For the increase of the stability and hardness of alginate beads, the addition of polyvinyl alcohol, polyvinylpyrrolidone, carboxymethylcellulose, $CaCO_3$, and $SrCl_2$ have been reported, but neither of these agents determined a significant increase in the mechanical strength. However, the integrity of the shape remains for at least 96 h, a fact that supports the use of these systems in the tertiary wastewater treatment.

Another example of the application of immobilized microalgae to remove the contaminants from wastewater has been reported in the treatment of industrial textile wastewater (Chu et al. 2008). Carrageenan and alginate were used as immobilization matrices. More recently, a new immobilized complex of microalgae-biochar was successfully used to remove cadmium from aqueous solution and it was demonstrated that the referred complex improved the adsorption capacity of Cd (II) (Shen et al. 2017).

Table 5.1 presents a few examples of the use of immobilized microalgae systems in the removal of several pollutants.

In addition to the pollutants already studied in wastewater, there are numerous environmental chemicals, including pesticides, polychlorinated biphenyls (PCBs), pharmaceuticals, etc. that could have an adverse impact on human health and wildlife (Mcdonald 2002, Xiong et al. 2018, Zhang et al. 2015). These pollutants can be associated with some medical conditions, such as breast and prostate cancer. Nonetheless, in the last decades type 2 diabetes and cardiovascular diseases have found to be affected by the pollutants and therefore microalgae bioremediation has gained a special interest in this field as well (Risk and Factor Collaboration 2008, Xiong et al. 2018, Zhang et al. 2015). There are studies that show that the partial or total removal of these pollutants is possible, thus allowing for a new treatment of wastewater (Gentili and Fick 2017). However, these studies were carried out with free microalgae and studies have not yet been reported, as far as we are aware, with immobilized microalgae for bioremediation of water contaminated with these pollutants.

TABLE 5.1 Removal of pollutants by immobilized microalgae.

Microalgae Specie	Immobilization System	Pollutant	Source	Refs
Chlamydomonasreinhardtii	Polysulfonenanofibrous web	Reactive dyes	Synthetic textile wastewater	(San Keskin et al. 2015)
Chlamydomonasreinhardtii	Ca alginate	Hg (II), Cd (II) and PB (II)	Aqueous solution	(Bayramoǧlu et al. 2006)
Chlorella pyrenoidosa	Sodium alginate algal beads	Phosphate, ammonium	Dairy effluent	(Yadavalli and Heggers 2013)
Chlorella sorokiniana	Loofa sponge-immobilized microalgae	Nickel (II)	Aqueous solution	(Akhtar et al. 2004)
***Chlorella* sp.**	Microalgae-biochar complex	Cadmium	Aqueous solution	(Shen et al. 2017)
Chlorella vulgaris	Alginate algal beads	Nitrogen and phosphorous	Synthetic wastewater	(Tam and Wong 2000)
Chlorella vulgaris	Twin-layer system	Nitrate phosphate, ammonium	Municipal wastewater Synthetic wastewater	(Shi et al. 2007)
Chlorella vulgaris	Alginate algal beads	Nitrate phosphate	Urban wastewater	(Ruiz-Marin et al. 2010, Whitton et al. 2016)
Chlorella vulgaris	Alginate algal/bacteria beads	Ammonium and phosphorus ions	Synthetic wastewater	(Luz et al. 2002)
Scenedesmus obliquus	Carrageenan bead	Nitrogen and phosphorous	Secondary effluent	(Laval and Louis 1985)
Scenedesmusobliquus, Scenedesmusacutus	SRB-microalgae beads	Copper	Mine wastewater	(Li et al. 2018)
Scenedesmus obliquus	Alginate algal beads	Nitrate phosphate	Urban wastewater	(Ruiz-Marin et al. 2010, Whitton et al. 2016)
Scenedesmusquadricauda	Calcium alginate/ polyvinyl alcohol composite hydrogel beads	Cu (II) and Cd (II)	Aqueous solutions	(Bayramoglu and Arica 2011)
Scenedesmusrubescens	Twin-layer system	Nitrate, phosphate, and ammonium	Municipal wastewater and synthetic wastewater	(Shi et al. 2007)

OMIC APPROACHES IN ENVIRONMENTAL BIOREMEDIATION

In response to environmental stress or disturbances, microalgae and cyanobacteria activate specific genes involved in tolerance to environmental changes. As a result of the gene activation, the cell produces proteins or a set of proteins. These are involved in pathways associated with the production of specific metabolites that protect them from stress or promote their adaptation. Lipids, carbohydrates, amino acids, carotenoids and other microalgal metabolites can change the molecular structure and concentration with the exposure to anthropogenic pollutants (Gauthier et al. 2020, Encarnação et al. 2012). The advances in genome sequence of microorganisms and bioinformatics have provided new insights into the complexity of these and many other biological functions. Genome sequences of thousands of microbes are available in public repositories. Omics technologies, that describe biological systems, are able to determine changes in the genome, transcriptome, proteome and metabolome. These omics techniques generate complex and multivariate data, which require computational resources; computational and public web platforms, such as the Integrated Microbial Genomes Atlas of Biosynthetic Gene Clusters (IMG-ABC) (https://img.jgi.doe.gov/abc/), the Atlas of Biosynthetic Gene Clusters (Blin et al. 2013) and the antiSMASH (Blin et al. 2017a, b, Medema et al. 2011, Weber et al. 2015) are used to predict possible products of genes whose DNA sequences are known. Genomics, transcriptomics, proteomics and metabolomics technologies rely on analytical chemistry techniques. These include advanced modern technologies such as UHPLC, NMR and various mass spectroscopy methodologies and, therefore, the products of such genes can be identified and characterized (Dupré et al. 2020). Omics approaches have been applied to the discovery of novel bioactive compounds and in drug discovery. More recently, these approaches have been used for environmental purposes, although limited research has been devoted to environmental omic approaches.

CONCLUDING REMARKS

There are several challenging technical aspects that have to be overcome, or improved, for the generalized implementation of microalgae systems to wastewater treatment plants. The optimization of the efficiency in the bioremediation process can be compromised, simply by a constantly changing medium. Wastewaters, by their nature, vary in the composition of nutrients and pollutants. Nitrates and phosphates, heavy metals, organic pollutants and several other parameters can affect the performance of the living organisms, such as pH and temperature. The stability of the polymeric matrix is another issue. Ensuring the stability of the immobilizing polymer in the presence of different chemicals and other organisms, such as bacteria, could be challenging and compromise the remediation efficiency. Another important aspect to mention in this technology is the fact that integrating a microalgae system to conventional wastewater system can be costly, depending on the chosen photobioreactor. Nonetheless, adding value to the biomass produced in wastewater treatment plants, in a biorefinery concept, can encourage the investment in the implementation of bioremediation using algae. The value-added products include lipids, sugars, proteins and pigments that can be used as feedstock for biofuels, biosurfactants and biopolymers, thus generating profit from the biomass produced. Another challenge, albeit indirect, is the improvement and advance in the downstream technologies. The upgrade in technology for extraction, separation, refining and purification of the products of interest may lower the costs and promote an increase in the production of biomass.

ACKNOWLEDGEMENTS

The authors acknowledge the Fundação para a Ciência e a Tecnologia (FCT), Portuguese Agency for Scientific Research, for the PhD research grants SFRH/BD/81385/2011. The authors are also grateful for support from The Coimbra Chemistry Centre, which is funded by the FCT, through the projects PEst-OE/QUI/UI0313/2014 and POCI-01-0145-FEDER-007630.

REFERENCES

Abu Sepian, N.R., N.H. Mat Yasin, N. Zainol, N.H. Rushan and A.L. Ahmad. 2017. Fatty acid profile from immobilised *Chlorella vulgaris* cells in different matrices. Environ. Technol. 40: 1110-1117.

Akhtar, N., J. Iqbal and M. Iqbal. 2004. Removal and recovery of nickel(II) from aqueous solution by loofa sponge-immobilized biomass of chlorella sorokiniana: Characterization studies. J. Hazard Mater. 108: 85-94.

Ali, Imran and V.K. Gupta. 2007. Advances in water treatment by adsorption technology. Nat. Protoc. 1: 2661-2667.

Atkins, P.W. 1978. Physical Chemistry. Oxford University Press.

Bayramoğlu, G., I. Tuzun, M. Yilmaz and M.Y. Arica. 2006. Biosorption of mercury(II), cadmium(II) and lead(II) ions from aqueous system by microalgae chlamydomonas reinhardtii immobilized in alginate beads. Int. J. Miner. Process. 81: 35-43.

Bayramoglu, Gulay and M. Yakup Arica. 2011. Preparation of a composite biosorbent using scenedesmus quadricauda biomass and alginate/polyvinyl alcohol for removal of Cu(II) and Cd(II) ions: Isotherms, kinetics and thermodynamic studies. Water Air Soil Poll. 221: 391-403.

Blin, K., M.H. Medema, D.K. Kazempour, M.A. Fischbach, R. Breitling, E. Takano, et al. 2013. AntiSMASH 2.0: A versatile platform for genome mining of secondary metabolite producers. Nucleic Acids Res. 41: W204-W212.

Blin, K., T. Wolf, M.G. Chevrette, X. Lu, C.J. Schwalen, S.A. Kautsar, et al. 2017. AntiSMASH 4.0: Improvements in chemistry prediction and gene cluster boundary identification. Nucleic Acids Res. 45: W36-W41.

Blin, K., M.H. Medema, R. Kottmann, S.Y. Lee and T. Weber. 2017. The AntiSMASH database, a comprehensive database of microbial secondary metabolite biosynthetic gene clusters. Nucleic Acids Res. 45: D555-D559.

Borowitzka, M.A. 2013. High-Value products from microalgae-their development and commercialisation. J. Appl. Phycol. 25: 743-756.

Chevalier, P. and J. de la Noüe. 1985. Wastewater nutrient removal with microalgae immobilized in carrageenan. Enzyme Microb. Technol. 7: 621-624.

Chu, W., Y. See and S. Phang. 2008. Use of immobilised *Chlorella vulgaris* for the removal of colour from textile dyes. J. Appl. Phycol. 21: 641-648.

Cohen, Y. 2001. Biofiltration – The treatment of fluids by microorganisms immobilized into the filter bedding material: A review. Bioresource Technol. 77(3): 257-274.

Cruz, I., Y. Bashan, G. Hernàndez-Carmona and L.E. De-Bashan. 2013. Biological deterioration of alginate beads containing immobilized microalgae and bacteria during tertiary wastewater treatment. Appl. Microbiol. Biotechnol. 97: 9847-9858.

Das, M. and A. Adholeya. 2015. Potential uses of immobilized bacteria, fungi, algae and their aggregates for treatment of organic and inorganic pollutants in wastewater. ACS Symposium Series 1206: 319-337.

De-Bashan, L.E. and Y. Bashan. 2008. Joint immobilization of plant growth-promoting bacteria and green microalgae in alginate beads as an experimental model for studying plant-bacterium interactions. Appl. Environ. Microbiol. 74(21): 6797-6802.

de-Bashan, L.E. and Y. Bashan. 2010. Immobilized microalgae for removing pollutants: Review of practical aspects. Bioresource Technol. 101: 1611-1627.

Dupré, C., H.D. Burrows, M.G. Campos, C. Delattre, T. Encarnação, M. Fauchon, et al. 2020. Microalgal Biomass of Industrial Interest: Methods of Characterization. In Handbook on Characterization of Biomass, Biowaste and Related By-Products. (pp. 537-639). Springer, Cham.

El-Sheekh, M.M., A.A. Farghl, H.R. Galaland and H.S. Bayoumi. 2016. Bioremediation of different types of polluted water using microalgae. Rend. Lincei 27: 401-410.

Elakkiya, M., D. Prabhakaran and M. Thirumarimurugan. 2016. Methods of cell immobilization and its applications. Int. J. Innov. Res. Sci. Eng. Technol. 5: 5429-5433.

Encarnação, T., H.D. Burrows, A.C. Pais and M.G. Campos. 2012. Effect of N and P on the uptake of magnesium and iron and on the production of carotenoids and chlorophyll by the microalgae *Nannochloropsis* sp. J. Agric. Sci. Technol. 2: 824-832.

Encarnação, T., A.C. Pais, M.G. Campos and H.D. Burrows. 2019. Endocrine disrupting chemicals: Impact on human health, wildlife and the environment. Sci. Prog. 102: 3-42.

Eroglu, E., S.M. Smith and C.L. Raston. 2015. Application of various immobilization techniques for algal bioprocesses. *In*: N. Moheimani, M. McHenry, K. de Boer, P. Bahri. (eds.). Biomass and Biofuels from Microalgae. Biofuel and Biorefinery Technologies, vol 2. Springer, Cham.

Gauthier, L., J. Tison-Rosebery, S. Morin and N. Mazzella. 2020. Metabolome response to anthropogenic contamination on microalgae: A review. Metabolomics. 16: 8.

Gentili, F.G. and J. Fick. 2017. Algal cultivation in urban wastewater: An efficient way to reduce pharmaceutical pollutants. J. Appl. Phycol. 29: 255-262.

González-Delgado, Á.D., A. Barajas-Solano and Y. Peralta-Ruiz. 2016. Microalgae immobilization using hydrogels for environmental applications: Study of transient photopolymerization. Chem. Eng. Trans. 47: 457-462.

Gressler, P., T. Bjerk, R. Schneider, M. Souza, E. Lobo, A. Zappe, et al. 2014. Cultivation of desmodesmus subspicatus in a tubular photobioreactor for bioremediation and microalgae oil production. Environ. Technol. 35: 209-219.

Halder, S. 2014. Bioremediation of heavy metals through fresh water microalgae : A review. Scholars Acad. J. Biosci. 2: 825-830.

Jia, H. and Q. Yuan. 2016. Removal of nitrogen from wastewater using microalgae and microalgae-bacteria consortia. Cogent Environ. Sci. 2: 1-15.

Joo, D.S., M.G. Cho, J.S. Lee, J.H. Park, J.K. Kwak, Y.H. Han, et al. 2001. New strategy for the cultivation of microalgae using microencapsulation. J. Microencapsul. 18: 567-576.

Kaparapu, J. 2017. Microalgal immobilization techniques. J. algal biomass Utln. 8: 64-70.

Kaparapu, J., M. Narasimha, R. Geddada and M.N. Rao. 2016. Applications of immobilized algae. J. Algal Biomass Utln. 7: 122-128.

Katchalski-Katzir, E. 1993. Immobilized enzymes: Learning from past Successes and failures. Trends Biotechnol. 11: 471-478.

Keskin, S., N. Oya, A. Celebioglu, T. Uyar and T. Tekinay. 2015. Microalgae immobilized by nanofibrous web for removal of reactive dyes from wastewater. Ind. Eng. Chem. Res. 54: 5802-5809.

Klein, J. and H. Ziehr. 1990. Immobilization of microbial cells by adsorption. J. Biotechnol. 16: 1-2.

Kumar, S. and A. Saramma. 2012. Nitrate and phosphate uptake by immobilized cells of gloeocapsa gelatinosa. J. Marine Biol. Ass. India 54: 119-122.

Leenen, E., V. Dos Santos, K. Grolle, J. Tramper and R. Wijffels. 1996. Characteristics of and selection criteria for dupport materials for cell immobilization in wastewater treatment. Water Res. 30: 2985-2896.

Li, Y., X. Yang, B. Geng and X. Liu. 2018. Effective bioremediation of Cu(II) contaminated waters with immobilized sulfate-reducing bacteria-microalgae beads in a continuous treatment system and mechanism analysis. J. Chem. Technol. Biotechnol. 93: 1453-1461.

Luz, E., M. Moreno, J. Hernandez and Y. Bashan. 2002. Removal of ammonium and phosphorus ions from synthetic wastewater by the microalgae *Chlorella vulgaris* coimmobilized in alginate beads with the microalgae growth-promoting bacterium azospirillum brasilense. Water Res. 36: 2941-2948.

Malik, N. 2002. Biotechnological potential of immobilized algae for wastewater N, P and metal removal: A review. BioMetals 15: 377-390.

Manojlovic, V., J. Djonlagic, B. Obradovic, V. Nedovic and B. Bugarski. 2006. Immobilization of cells by electrostatic droplet generation: A model system for potential application in medicine. Int. J. Nanomedicine 1: 163-171.

Martins, S., C. Martins, L. Fiúza and S. Santaella. 2013. Immobilization of microbial cells: A promising tool for treatment of toxic pollutants in industrial wastewater. Afr. J. Biotechnol. 12: 4412-4418.

Mcdonald, T.A. 2002. A perspective on the potential health risks of PBDEs. Chemosphere 46: 745-755.

Medema, M.H., K. Blin, P. Cimermancic, V. de Jager, P. Zakrzewski, M.A. Fischbach, et al. 2011. AntiSMASH: Rapid identification, annotation and analysis of secondary metabolite biosynthesis gene clusters in bacterial and fungal genome sequences. Nucleic Acids Res. 39: W339-W346.

Moreno-Garrido, I. 2008. Microalgae immobilization: Current techniques and uses. Bioresour. Technol. 99: 3949-3964.

Moreno-garrido, I. 2013. Immobilization of enzymes and cells. pp. 327-347. *In*: J.M. Guisan. (ed.). Microalgal Immobilization Methods. Humana Press, Totowa, NJ.

Narasimhan, A.M. 2010. Microalgal bioremediation of nutrients in wastewater and carbon dioxide in flue Gas. Masters Teses. 4779.

NCD RiskFactor Collaboration. 2008. Worldwide trends in diabetes since 1980: A pooled analysis of 751 population-based studies with 4.4 million participants. The Lancet 387: 1513-1530.

Ng, F., S. Phang, V. Periasamy, K. Yunus and A. Ficher. 2017. Enhancement of power output by using alginate immobilized algae in biophotovoltaic devices. Sci. Rep. 7: 1-8.

Nussinovitch, A. 2010. Bead formation, strengthening and modification. pp. 27-52. *In*: Polymer Macro- and Micro-Gel Beads: Fundamentals and Applications. Springer.

O'Reilly, A.M. and J. A. Scott. 1995. Defined coimmobilization of mixed microorganism cultures. Enzyme Microb. Technol. 17: 636-646.

Priyadarshani, I., D. Sahu and R. Biswajit. 2012. Microalgal bioremediation: Current practices and perspectives. J. Biochem. Technol. 3: 299-304.

Pulz, O. and W. Gross. 2004. Valuable products from biotechnology of microalgae. Appl. Microbiol. Biotechnol. 65: 635-648.

Ramírez, M.E., Y.H. Vélez, L. Rendón and E. Alzate. 2018. Potential of microalgae in the bioremediation of water with chloride content. Braz. J. Biol. 78: 1-5.

Rengaraj, S. and S. Moon. 2002. Kinetics of adsorption of Co(II) removal from water and wastewater by ion exchange resins. Water R. 36: 1783-1793.

Riva, F., S. Castiglioni, E. Fattore, A. Manenti, E. Davoli and E. Zuccato. 2018. Monitoring emerging contaminants in the drinking water of milan and assessment of the human risk. Int. J. Hyg. Environ. Health. 221: 451-457.

Ruiz-Marin, A., L. Mendoza-Espinosa and T. Stephenson. 2010. Growth and nutrient removal in free and immobilized green algae in batch and semi-continuous cultures treating real wastewater. Bioresour. Technol. 101: 58-64.

Salgueiro, J., L. Perez, R. Maceiras, A. Sanchez and A. Cancela. 2016. Bioremediation of wastewater using *Chlorella vulgaris* microalgae: Phosphorus and organic Matter. Int. J. Environ. Res. 10: 465-470.

Salim, S., R. Bosma, M. Vermuë and R.H. Wijffels. 2011. Harvesting of microalgae by bio-flocculation. J. Appl. Phycol. 23: 849-855.

Selvi, K. and K. Kadirvelu. 2001. Removal of Cr(VI) from aqueous solution by adsorption onto activated carbon. Bioresour. Technol. 80: 89-91.

Shen, Y., H. Li, W. Zhu, S. Ho, W. Yuan, J. Chen, et al. 2017. Microalgal-biochar immobilized complex: A novel efficient biosorbent for cadmium removal from aqueous solution. Bioresour. Technol. 244: 1031-1038.

Shi, J., B. Podola and M. Melkonian. 2007. Removal of nitrogen and phosphorus from wastewater using microalgae immobilized on twin layers: An experimental study. J. Appl. Phycol. 19: 417-423.

Snyder, S.A., P. Westerhoff, Y. Yoon and D.L. Sedlak. 2003. Pharmaceuticals, personal care products and endocrine disruptors in water: Implications for the water industry. Environ. Eng. Sci. 20(5): 449-469.

Soo, C., C. Chen, O. Bojo and Y. Siang Hii. 2017. Feasibility of marine microalgae immobilization in alginate bead for marine water treatment: Bead stability, cell growth and ammonia removal. Int. J. Polym. Sci. 2017: 1-7.

Talukder, M., P. Das and J. Wu. 2014. Immobilization of microalgae on exogenous fungal mycelium: A promising separation method to harvest both marine and freshwater microalgae. Biochem. Eng. J. 91: 53-57.

Tam, N.F.Y. and Y.S. Wong. 2000. Effect of immobilized microalgal bead concentrations on wastewater nutrient removal. Environ. Pollut. 107: 145-151.

U.S. DOE. 1985. Review and Evaluation of Immobilized Algae Systems for the Production of Fuels from Microalgae. Alexandria (VA): Solar Energy Research Institute; 1985 Nov. Report No. SERI/STR-231–2798. Prepared for the U.S. Department of Energy.

Valuev, L., V. Chupov and N. Plate. 1994. Covalent immobilization of microorganisms in polymeric hydrogels. J. Biomater. Sci. Polym. Ed. 5: 37-48.

Vandamme, D., A. Beuckels, G. Markou, I. Foubert and K. Muylaert. 2015. Reversible flocculation of microalgae using magnesium hydroxide. Bioenergy Res. 8: 716-725.

Vasilieva, S.G., E. Lobakova, A. Lukyanov and A.E. Solovchenko. 2016. Immobilized microalgae in biotechnology. Moscow Univ. Biol. Sci. Bull. 71: 170-176.

Verwey, E.J.W. 1947. Theory of the stability of lyophobic colloids. J. Phys. Coll. Chem. 51: 631-636.

Verwey, E.J.W. and J.Th.G. Overbeek. 1955. Theory of the stability of lyophobic colloids. J. Coll. Sci. 10: 224-225.

Wang, S., J. Liu, C. Li and B. Chung. 2018. Efficiency of nannochloropsis oculata and bacillus polymyxa symbiotic composite at ammonium and phosphate removal from synthetic wastewater. Environ. Technol. 0: 1-10.

Weber, T., K. Blin, S. Duddela, D. Krug, H. Kim, R. Bruccoleri, et al. 2015. AntiSMASH 3.0: A comprehensive resource for the genome mining of biosynthetic gene clusters. Nucleic Acids Res. 43: W237-W243.

Whitton, R., A. Mével, M. Pidou, F. Ometto, R. Villa and B. Jefferson. 2016. Influence of microalgal N and P composition on wastewater nutrient remediation. Water Res. 91: 371-378.

Willaert, R. and G. Baron. 1996. Gel entrapment and micro-encapsulation: Methods, applications and engineering principles. Rev. Chem. Eng. 12: 1-205.

Xiong, Jiu Qiang, M. Kurade and B. Jeon. 2018. Can microalgae remove pharmaceutical contaminants from water? Trends Biotechnol. 361: 30-44.

Yadavalli, R. and G. Heggers. 2013. Two stage treatment of dairy effluent using immobilized chlorella pyrenoidosa. J. Environ. Health Sci. Eng. 1: 36.

Ying L., M.H. Rafailovich, R. Malal, D. Cohn and D. Chidambaram. 2009. Engineering of Bio-Hybrid Materials by Electrospinning Polymer-Microbe Fibers. PNAS 106: 14201-14206.

Zeraatkar, A., H. Ahmadzadeh, A. Talebi, N. Moheimani and M. McHenry. 2016. Potential use of algae for heavy metal bioremediation, a critical review. J. Environ. Manage. 181: 817-831.

Zhang, J., Y. Huang, X. Wang, K. Lin and K. Wu. 2015. Environmental polychlorinated biphenyl exposure and breast cancer risk: A meta- analysis of observational studies. PLOS One 10: 1-18.

Zur, J., D. Wojcieszyńska and U. Guzik. 2016. Metabolic responses of bacterial cells to immobilization. Molecules 21: 958.

6

Bioremediation Markers in Marine Environment

Ananya Mitra[1], Neelam M. Nathani[2] and Chandrashekar Mootapally[3,*]

INTRODUCTION

Bioaccumulation of various anthropogenic pollutants has raised a serious threat concerning the sustenance of marine ecosystem worldwide. The persistence of pollutants such as heavy metals, radioactive wastes, polynuclear aromatic wastes, and other organic pollutants poses a significant danger to human life by getting incorporated into the tissues of marine species and amplifying through their food chain (Sarkar 2006). With an increase in tourism near coastal areas of the world, and the continuous development around it to promote the economy due to tourism, the harm mentioned above to marine life is inevitable. But in order to sustain the outcome, it is also imperative to create a continuous check on minimizing the dire damage and develop revolutionary techniques to assess and cure the marine health. Bioremediation of marine environment has been the need of the hour, which requires huge developments and bringing them into regular practice. Prior to that, it is also important to assess the impact of each pollutant in the environment and thereby accurately formulate the necessary measures to be taken for its remediation. Therefore, biomarkers have gained much attention due to their capability for rapid detection and monitoring of pollutants in a particular environment.

Biomarkers possess a unique order of molecular structures, which give it a characteristic sensitivity for effective measurements of the quality of the environment. They can be classified as exposure markers, effect markers and markers of susceptibility. Major importance of biomarkers resides in their property to be measurable using different biochemical and molecular approaches. Biomarkers such as genes, proteins, metabolites, and other molecules are being used since half a century by other scientific areas for diagnosis and prognosis of diseases or for measuring the response process in the functioning of a living cell. However, its importance

[1] SRL Diagnostics, Dr. Phadke Lab, Mahin West, Mumbai - 400016, Maharashtra, India.
[2] Department of Life Sciences, Maharaja Krishnakumarsinhji Bhavnagar University, Bhavnagar - 364002, Gujarat, India.
[3] Department of Marine Science, Maharaja Krishnakumarsinhji Bhavnagar University, Bhavnagar - 364002, Gujarat, India.
[*] Corresponding Author: chandu.avi@gmail.com

has been thriving since the 21st century due to the development of technologies that can exploit the use of biomarkers for the molecular level measurements on large scale. In today's era of omics, these developments have motivated the discovery and validation of more biomarkers in the interest of environmental biotechnology (Michel and Soto 2016).

Considering the usefulness of biomarkers in assessment of environmental health risks, it is paramount that much focus needs to be given on the use of relevant markers and the validation of the same. Now the question arises as to what it means by a relevant marker? A relevant biomarker is the one which holds appropriate characteristic to guide the decision makers to track the pathway by which a xenobiotic component takes in the environmental quality disruption process. This is important in order to answer many other questions pertaining to the public health and hazards which otherwise could not be obtained by mere statistical reviews of records obtained from different approaches collected by observation of different events. Relevance also depends on the grounds of the question based on which biomarkers are selected, if they are relevant enough in guiding the researcher towards the desired information (Muscat 1996).

Moreover, it should be kept in mind that the concentration of pollutants in the environment varies from place to place in the same environment and so will the impact from one individual species to another and also within the population of the same species. Therefore, the choice of biomarkers is crucial and strategies pertaining to the correct measurement are to be decided, keeping in mind the variation in the particular environment. However, during a prolonged exposure these anthropogenic chemicals in the environment tend to interfere in the host physiological processes like growth and reproduction (Michel and Soto 2016). For example, in a study conducted at the Kandla Port situated on the west coast of India, it was found that the growth of phytoplankton *Chaetoceros tenuissimus* was highly affected due to the genotoxic effect of cadmium. Exposure biomarkers highlight the occurrence and distribution of the xenobiotic component or its metabolite through different biological levels of the organism and thus can be tracked on genetic, cellular and physiological parameters (Garban and Section 2006). This is indicative of the relationship between the chemical structure and the biological activity of an organism and thus it is evident that the impact of xenobiotic interference is injurious at the molecular level. Therefore, by quantifying the generalized covalent binding between the macromolecules to the reactive species help in measuring the absorbed dose delivered to the target cell whereas the number of DNA adducts will help to measure the dose delivered to the target organelle or macromolecule (Anderson et al. 1994). This is where the importance of determining the validity of a biomarker comes into picture. Validity is the approximate value of a range of characteristics that is responsible for producing a true or false information (Dor et al. 1999). In other words, the validation of biomarkers is the process of designing an assay where the biomarkers may function towards producing actual results by eliminating factors that may lead to false interpretation. The validity of biomarkers includes three categories: validity of measurement, external validity and internal study validity (Schulte and Perera 1993). Measurement validity is the degree of accuracy with which the biomarker is expected to perform. External validity refers to the applicability of the finding in a different set of population. Internal study validity refers to the degree of confidence on the inferences of the study and is free from any confounding factors.

Characteristics of a Valid Biomarker

The validation of biomarkers is widely described by WHO (1975). It generally involves knowing the history, understanding the temporal and biological correlation and pharmacokinetics, response from dosage and studying the probable causes of errors (Schulte and Talaska 1995). Following are few examples of errors as listed by White (1997) that can create a conflict in validity of a biomarker that can occur in laboratory.

- The designed method may not cover all the factors the subject is exposed to which is responsible for the consequent effect.
- The designed method may include factors other than the ones responsible for the consequent effect and lead to confounding circumstances.
- Missing out minor details in the protocol such as the method of collection, time of collection, handling and storage of specimen etc.
- Failure in maintaining the periodic standardization of instrument during the date collection procedure.
- Failure in maintaining the uniform method or handling techniques for repeated sample collection.
- Variation in maintenance of time duration between sample collection and sample processing.
- Techniques used may also vary among different technicians of laboratory.
- Different batches of chemicals or variation in calibration of instrument used between different batches of sample processing.

Capabilities of a valid biomarker have been listed by Mayeux (2004) as follows:

- Tracing the order of events taking place between exposure and consequent effect
- Detection of dose response
- Detection of events occurred in past in nature
- Description of the mechanism by which a response is generated as a result of exposure to certain factor
- Leads to accurate classification of risk factors and causative effect
- Establishment of variability and effect modification

Frequently used Biomarkers in Marine Bioremediation

1. The most common pollutant in marine environment are the hydrocarbon compounds which has drawn mass attention. Microorganisms degrading isopropenoid compounds such as pristane and phytane degrade branched alkanes at a slower rate than it degrades the straight chain hydrocarbons

despite being subjected to the same physical and chemical removal mechanism as of pristane and phytane due to which the ratio of n-C17 : pristane and nC18 : phytane have been used as an important evidence to determine the extent of biodegradation. Later, it was found that these compounds resist degradation for an initial short stretch of time and then suddenly peaks up rapidly over a longer period of time (Prince 1993, Bragg et al. 1994).

2. Cytochrome P4501A (CYP1A) has been proven to be an effective biomarker in detection of contaminants such as dioxins, furans, polychlorinated biphenyls and polycyclic aromatic hydrocarbons (Fernando et al. 2005), wherein the induction of CYP1A is enhanced as a result of aryl hydrocarbons receptor from cytosol when the organism is exposed to such pollutants. Therefore, this phenomenon is used as a biomarker for monitoring pollutants in fishes. The concept of using such biomarkers has assisted in application of ethoxyresorufin dealkylation (EROD) in bivalves (Binelli et al. 2006). For example, when Zebra mussel (*Dreissena polymorpha*) was exposed to 100 ng/l of PCB mixture of Arochlor 1260 and dioxin-like CB-126, a significant activity could be observed of ethoxyresorufin dealkylation. Although this underwent a competitive inhibition after an exposure above 48 hours, the CYP1A biomarker could quantify different levels of pollutants by AhR-binding PAHs and PCBs (Behrens and Segner 2005).

3. Pigments can be found in all photosynthetic organisms, which are responsible for harvesting light for the generation of energy and for photoprotection. There are three major kinds of pigments present in plants and algae. The first among them is a porphyrin containing chlorophyll which is a greenish pigment and is present in plants, algae and cyanobacteria. Caretenoids are usually red, orange or yellowish colored pigment that is responsible for their characteristic color in some plants and algae. Caretenoids are composed of two small six carbon rings that are connected to each other by a chain of carbon atoms. The third kind of pigment is the fluorescent pigments phycocyanin and phycoerythrin, which can be used as chemical tags when subjected to a particular wavelength and have been used extensively for tagging tumor cells in cancer research. The unique distribution of pigments in plants and algae specific to their taxonomy makes it a reliable biomarker. They have been used as antioxidants in assessment of the rates at which the key oceanographic processes functions and for characterizing the marine organic matter. They are also employed as an alternative to microscopical counting process to determine the structure of phytoplankton community as demonstrated in a study conducted by Jeffrey and Vesk (1997).

   Major phytoplankton pigments used as standards are well described by Michel and Soto (2016) in the form of the chromatogram and these include different types of chlorophyll pigments, carotenes, violaxanthin, alloxanthin, and other compounds detected through the broad range of retention time.

4. DNA integrity can be affected due to the presence of exogenous genotoxic agents owing to the formation of DNA adducts, breakage of strands or loss of

methylation. Members of Polyaromatic hydrocarbons (PAH) such as Benzo(a) pyrene (BaP) undergoes oxidation and forms diolepoxide (BaPDE), which is a chemically reactive compound which reacts with the DNA to form stable or unstable adducts with DNA which eventually causes a transformation of the cell and interferes in the general cellular mechanisms (Everaarts and Sarkar 1996). A direct chemical damage caused by ionizing radiation, oxidation-reduction reaction or photoreaction can cause single strand breaks (Michel and Soto 2016). Based on measurement of DNA strand breaks in marine snails (*Planaxis sulcatus*), impact of pollution attributed to petroleum hydrocarbon contamination at different sites was estimated which was evident from a remarkable decrease in DNA integrity.

5. Metallothioneins (MTs) are cysteine rich, low molecular weight protein present in cytosol and have ubiquitous occurrence. The sulfur atom of the cysteine residue of metallothioneins binds with the toxic metal ions and causes its inactivation (Michel and Soto 2016). The concentration levels of Metallothioneins is observed to correlate with the accumulation fraction of toxic metal ions like copper or cadmium in the tissues of organisms (Roesijadi 2000). The metal handling capability owing to the presence of metallothioneins in molluscs bivalve species is employed for measuring the level of toxic metal ions and thus has proven to be an effective biomarker in monitoring the metallic pollutant in marine habitats (Michel and Soto 2016). Metallothioneins in aquatic animals are also used as biomarkers in estimation of oxidative stress. Since Metallothioneins are metal chelating agents, their role is significant in metabolism and detoxification of excess metallic components in cell by binding to its free ion and scavenging of oxygen free radical (Andrew 2000). In recent years, extensive researches have been conducted to employ free radical reaction and ROS in physiological processes of living organism to monitor oxidative stress and the resultant oxidative damage to DNA, proteins, lipids, antioxidants and enzymatic or non-enzymatic defense mechanism of an organism are used as biomarkers for monitoring the contaminants in the pollutant sites. Moreover, the ubiquitous nature of Metallothioneins and its unique structure in eukaryotes gives it metal binding and redox capabilities in such eukaryotes. The thiol group of the cysteine residues can bind to heavy metals, which makes it an important tool in estimation of heavy metal pollutants. In many vertebrates including many species of fishes and invertebrates, which mainly includes crustaceans and molluscs, the accumulation of metallothioneins can be attributed to the induction caused by the presence of essential metals such as copper and zinc and non-essential metals like cadmium, silver and mercury (Roeva et al. 1999, Isani et al. 2000). Effective induction of metallothioneins have been observed in more than fifty different species which includes oyster *Crassostera virginica* and mussel (*M. galloprovincialis*, *Mullus barbatus*) (Roesijadi 1997, Petrović et al. 2001, Lionetto et al. 2003). Exposure of aquatic animals to such metals can cause adverse biochemical changes in them and should be considered as an early warning sign for an underlying fatal situation.

6. Measurement of Acetylcholinase (AChE) inhibition has been one of the first validated biomarkers in effect since the past two decades and has been useful in monitoring the concentration of organophosphate and carbamate compounds in the environment. These compounds are widely used in the pesticides and have become the most commonly used pesticides in order to replace the organochlorine pesticides. These pesticides persist in the environment for a long time and get introduced to the ground water by seeping into the soil and contaminate the marine shoreline, posing a toxic threat to the marine flora and fauna during long exposure. Organophosphorus and carbamate compounds bind to the esteratic site of AChE with variable affinity by phosphorylation or decarbamylation reaction and causes inactivation of the enzyme (Lionetto et al. 2013). Organophosphorus compounds are considered to exhibit irreversible inhibition of AChE owing to the excessive time required to release the enzyme from the inhibition and in synthesis of new AChE. A similar activity is observed under the influence of other chemical pesticides families such as pyrethroids (Reddy and Philip 1994), triazines (Davies and Cook 1993) and paraquat (Szabo et al. 1992). Moreover, instances of AchE inhibition have been demonstrated by some potential metallic ions, such as Hg^{2+}, Cd^{2+}, Cu^{2+}, and Pb^{2+} *in vivo* by Lionetto et al. (2013), Ademuyiwa et al. (2007) and Frasco (2005). Due to recent increase in technologies employing nanomaterials in various applications such as biomedicines, electronic and material science etc., there has been increase in the studies pertaining to the probable toxic consequences. In a study, Wang et al. (2009) demonstrated adsorption of nanoparticles with Acetylcholinase enzyme and inhibiting its activity *in vitro* using different classes of nanoparticles widely used such as metals, oxides, and carbon nanotubes such as SiO_2, TiO_2, Al_2O_3, aluminum, copper, carbon-coated copper, multi-walled carbon nanotubes, and single-walled carbon nanotubes. A dose response inhibition of AChE activity was observed with IC50 values of 4, 17, 156 and 96 mg/L in response to Cu, Cu-multiwalled carbon nanotubes, and single-walled carbon. All these studies are suggestive of the affinity of AChE towards various contaminants other than organophosphates and carbamates that makes it a reliable biomarker in marine environmental monitoring for various pollutants.

Applications of Marine Biomarkers in India

According to the first ever United Nations system wide evaluation of global water quality, India has been ranked 120 out of 122 countries of the world for its water quality and 133 out of 180 countries for poor availability of water viz. 1,880 cubic meter per person annually. Therefore, it becomes the need of the hour to employ effective measures to monitor and treat the available supply of water for the sustenance of ever-increasing population of the country. Many studies on the use of molecular biomarkers have been encouraged in recent years for its effectiveness in contemporary sciences and its known reliability. According to Sarkar (2006), studying the DNA integrity in marine snails (*Planaxis sulcatus*) of the Arabian

Sea along the Goan coast provides a remarkable information on the measure of pollution in the coastal regions. Also, the seasonal variation in the AChE activity in the marine snails (*Cronia contracta)* gives an indication of the presence of neurotoxic pollutants at the coastal lines of Goa (Gaitonde et al. 2006). In a study involving the freshwater prawns of Kaveri river basin on the East coast of India, enhanced activity of cytochrome P450-dependent mixed function oxygenase and the conjugating enzymes glutathione-S-transferase in different tissues as a part of their detoxification mechanism threw a light on their potential use as biomarker of stress in early detection of oil pollution in marine environment (Arun et al. 2006).

An extensive study of biological responses in green mussels to evaluate the xenobiotic contaminants was performed by Sasikumar et al. (2006) along the southwest coast of India, taking into consideration the resultant biotransformation due to the presence of such pollutants. These responses were studied based on sister chromatid exchange (SCE), chromosomal aberration, micronucleus test, hemic neoplasia (HN), chromtest (Ames test) and comet assay. DNA damage caused by varying concentration of smoke has been shown to cause disruption of male gonads in the marine mussels (*Perna viridis*) in a comet assay for a period of 16 days, whereas tobacco extracts of non-smoked cigars were found to induce deformity in their nuclei. Moreover, genotoxic effects of cadmium on phytoplankton (*Chaetoceros tenuissiums*) near the Kandla Port was estimated by Comet Assay to analyze the load of pollutants along the west coast of India, suggesting that the cadmium concentration has a severe effect on the growth of diatoms.

Role of Biomarkers in the Assessment of Efficacy of Marine Bioremediation

Advancement in bioremediation in the past few decades has also given rise to the question of whether we are actually getting rid of the target contaminant, or are we only giving rise to another threat waiting to take a hideous form. Thus, it is equally important to monitor the resultant product of bioremediation as bioremediation itself. The environmentally transformed product of DDT [1,1,1-trichloro-2,2-bis(p-chlorophenyl) ethane] is DDE [1,1-dichloro-2,2-bis(p-chlorophenyl)ethylene], which is known to be a more potent antagonist to androgen receptor than its parent compound (Kelce et al. 1997). Moreover, the microbial reductive dechlorinated products of polychlorinated biphenyls (PCBs) are found to cause more cases of uterine contraction *in vitro* than the parent polychlorinated biphenyls mixtures (Bae et al. 2001). Biomarkers have been developed to investigate the stable transformation product of bioremediation. In an evaluation study by Ganey and Boyd (2005), they selected bioassays based on the existing information regarding the degradation mechanisms in the initial part of the examination. For instance, in case of PCBs the effect on cytochrome P450 induction is taken into consideration due to the known phenomenon of higher P4501A stimulation in presence of PCBs. Likewise, some chemicals are known to interfere in the intracellular signaling pathways. For example, in case of polycyclic aromatic hydrocarbons (Burdick et al. 2003, Patten

Hitt et al. 2002), it is preferred to examine the activation of mitogen-activated protein kinases or for any change in neutrophil function (Ganey and Boyd 2005).

Biomarkers are also developed in order to monitor the efficacy and durability of specific bacteria used in bioremediation as an inoculum. The luc gene responsible for encoding of the luciferase and the gfp gene responsible for encoding the green fluorescence protein in fireflies are used to tag different bacteria used in the bioremediation of sites contaminated with gasoline and chlorophenols. The activity of bacteria is measured by measuring the luciferase activity obtained from the cell extracts. Similarly, the gfp gene is involved in monitoring bacteria employed in the degradation of chlorophenol by evaluating the fluorescence produced by the gfp gene (Jansson et al. 2000). In a ground water aquifer contaminated with 1,2 dichloroethane (1,2-DCA), *Dehalobacter* sp. and *Desulfitobacterium* sp. were found to be involved in the dehalorespiration of 1,2-DCA and a new gene cluster encoding the 1,2-DCA reductive dehalogenase were identified and described by Marzorati et al. (2010). In another study conducted by Brenner et al. (2006), the efficacy of the biodegradation process of tetrabromobisphenol A and tribromophenol was assessed by using indigenous bacteria in an activated sludge. A mixture of 40% TBBPA, 20% 2,4,6-tribromophenol (TBP), and 8% 2,4-dibromophenol, apart from some other materials in the waste mixture, was produced. Evaluation of bacterial multicomponent monooxygenase gene targets in *Pseudonocardia dioxanivorans* CB1190 was performed and their use as potential biomarkers was illustrated using industrial activated sludge samples. According to that, the presence of gene responsible for dioxane monooxygenase, propane monooxygenase, alcohol dehydrogenase, and aldehyde dehydrogenase are indicative of 1,4-dioxane biotransformation and have promising application in detecting the efficacy of the same. However, the method highlighted a limitation of gene abundance to delineate the actual degradation process. A consortium activity of *Pseudonocardia dioxanivorans* CB1190 and mixed communities in wastewater associated with the removal rate of 1,4-dioxane was revealed by a time course gene expression analysis of dioxane and propane monooxygenase. An upregulation of the transcripts of alcohol dehydrogenase and aldehyde dehydrogenase genes occurs during the biodegradation, which is suggestive of a correlation of these genes with the concentration of 1,4-dioxane (Gedalanga et al. 2014).

CONCLUSIONS

The use of biomarkers has proven its efficiency in various fields of science and has been a boon in detection and identification of many deleterious phenomenon in causing unacceptable changes. The potentials of biomarkers are still not fully utilized in its practical application in determining the pollutant levels and achieving its effective biodegradation. The application of the techniques demonstrated in the case studies demonstrated here can improve the understanding of the mechanisms of each phenomenon involved in degradation process of any contaminant and thus provide an insight to efficiently formulate a bioremediation strategy for the same contaminant. Moreover, it is also important to monitor the biotransformed product to ensure complete detoxification.

REFERENCES

Ademuyiwa, O.U.R.N., R.N. Ugbaja, S.O. Rotimi, E. Abam, B.S. Okediran, O.A. Dosumu, et al. 2007. Erythrocyte acetylcholinesterase activity as a surrogate indicator of lead-induced neurotoxicity in occupational lead exposure in Abeokuta, Nigeria. Environ. Toxicol. Pharmacol. 24(2): 183-188.

Anderson, S., W. Sadinski, L. Shugart, P. Brussard, M. Depledge, T. Ford, et al. 1994. Genetic and molecular ecotoxicology: A research framework. Environ. Health Perspect. 102(12): 3-8.

Andrews, G.K. 2000. Regulation of metallothionein gene expression by oxidative stress and metal ions. Biochem. Pharmacol. 59(1): 95-104.

Arun, S., A. Rajendran and P. Subramanian. 2006. Subcellular/tissue distribution and responses to oil exposure of the cytochrome P450-dependent monooxygenase system and glutathione S-transferase in freshwater prawns (*Macrobrachium malcolmsonii, M. lamarrei lamarrei*). Ecotoxicology 15(4): 341-346.

Bae, J., M.A. Mousa, J.F. Quensen 3rd, S.A. Boyd and R. Loch-Caruso. 2001. Stimulation of contraction of pregnant rat uterus *in vitro* by non-dechlorinated and microbially dechlorinated mixtures of polychlorinated biphenyls. Environ. Health Perspect. 109(3): 275-282.

Behrens, A. and H. Segner. 2005. Cytochrome P4501A induction in brown trout exposed to small streams of an urbanised area: Results of a five-year-study. Environ. Pollut. 136(2): 231-242.

Binelli, A., F. Ricciardi, C. Riva and A. Provini. 2006. New evidences for old biomarkers: Effects of several xenobiotics on EROD and AChE activities in Zebra mussel (*Dreissenapolymorpha*). Chemosphere 62: 510-519.

Brenner, A., I. Mukmenev, A. Abeliovich and A. Kushmaro. 2006. Biodegradability of tetrabromobisphenol A and tribromophenol by activated sludge. Ecotoxicology 15(4): 399-402.

Burdick, A.D., J.W. Davis 2nd, K.F. Liu, L.G. Hudson, H. Shi, M.L. Monske, et al. 2003. Benzo(a)pyrene quinones increase cell proliferation, generate reactive oxygen species and transactivate the epidermal growth factor receptor in breast epithelial cells. Cancer Res. 63: 7825-7833.

Costa, L.G. 1992. Effect of neurotoxicants on brain neuro-chemistry. Neurotoxicology 101-123.

Curseu, D., D. Sirbu, M. Popa, A. Ionutas and E. Czuczi. 2007. The Biologic Markers as useful tools in Epidemiological Research. *In*: 1st International Conference on Advancements of Medicine and Health Care through Technology. Retrieved from http://ie. utcluj. ro/files/acta/2007/Number4/Papers P (Vol. 50308).

Davies, P.E. and L.S.J. Cook. 1993. Catastrophic macroinvertebrate drift and sublethal effects on brown trout, Salmo trutta, caused by cypermethrin spraying on a Tasmanian stream. Aquat. Toxicol. 27(3-4): 201-224.

Desai, S.R., X.N. Verlecar and U. Goswami. 2006. Genotoxicity of cadmium in marine diatom *Chaetoceros tenuissimus* using the alkaline Comet assay. Ecotoxicology 15(4): 359-363.

Dor, F., W. Dab, P. Empereur-Bissonnet and D. Zmirou. 1999. Validity of biomarkers in environmental health studies: the case of PAHs and benzene. Crit. Rev. Toxicol. 29(2): 129-168.

Everaarts, J.M. and A. Sarkar. 1996. DNA damage as a biomarker of marine pollution: strand breaks in seastars (Asterias rubens) from the North Sea. Water Sci. Technol. 34(7-8): 157-162.

Fernando, R., L. Ferrari and A. Salibián. 2005. Biomarkers of a native fish species (Cnesterodon decemmaculatus) application to the water toxicity assessment of a peri-urban polluted river of Argentina. Chemosphere 59(4): 577-583.

Frasco, M.F., D. Fournier, F. Carvalho and L. Guilhermino. 2005. Do metals inhibit acetylcholinesterase (AChE)? Implementation of assay conditions for the use of AChE activity as a biomarker of metal toxicity. Biomarkers 10(5): 360-375.

Gaitonde, D., A. Sarkar, S. Kaisary, C.D. Silva, C. Dias, D.P. Rao, et al. 2006. Acetylcholinesterase activities in marine snail (Cronia contracta) as a biomarker of neurotoxic contaminants along the Goa coast, West coast of India. Ecotoxicology 15(4): 353-358.

Ganey, P.E. and S.A. Boyd. 2005. An approach to evaluation of the effect of bioremediation on biological activity of environmental contaminants: Dechlorination of polychlorinated biphenyls. Environ. Health Perspect. 113(2): 180-185. doi: 10.1289/ehp.6935.

Ganguly, A. and U. Goswami. 2006. DNA damage in male gonad cells of Green mussel (Perna viridis) upon exposure to tobacco products. Ecotoxicology 15(4): 365-369.

Garban, Z. and F.C. Section. 2006. Biomarkers: Theoretical aspects and applicative peculiarities note II. nutritional biomarkers. J. Agroaliment. Processes Technol. XII(2): 349-356.

Jansson, J.K., K. Björklöf, A.M. Elvang and K.S. Jørgensen. 2000. Biomarkers for monitoring efficacy of bioremediation by microbial inoculants. Environ. Pollut. 107(2): 217-223.

Lionetto, M.G., R. Caricato, A. Calisi, M.E. Giordano and T. Schettino. 2013. Acetylcholinesterase as a biomarker in environmental and occupational medicine: New insights and future perspectives. Biomed. Res. Int. 2013: 1-8. doi: 10.1155/2013/321213.

Marzorati, M., A. Balloi, F. De Ferra and D. Daffonchio. 2010. Identification of molecular markers to follow up the bioremediation of sites contaminated with chlorinated compounds. In *Metagenomics* (pp. 219-234). Humana Press, Totowa, NJ.

Mayeux, R. 2004. Biomarkers: Potential uses and limitations. NeuroRx, 1(2), pp. 182-188. doi: 10.1602/neurorx.1.2.182.

Michel J.P. and J.O. Soto. 2016. Modern approaches into biochemical and molecular biomarkers: Key roles in environmental biotechnology. J. Biotechnol. Biomater. 06(01): 1-8. doi: 10.4172/2155-952x.1000216.

Sarkar, A. 2006. Biomarkers of marine pollution and microbial degradation of pollutants. Special issue dedicated to Dr. Simao Nascimento de Sousa and Proceedings of the International Workshop on Marine Pollution and Ecotoxicology. February 25-26, 2004. Dona Paula, Goa, *In*: Ecotoxicology (London, England), 15(4), pp. 331-402. doi: 10.1007/s10646-006-0073-5.

Schulte, P.A. and F.P. Perera. 1993. Validation. pp. 79-107. *In*: P.A. Schulte, F.P. Perera. (eds.). Molecular Epidemiology: Principles and Practices. San Diego: Academic Press.

Schulte, P.A. and G. Talaska. 1995. Validity criteria for the use of biological markers of exposure to chemical agents in environmental epidemiology. Toxicology 101(1-2): 73-88.

Szabo, A., J. Nemcsok, B. Asztalos, Z. Rakonczay and P. Kasa. 1992. The effect of pesticides on carp (Cyprinus carpio L). Acetylcholinesterase and its biochemical characterization. Ecotoxicol. Environ. Saf. 23(1): 39-45.

Wang, Z., J. Zhao, F. Li, D. Gao and B. Xing. 2009. Adsorption and inhibition of acetylcholinesterase by different nanoparticles. Chemosphere 77(1): 67-73.

White, E. 1997. Effects of biomarker measurement error on epidemiological studies. IARC Scientific Publications (142): 73-93.

7

Antarctic Marine Fungi and Their Potential Application in Bioremediation

Mayara Baptistucci Ogaki

INTRODUCTION

Since the beginning of the Antarctica occupation, the continent has been impacted by human disturbance with fishing, whaling, waste and fuels used by vessels and vehicles in scientific stations or by tourism (Bargagli 2008). Even Antarctica, still largely preserved when compared with another places around the world, is not entirely protected from local anthropogenic contamination and global environmental changes (Aronson et al. 2011).

In accordance with the environmental protocol in the Antarctic Treaty, the recovery of contaminated areas should not be carried out in this sensitive environment by physical-chemical disruptive cleaning methods or treatments with exogenous agents that cause ecological impact in Antarctic biological communities (Bargagli 2008, Flocco et al. 2019, Martorell et al. 2019). Thereby, bioremediation processes using indigenous or native microbial communities, including endogenous fungi, become an interesting tool in protecting Antarctic areas (Martorell et al. 2019).

Among the fungi used in bioremediation processes, marine-derived fungi are commonly used for the production of secondary metabolites, biosurfactants, enzymes, polysaccharides and polyunsaturated fatty acids (Damare et al. 2012). The ability of these microorganisms to survive in extreme temperatures, salinity and pressure demonstrates their potential in biotechnological processes (Rosa et al. 2019).

This chapter provides an overview of the local and global anthropogenic impacts in Antarctica, the principal contaminants in Antarctic marine habitats and some characteristics and applications of Antarctic fungi, mainly marine fungi, as native species in bioremediation methods.

Antarctica and the Anthropogenic Activities

Antarctica is one of the most inhospitable and unexplored regions on Earth, spanning 13.8 million km^2, and generally seen as a remote and hostile place, cold, windy and

Microbiology Department of Universidade Federal de Minas Gerais, Belo Horizonte, Brazil.
E-mail: mayaraogaki@gmail.com

dry, which grants extreme conditions of survival for sensitive and unique species, which are generally composed by simple trophic chains, including a variety of microorganisms such as Bacteria, Archaea and Fungi (Rosa et al. 2019).

Antarctic continent is protected from the entry of water and air masses by natural barriers, being geographically isolated by the Antarctic Circumpolar Current (ACC) and thermally by the strong catabatic winds that descend from the central plateau towards the coast (Wynn-Williams 1990, Simões 2011). Even with the thermic and physical isolation, this cold desert, which remained untouched for a long time by human disturbance, is no longer protected from the impact of global and local anthropogenic activities (Bargagli 2008).

Global Impacts

Since 1961, temperatures are monitored by scientists in Antarctica and warming has not been so uniform around the continent. Many continental regions have not exhibited significant change over the past century (Convey and Peck 2019), but in contrast, in the recent decades, the rapid increase of temperature recorded specially in Antarctic Peninsula indicates a correlation between global warming and the impact of anthropogenic activities in the Antarctic continent (Bargagli 2008).

In Antarctic Peninsula, the average annual temperature ranges from about −10°C, contrasting to −60°C in the interior on the Continental Antarctica (WMO 2020). According to World Meteorological Organization - WMO (2020), the Antarctic Peninsula is among the fastest warming regions of the planet, their temperatures have risen 3°C over the last 50 years. In 2020, at first researchers logged a record on temperatures of 20.75°C in Seymour Island area. These records allow scientists to monitor climate, weather patterns and rising sea levels on the continent over time, all changes that influence the equilibrium of Antarctic environments.

The increase on annual temperatures cause alteration in the extent of sea ice and ocean acidification. Polar ecosystems are especially sensitive to these two aspects of global change, specifically in marine environments, where the deleterious effects of warming and acidification occur sooner than one might assume from "longer-term" (Aronson et al. 2011). To exempify these changes, according to Clarke et al. (2006), the winter sea ice in the Bellingshausen and Amundsen seas has been decreasing 10% in extent per decade, and since 1950s there are records that their surface waters and the deep water of the ACC have warmed, and this continuous heating may induce sublethal effects which influence ecological interactions and the operation of food chains in Antarctica.

An endemic terrestrial and marine biota was established in the Antarctic continent, surviving to the selective pressure imposed by the limiting cold desert conditions. However, this unique biota is sensitive to environmental imbalance, for being subjected to such specific conditions of cold, light, water and nutrients availabilty.

Over past decades, the climate change caused perceptible changes on biotic patterns that are known to have occurred during past climatic shifts (Aronson et al. 2011). The increase in temperature and environmental degradation furthers the process of habitat loss, which can extinguish entire populations, leading many endemic species to the brink of extinction.

The competition with non-native species, mainly for microorganisms, that excel at milder temperatures also influences the endemic species population. Uncontrolled introduction of foreign microrganisms combined with the increase in temperatures could make these species capable of colonizing and/or increasing their population size, which may present an even greater threat than climate change itself.

Local Impacts

In the last 50 years, overfishing, tourism, and the development of scientific research have intensified the human presence in Antartica. The expeditions and the current human activities have affected local terrestrial and marine ecosystems, disrupting habitats, fauna and flora, through chemical contamination derived from human waste (including open air incineration executed prior to regulations), dust and particulate matter produced during the construction of scientific stations, from the combustion of fossil fuels and/or accidental oil spills (derived from the transit of vehicles, pedestrians and the production of energy) that brought some Antarctic marine species to the brink of extinction (Bargali 2008, Flocco et al. 2019, Frenot et al. 2005).

The human occupation in Antarctica is restricted to scientific and logistic crews that temporarily live in research stations or tourists (Flocco et al. 2019). However, the anthropogenic influence is not recent in Antarctica, principally in Maritime Antarctica. Although the first adventurous explorations and hunting expeditions began on 18th century (Flocco et al. 2019), the effective human presence began in the early 1900 for hunting, fishing and whale exploration (Bargagli 2008).

Fishing is now temporarily closed to exploitation in the Antarctic Atlantic, being controlled by the Convention on the Conservation of Antarctic Marine Living Resources (CCAMLR), while whales and seals are currently protected by agreements from International Whale Commission (IWC), Japan being the only country that still exploits whales in the Southern Ocean (Aronson et al. 2011).

Past expeditions left their footprint in Antarctica and the continent still suffers from some current activities as illegal fishing and whaling and the disposal of sewage and waste generated by human occupation and vessels. The Antarctic Treaty regulates the activities in the continent, prohibiting the militarization and exploitation of natural resources (Convey et al. 2012); thus, the human presence in Antarctica is currently restricted to people at research stations and regulated tourist visits (Flocco et al. 2019), and both are recent, dating from the early 1950s (Tin et al. 2009) and expanding constantly since then (Frenot et al. 2005, IAATO 2020).

Tourism has incredibly intensified the human occupation in Antarctica over the past two decades. During 1989-1990, the National Science Foundation of the United States started recording tourist visitation data, years after the International Association of Antarctic Tour Operators (IAATO) began compilling this data, including annual total number of visitors and visitors ashore (Bender et al. 2016).

According to IAATO, the number of annual visits in Antarctica was on average less than 1,000 tourists from 1965 to 1990, increasing for more than 33,000 per year in 2007-2008 (Aronson et al. 2011), and by the last records 44,600 land visitants per year in 2018-2019 (IAATO 2020). The monitoring of these human impacts has been little addressed, although some initiatives of comprehensive and systematic monitoring programs have been called (Lee and Hughes 2010).

Tourist visitation may impact entire ecosystems and wildlife, such as seabirds, seals, shallow invertebrate-dominated ecosystems, soil and some abiotic elements (Bender et al. 2016). However, the environmental impacts of tourism activities are generally transient, due to the visitors' flow that occur temporarily on the continent. Ship, air or landborne visitors require some infrastructure, and exert pressure on local biological communities, damaging heritage sites, due to the repeatedly visited shore-based attractions in coastal areas (Kriwoken and Rootes 2000).

Thereby, the nature of the tourism impacts may include: disturbance and introduction of diseases in colonies of birds and other marine animals, inadequate disposal of sewage and waste, water pollution, oil spills, collection of souvenirs, introduction of non-native species, incineration of ships and disruption of scientific research activities (Aronson et al. 2011, Kriwoken and Rootes 2000).

Even with the increase in transitory anthropogenic activities, the sources of local pollution in Antarctica are mainly concentrated in the ~75 research stations (Aronson et al. 2011). The construction and occupation of scientific stations began in 1957-1958 with the modern navigation in Antarctica, and most of them are distributed in coastal ice-free areas, adjacent to shallow water, sediments or other marine habitats, which are locally influenced by environmental impacts, as the displacement of fauna from the site with the station construction and the emission of pollutants into the marine and terrestrial environment (Aronson et al. 2011), since the stations are replenished by air or naval transport that use fuels.

Until the 1980s, waste produced from Antarctic stations was locally dumped in landfills, discarded at sea or ice or burned in open air, and included the discard of batteries, food waste, building materials, laboratory chemicals, sewage, shipping containers, fuel drums, and waste oil and lubricants (Aronson et al. 2011, Crockett and White 2003, Martorell et al. 2019). The current waste management practices limit, with the exception of the sewage, that the waste must be removed from Antarctica, open burning is prohibited and each nation is responsible for cleaning up its old waste dumping sites. But some contaminants that resulted from past or improper practices, accidents and fuel spills are still released in the marine environment (Tin et al. 2009, Riddle 2009).

Even with the measures implemented by the Treaty, heavy metals, polyaromatic hydrocarbons (PAHs), and persistent organic pollutants (POPs) are still detectable near some research stations and may leach into the marine environment, particularly during the summer when the pollutants can be mobilized by meltwater (Aronson et al. 2011).

In Antarctic stations and vessels, the sewage of bathrooms, kitchens and laundries may contain (1) human waste with microorganisms (pathogenic bacteria, viruses and fungi) and antibiotics; (2) organic material, such as food waste and toilet paper; and (3) chemicals, such as detergents, hydrocarbons and heavy metals (Martorell et al. 2019). These sewage components can cause increase in microbial biomass, reducing oxygen levels in seawater; suspended solids can bury sessile invertebrates, interfering with feeding activities in the marine environment, and the impact of toxic chemicals (Martorell et al. 2019) that are persistent can accumulate in sediments and also in marine organisms. Because many polar organisms grow slowly, with longer initial stages of development and therefore are long-lived, they accumulate more contaminants over time (Snape et al. 2003). In addition, all these above-mentioned contaminants can be transferred to marine environments through mass transfer in the atmosphere and oceans; therefore, they are not restricted to contamination sites affecting ecosystems on a large scale.

Antarctic Marine Ecosystems

The Southern Ocean or Antarctic Ocean is a deep marine system that starts from 60°s encircling Antarctica continent in a 2,500-km-wide, semi-closed belt (Bargagli 2008, Rosa et al. 2019). The water circulation of this large ocean is an important component of the world climate system (Donohue et al. 2016). Therefore, the flows and dynamics of the Antarctic marine systems become significant on a global scale.

Antarctic marine ecosystems were protected from human influences by the physical barrier created by the cold sea temperatures at high Southern latitudes and the Antarctic Circumpolar Current (ACC). The southern edge of the ACC, called Polar Front, creates a barrier between sub-Antarctic waters and colder waters surrounding the Antarctic continent, promoting its thermal insulation (Mays et al. 2019).

The ACC is the largest ocean current with the dominant oceanographic feature of the Southern Ocean. This current has dense waters with intense flow that practically does not attenuate throughout the entire water column; downing the continental Antarctica, it spreads through upper and deep layers, connecting the major ocean basins and ventilating a large proportion of the global ocean (Donohue et al. 2016, Rintoul and Garabato 2013).

ACC circulation acts to facilitate dispersal within the Southern Ocean and supports biological activity at its boundaries where upwelling of nutrients occurs (Mays et al. 2019). The resurgence of deep waters of Southern Ocean, rich in nitrates, nitrites and phosphates, added to the high concentrations of oxygen, make this marine environment very favorable to the development of life, the Ocean being considered as one of the quantitatively richest communities in the world, despite the low temperatures of its freezing water bodies (about –1.9°C) (Bargagli 2008, Schell 1965).

Antarctic marine ecosystems are composed by seemingly simple food webs, from primary producers characterized by phytoplankton and algas blooms, which sustain large communities of consumers as pelagic tunicates, salps, copepods, gammarids, hyperiids, chaetognaths and, above all, krill (*Euphausia superba*) which provides the food base for cephalopods, fishes, seals, whales, penguins and other seabirds (Bargagli 2008, Jacob 2007).

However, these systems consist of many subsystems which make it more complex, as a greater number of feeding links increases the complexity of the food web (Jacob 2007). Especially leading with microorganisms that play a lage role in marine ecosystems as photosynthetic primary producers, consumers (Glöckner et al. 2012) and decomposers. They are responsible to balance organic matter and to produce oxygen, influencing the fitness responses of wildlife in the sea and oceans. Different substrates of the Southern Ocean, like water, sea ice, rocks, sediments, seaweeds, invertebrates, and vertebrates may constitute promising habitats that shelter microorganisms (virus, archaea, bacteria, and fungi) (Rosa et al. 2019).

Marine ecosystems and their microhabitats have immediate impact on local and global scale with the advance of human activities and occupation, such as overfishing, pollution, introduction of non-native species, rising sea temperatures, altered sea ice (and iceberg scour in benthic habitats) and ocean acidification (Aronson et al. 2011, Barnes 2017, Convey and Peck 2019). Therefore, all these fluctuations occurring in inert substrates, biological communities and organic matter interfere with the composition and biological activity of the perennial and/or transitory microbiota that is deeply related to all levels of the Antartic marine food chain and, consequently, to the cycling of nutrients.

Antarctic Marine Fungi

In salt-waters, saprophytic fungi act on decaying organic matter and decomposing of wood, algae and calcareous materials, chitin and keratin of marine animals (Arenz et al. 2006, Raghukumar 2006). Moreover, fungi could be associated with other organisms as parasites/pathogens: in an obligate (Fraser and Waters 2013), a facultative (Gachon, 2010, Heesch et al. 2008), or an opportunistic association (Barrero-Canosa et al. 2013), or as endobiontic symbionts in a fungus-host equilibriumin host tissue, without causing any external symptoms (Raghukumar 2017). For example, marine fungi isolated from macroalgae as endophyte hosts (Schoenrock et al. 2015) or associated with animals like corals and marine sponges (Maldonado et al. 2005, Yarden 2014).

In Antarctic marine environments, a variety of substrata may harbor cosmopolitan or native fungi (Table 7.1), and many species were identified in association with driftwood (Pugh and Jones 1986), algae (Duarte et al. 2016, Furbino et al. 2014 2018, Godinho et al. 2013, Loque et al. 2010), marine invertebrates (Cui et al. 2016, Godinho et al. 2019, Henríquez et al. 2014, Poveda et al. 2018, Vaca et al. 2013), coastal waters (Grasso et al. 1997), seawater (Vaz et al. 2011) and deep sea (Bass et al. 2007, Gonçalves et al. 2017, Lopez-Garcia et al. 2001), and shallow (Gonçalves et al. 2015, Vaz et al. 2011, Wentzel et al. 2019) and deep sediments (Gonçalves et al. 2013, Ogaki et al. 2019).

TABLE 7.1 Marine Antarctic fungi from different substrata.

Substrate	Fungal Taxa	Region	Reference
Plants and Algae			
Wooden baits	*Trichocladium achrasporum*, *T. lignincola*, *T. constrictum* and *Phoma* sp.	Penguin Bay (Adelie Cove), South Georgia	Grasso et al. (1997)
Macroalgae[a]	*Antarctomyces pellizariae*, *A. psychrotrophicus*, *Aspergillus* sp., *A. conicus*, *A. protuberus*, *A. tabacinus*, *A. terreus*, *Aureobasidium pullulans*, *Beauveria bassiana*, *Cadophora malorum*, *Candida sake*, *Chaetomium* sp., *Cladosporium* sp., *C. lignicola*, *C. tenuissimum*, *Coprinellus radians*, *Cordyciptaceae* sp., *Cryptococcus* sp., *C. adeliensis*, *C. albidosimilis*, *C. laurentii*, *C. magnus*, *Cystofilobasidium infirmominiatum*, *Debaryomyceshansenii*, *Dioszegia athyri*, *D. xingshanensis*, *Dipodascus australiensis*, *Doratomyces* sp., *Engyodontium* sp., *Eurotium herbariorum*, *E. repens*, *Fusarium* sp., *Geomyces* sp., *G. luteus*, *Glaciozyma litorale*, *G. martinii*, *Guehomyces pullulans*, *Helotiales* sp., *Holtermanniella festucosa*, *H. nyarrowii*, *Hyaloscyphaceae* sp., *Lecanicillium* sp., *Leucosporidiium fragararium*, *L. muscorum*, *Metschnikowia australis*, *Meyerozyma caribbica*, *M. guilliermondii*, *Mortierella* sp., *M. antarctica*, *Mrakia* sp., *Mycoarthriscf. corallines*, *Oidiodendron* sp., *O. truncatum*, *Penicillium* sp., *P. chrysogenum*, *P. citrinum*, *P. crustosum*, *P. discolor*, *P. spinulosum*, *Penicillium steckii*, *Phaeosphaeria herpotrichoides*, *Phoma* sp., *Pseudogymnoascus* sp., *P. destructans*, *P. pannorum*, *Pseudozyma* sp., *P. tsukubaensis*, *Rhodotorula glacialis*, *R. laryngis*, *R. marina*, *R. minuta*, *R. mucilaginosa*, *Sporidiobolus pararoseus*, *Thelebolus globosus*, *Tilletiopsis washingtonensis*, *Ustilaginaceae* sp., *Verticillum* sp., *Vishniacozyma carnescens*, *V. victoriae* and *Yamadazyma mexicana*	Deception, Elephant, Half Moon, King George, Nelson and Robert Islands	Duarte et al. (2016), Furbino et al. (2014, 2018), Godinho et al. (2013), Loque et al. (2010)

Contd.

TABLE 7.1 Contd.

Substrate	Fungal Taxa	Region	Reference
Invertebrates			
Sponge[b]	*Acremonium* sp., *Aureobasidium pullulans*, *Bullera pseudoalba*, *Cladosporium* sp., *Cryptococcus laurentii*, *Cystofilobasidium infirmominiatum*, *Debaryomyces hansenii*, *Epicoccum* sp., *Leucosporidium* sp., *L. creatinivorum*, *Metschnikowia australis*, *Penicillium polonicum*, *Penicillium* sp., *Phoma* sp., *Pseudgymnoascus* sp., *P. pannorum*, *Rhodotorula pinicola*, *R. muscilaginosa*, *Thelebolus* sp., *Trichocladium* sp. and *Pseudeurotium* sp.	King George Island	Duarte et al. (2013), Henriquez et al. (2014), Poveda et al. (2018), Vaca et al. 2013)
Mollusca[c]	*A. psychrotrophicus*, *Aspergillus* sp., *Candida* sp., *C. spencermartinsiae*, *C. zeylanoides*, *Cladosporium halotolerans*, *Clavispora lusitaniae*, *Cladosporium* sp., *Cryptococcus laurentii*, *Debaryomyces hansenii*, *Galactomyces* sp., *G. martini*, *H. festucosa*, *Meyerozyma guilliermondii*, *Mortierella* sp., *Mrakia frigida*, *Penicillium* sp., *Polypaecilum botryoides*, *P. destructans*, *P. verrucosus*, *P. pannorum*, *R. mucilaginosa*, *Thelebolus balaustiformis*, *T. globosus*, *Tolypocladium tundrense*, *V. Victori* and *Wickerhamomyces anomalus*	Deception, King George and Livingston Islands; and Hope Bay	Duarte et al. (2013), Godinho et al. (2019)
Nemertea[d]	*Cladosporium* sp., *Didymella longicolla*, *Glaciozyma martinii*, *Metschnikowia australis*, *Mollisia* sp., *Mortierella* sp., *Mrakia* sp., *Penicillium* sp., *P. brevicompactum*, and *Pestalotiopsis kenyana*	Deception and King George Islands; Dorian, Hope and Orne Bay	Godinho et al. (2019)
Cnidaria[e]	*M. australis*	Robert and Snow Islands, Sissini, Greenwich, Dorian Bay	Godinho et al. (2019)
Oligochaeta[f]	*Aspergillus* sp., *Cystobasidium slooffiae*, *Didymella coffeae-arabicae*, *Letendraea* sp., *M. australis*, *Nothophoma macrospora*, *Penicillium* sp., *Penicilliums wiecickii*, *Pestalotiopsis kenyana*, *Phoma* sp., *P. destructans*, *Rhodotorula glutinis*, *R. mucilaginosa* and *T. globosus*	Maxwell Bay, King George Island	Godinho et al. (2019), Herrera et al. (2017)

Contd.

TABLE 7.1 Contd.

Substrate	Fungal Taxa	Region	Reference
Annelida[g]	*M. australis*		Godinho et al. (2019),
Echinodermata[h]	*C. adeliensis*, *C. albidosimilis*, *C. laurentii*, *Cystofilobasidium infirmominiatum*, *Guehomyces pullulans*, *M. australis*, *M. guilliermondii*, *Penicillium* sp., *P. destructans* and *R. mucilaginosa*	Deception and King George Islands	Duarte et al. (2013), Godinho et al. (2019)
Platyhelmintes[i]	*Cladosporium* sp., *D. hansenii*, *Penicillium* sp. and *Pseudogymnoascus* sp.	Snow Island and Dorian Bay - Antarctica Peninsula	Godinho et al. (2019)
Arthropoda[j]	*Aspergillus* sp., *Cladosporium* sp., *D. hansenii*, *Meyerozyma* sp., *M. guilliermondii*, *P. appendiculatus*, *P. destructans*, *P. pannorum*, *P. verrucosus*, *Penicillium* sp., *Septoriachromolaenae,* and *Talaromyces* sp.	King George, Livingston, Robert and Snow Islands and Antarctic Ocean	Cui et al. (2016), Duarte et al. (2013), Godinho et al. (2019)
Chordata[g]	*Antarctomyces psychrotrophicus*, *Candida sake*, *Cystofilobasidium* capitatum, *C. infirmominiatum*, *M. australis*, *R. mucilaginosa* and *W. anomalus*	Deception and King George Islands	Duarte et al. (2013), Godinho et al. (2019)
Sediments			
Shallow sediment	*Cadophora* sp., *Cladosporium* sp., *Cryptococcus* sp., *C. adeliensis*, *Cystobasidium* sp., *Debaryomyces hansenii*, *Geomyces* sp., *Glaciozyma* sp., *Guehomyces pullulans*, *Holtermanniella* sp., *Leucosporidium scottii*, *L. muscorum*, *Metschnikowia* sp., *M. australis*, *Meyerozyma* sp., *M. guilliermondii*, *Mrakia* sp., *Paraconiothyrium* sp., *Penicillium* sp., *Pestalotiopsis* sp., *Phenoliferia* sp., *Pseudocercosporella* sp., *Pseudogymnoascus* sp., *Rhodotorula* sp., *R. glacialis*, *Simplicillium lamellicola*, *Toxicocladosporium* sp. and *V. victoriae*	Deception and King George Island	Duarte et al. (2013), Gonçalves et al. (2015), Song et al. (2010), Vaz et al. (2011), Wentzel et al. (2019)
Deep sediments	*Acremonium fusidioides*, *P. allii-sativi*, *P. chrysogenum*, *P. palitans*, *P. solitum* and *P. verrucosus*	Maxwell Bay, Shetland Islands sea and Antarctic Ocean	Gonçaves et al. (2013), Ogaki et al. (2019)
Water			
Water column (different depth)	Uncultured fungi	Drake Passage	Bass et al. (2007)

Contd.

TABLE 7.1 Contd.

Substrate	Fungal Taxa	Region	Reference
Seawater	*Candida spencermartinsiae, C. zeylanoides, Leucosporidium scottii, L. creatinivorum* and *M. australis*	King George Island	Vaz et al. (2011)
Deep Seawater	*Acremonium* sp., *Aspergillus pseudoglaucus, Cladosporium sphaerospermum, Cystobasidium slooffiae, Exophiala xenobiotica, Glaciozyma antarctica, Graphium rubrum, Lecanicillium attenuatum, M. australis, P. chrysogenum, Penicillium citreosulfuratum, Purpureocillium lilacinum* and *Simplicillium aogashimaense*	Bransfieldand Gerlach Straits	Gonçalves et al. (2017)

Species or genera of: [a]Macroalgae: *Acrosiphonia arcta, Adenocystis utricularis, Ascoseira mirabilis, Curdiea racovitzae, Cystosphaera jacquinotii, Desmarestia anceps, Desmarestia menziesii, Georgiella confluens, Gigartina skottsbergii, Himantothallus grandifolius, Iridaea cordata, Monostroma hariotii, Palmaria decipiens, Phaeurus antarcticus, Pyropia endiviifolia and Ulva intestinalis.* [b]Sponge: *Dendrilla* sp., *Tedania* sp., *Hymeniacidon* sp., *Poecilosclerida* sp. [c]Mollusca: *Laevilacunaria antarctica and Nacella concinna.* [d]Nemertea: *Antarctonemertes valid.* [e]Cnidaria: *Halyclystus antarticu.* [f]Oligochaeta: *Grania* sp. and *Lumbricillus* sp. [g]Annelida: *Magelonidae* sp. [h]Echinodermata: *Ophiuroidea* sp., and two specimens no identified of star and urchin[i] Platyhelmintes: *Trepaxonemata* sp. [j]Arthropoda (Crustacea): *Tigriopus kingsejongensis, Euphausia superba* (krill). [k]Chordata: *Ascidia* sp. and *Salpa* sp.

Antarctic Fungi and Strategies in Bioremediation

The recovery of environments impacted with hydrocarbons, heavy metals, spilled oil and other contaminants is often done by physical-chemical methods, but in sensitive environments such as Antarctica these chemicals may afflict biological communities (de Jesus et al. 2015, Snape et al. 2003, Yang et al. 2009). Thereby, bioremediation process using native fungi has great potential in the treatment of impacted areas, due to its environmental compatibility and cost-benefits.

Antarctic fungi face several stresses simultaneously and they adopt different strategies to address these stresses (Ruisi et al. 2007). The factors that influence Antarctic mycobiota dynamics, adaptation to stress, biotechnological applications, and the advantages of their use in bioremediation are summarized in Figure 7.1.

Bioremediation in Polar Environments

Bioremediation is an economically effective cleanup system that speeds up the natural biodegradation using microorganisms or its components (such as enzymes) to reduce, eliminate, or transform toxic compounds (Margesin 2014, Martorell et al. 2019, Tyagi et al. 2011, Yang et al. 2009). Some methods, like bioaugmentation or biostimulation, are mainly based on the ability of these microorganisms to use organic compounds derived from contaminants as their carbon and energy source,

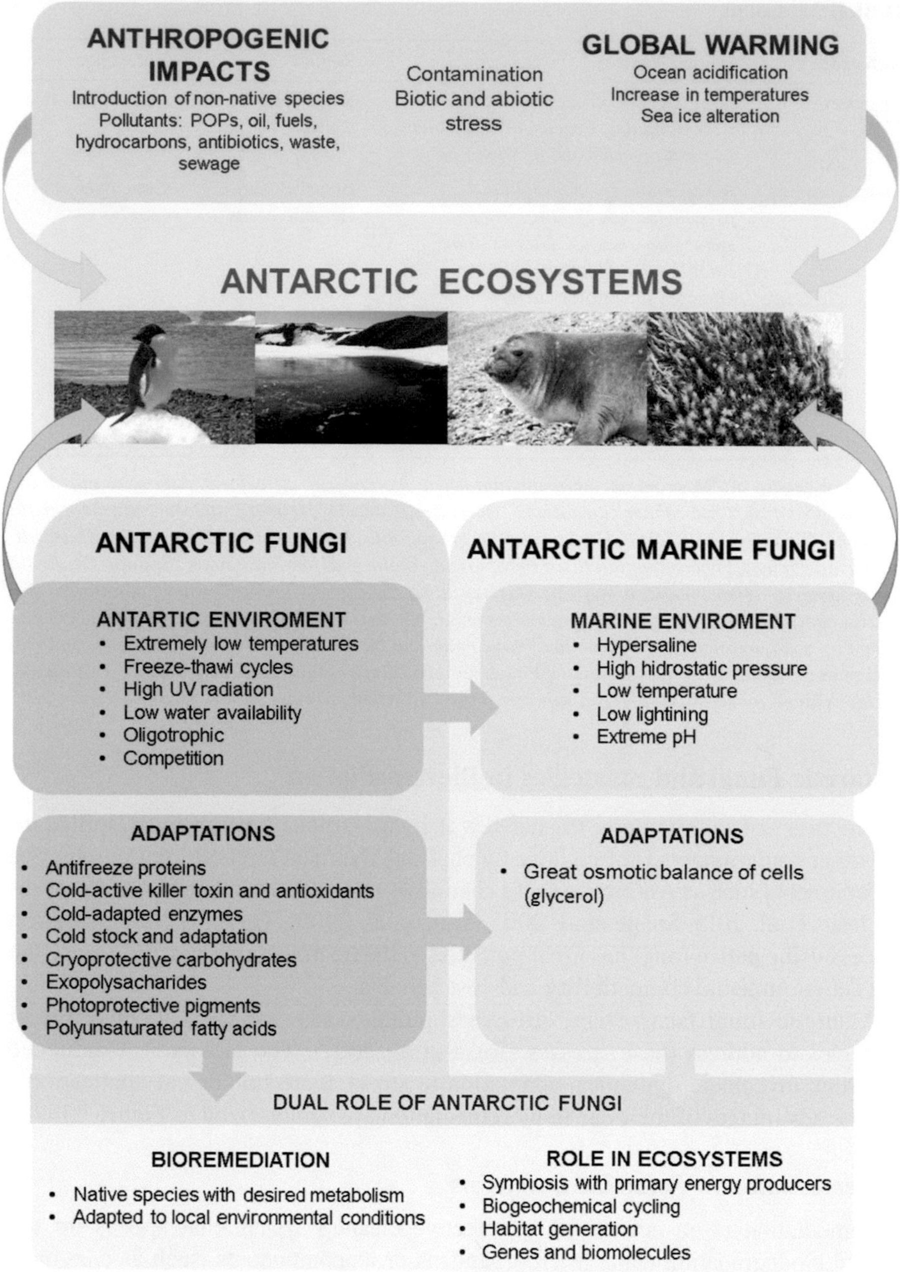

ANTHROPOGENIC IMPACTS
Introduction of non-native species
Pollutants: POPs, oil, fuels, hydrocarbons, antibiotics, waste, sewage
Contamination
Biotic and abiotic stress
GLOBAL WARMING
Ocean acidification
Increase in temperatures
Sea ice alteration
ANTARCTIC ECOSYSTEMS
ANTARCTIC FUNGI
ANTARCTIC MARINE FUNGI
ANTARTIC ENVIROMENT
• Extremely low temperatures
• Freeze-thawi cycles
• High UV radiation
• Low water availability
• Oligotrophic
• Competition
MARINE ENVIROMENT
• Hypersaline
• High hidrostatic pressure
• Low temperature
• Low lightining
• Extreme pH
ADAPTATIONS
• Antifreeze proteins
• Cold-active killer toxin and antioxidants
• Cold-adapted enzymes
• Cold stock and adaptation
• Cryoprotective carbohydrates
• Exopolysacharides
• Photoprotective pigments
• Polyunsaturated fatty acids
ADAPTATIONS
• Great osmotic balance of cells (glycerol)
DUAL ROLE OF ANTARCTIC FUNGI
BIOREMEDIATION
• Native species with desired metabolism
• Adapted to local environmental conditions
ROLE IN ECOSYSTEMS
• Symbiosis with primary energy producers
• Biogeochemical cycling
• Habitat generation
• Genes and biomolecules

FIGURE 7.1 Scheme illustrating the local and global impacts influencing the biotic and abiotic dynamics of Antarctica; the main abiotic factors of the Antarctic ecosystems and marine environments; the role and adaptations of fungi in these environments; and the advantages of using native mycobiota in bioremediation.

mineralizing them in CO_2 and H_2O or transforming in less harmful compounds (Martorell et al. 2019), and it could be alternatives in sensitive environments.

In cold systems, physical characteristics as freezing, marked seasonality, low available water and limited soil development, influence the stability and degradation of contaminants, in addition to the susceptibility of the biota to contaminants (Snape et al. 2003).

Considering this balance, bioremediation processes in polar regions require evaluation of the biotic and abiotic dynamics, the pollution dissipation processes under low temperature and fluctuations in environmental conditions. To design ideal decontamination strategies, prior assessment of contaminated site (ice, permafrost, soil, sediments) and the nature of the pollutants, harnessing active cold-adapted microorganisms with genetic catabolic capacity under these extreme conditions, should be considered (Flocco et al. 2019).

Antarctic Marine Fungi and their Biological Application

In accordance with the Antarctic treaty, the introduction of non-native microrganisms is not permitted and some endemic microrganisms have already been evaluated in bioestimulation and bioaugumentation assays (Cury et al. 2011, Gerginova et al. 2013, Martínez Álvarez et al. 2015), or in a combination of both methods, mainly in soil contaminated with fuel and heavy metals that are common pollutants in marine habitats.

Antarctic microorganisms make vital contributions to the functioning of ecosystem, mainly bacteria and fungi, which could also be applied in different bioremediation methods. Particularly, Antarctic fungi with bioremediation capacities have been reported at a fast pace during the last 20 years or more (Martorell et al. 2019). Some studies mentioned: (1) production of biosurfactants by *Candida glaebosa* (Bueno et al. 2019); (2) degradation of phenol contamination by *Aspergillus fumigatus* (Gerginova et al. 2013), both recovered from Antarctic soil samples; (3) non-identified yeasts obtained from soil capable of degrading phenol with heavy metal tolerance (Fernández et al. 2017) (4) non-ligninolytic fungi species with capacity to tolerate or degrade oil hydrocarbons (Hughes and Bridge 2010) and *Phialopora* sp. reducing concentrations of oil hydrocarbons in Antarctic soils (Aislabie et al. 2001), and (5) many species capable of producing enzymes with application in bioremediation, such as *Goffeauzyma gilvescens*, *Phenoliferia glacialis*, *Cystobasidium laryngis*, *Cystobasidium palidum*, *Dioszegia hungarica*, *Guehomyces pullulans*, *Holtermanniella* sp., *Leucosporidium creatinivorum*, *Leucosporidium scotti*, *Leuconeurospora* sp., *Protomyces inouyei* and others (Carrasco et al. 2012, Martorell et al. 2017, 2019, Rovati et al. 2013, Troncoso et al. 2017, Vaca et al. 2013, Vaz et al. 2011).

In contrast to these reports, fungi from Antarctic marine substrates were less explored and most of the studies involved the great potential of their extracellular enzymes in bioremediation. Some species are reported as producers of lipases, stearases, xylanases, galactosidase and proteases, for example: *Cryptococcusa deliensi*, *Cryptococcus gastricus*, *Leucosporidium creatinivorum*, *Glaciozyma litoralis*, *Nadsonia commutata*, *Rhodotorula mucilaginosa*, *Guehomyces pullulans*,

Mrakiafrigida, *Pichiacaribbica*, *Pseudogymnoascus* sp., *Metschnikowia* sp. and others (Duarte et al. 2013, 2018, Furbino et al. 2018, Song et al. 2010).

Antarctic marine microhabitats are characterized by hypersalinity, high hydrostatic pressure, low temperatures, oligotrophic and low lighting conditions and pH extremes. All these abiotic factors exert strong selective pressure inducing some adaptative responses and a great metabolic diversity in the local mycobiota, which contribute to the differences between the metabolites generated by marine microorganisms and their terrestrial homologous. Accordingly, Antarctic marine fungi become interesting to scientific researches and biotechnological applications.

Marine-derived fungi havc been described as producers of biosurfactants, extracellular enzymes, secondary metabolites, polysaccharides and polyunsaturated fatty acids with potential application in bioremediation (Bonugli-Santos et al. 2015, Damare et al. 2012). Some studies reported many taxa of marine-derived fungi with capacity to biotransform persistent organic polutantants (POPs) (Gao et al. 2013, Vacondio et al. 2015), phenolics compouds (Divya et al. 2014), polycyclic aromatic hydrocarbons (PAHs) (Vieira et al. 2018), crude oil (Hickey 2013), hydrocarbons (Raikar et al. 2001, Passarini et al. 2011) and to bioaccumulate heavy metals (Raghukumar 2008).

Some extracellular and intracellular fungal enzymes are interesting in pollutant-degrading in bioremediation strategies (Deshmukh et al. 2016). Reports relating to the enzyme production by marine-derived fungi began in 1980s, becoming frequent after 1999-2000 (Velmurugan and Lee 2012), and many marine strains have been reported to be producers of hydrolytic and/or oxidative enzymes such as alginatelyase, amylase, cellulase, chitinase, glucosidase, inulinase, keratinase, ligninase, lipase, nuclease, phytase, protease, and xylanase (Bonugli-Santos et al. 2015).

Duarte et al. (2013) evaluated yeasts obtained from different marine substrates for production of protease, xylanase and lipase in screening assays. The majority of isolates were lipase positive. In another study showing the production of similar enzymes, Wentzel et al. (2019) obtained many yeasts and filamentous fungi from Antarctic marine sediments with lipase and protease activities. The best results for lipase were presented by the taxa *Pseudogymnoascus* sp. and *Metschnikowia* sp. for protease. Both genera are common in other marine substrata of Antartica (Table 7.1). Lipases and sterases were promising enzymes in treating some kinds of phenol-polluted industrial wastewater containing heavy metals, such as effluents from petroleum refineries in cold environments (Martorell et al. 2019). Proteases may be also used in the wastewater treatment of detergent, food and textile industries.

Duarte et al. (2018) evaluated the production of ligninolytic enzymes by fungi from marine and terrestrial samples. Among the selected fungi, the strains *Cadophora luteoolivacea* and *Cadophora malorum* showed the highest laccase activity at 15°C, both recovered from marine sediment and sea star, respectively. In addition, the best results for xylanase were presented by unidentified fungus and *Penicillium* sp. recovered from marine invertebrates. Laccases degrade lignin and are used to decolorize and detoxify the industrial effluents, helping in wastewater treatment, as well as to act in highly recalcitrant environmental pollutants, and they can be effectively used in industrial and in xenobiotic degradation and bioremediation

(Viswanath et al. 2014). Xylanase has cellulolytic and hemycellulolytic actvities and laccases may be used in the textile and paper industies (Mishra and Thakur 2015).

Considering all results above, these positive activities of Antarctic marine isolates indicate the great potential of these fungi as metabolic fabrics capable of producing numerous compouds with application in bioremediation and other biotechnological assays.

CONCLUSIONS

Over the past 50 years, overfishing, tourism and the development of scientific research have intensified the human presence in Antarctica. These human activities affect local terrestrial and marine ecosystems with the deposition of residues and other contaminants which interfere in the balance of Antarctic biotic and abiotic systems. The recovery of environments impacted by hydrocarbons, heavy metals, spilled oil and other contaminants is usually done by physical-chemical methods, but these methods may not be applied in these sensitive ecosystems in accordance with Antarctic Treaty. Considering this, some bioremediation processes using native Antarctic mycobiota has become interesting in recovering polar environments. Many extremophilic and marine-derived fungi demonstrate their great potential in bioremediation and may be considered metabolic fabrics capable of producing numerous compounds for biotechnological purposes, or may be explored in the recovery of ecosystems and by industry in the treatment of effluents.

REFERENCES

Aislabie, J., R. Fraser, S. Duncan and R.L. Farrell. 2001. Effects of oil spills on microbial heterotrophs in Antarctic soils. Polar Biol. 24: 308-313.

Arenz, B.E., B.W. Held, J.A. Jurgcs, R.L. Farrell and R.A. Blanchette. 2006. Fungal diversity in soils and historic wood from the Ross sea region of Antarctica. Soil Biol. Biochem. 38: 3057-3064.

Aronson, R.B., S. Thatje, J.B. McClintockand K.A. Hughes. 2011. Anthropogenic impacts on marine ecosystems in Antarctica. Ann. NY Acad. Sci. 1223: 82-107.

Bargagli, R. 2008. Environmental contamination in Antarctic ecosystems. Sci. Total Environ. 400: 212-226.

Barnes, D.K.A. 2017. Iceberg killing fields limit huge potential for benthic blue carbono in Antarctic shallows. Glob. Chang. Biol. 23: 2649-2659.

Barrero-Canosa, J., L.F. Dueñas and J.A. Sánchez. 2013. Isolation of potential fungal pathogens in gorgonian corals at the Tropical Eastern Pacific. Coral Reefs. 32: 35-41.

Bass, D., A. Howe, N. Brown, H. Barton, M. Demidova, H. Michelle, et al. 2007. Yeast forms dominate fungal diversity in the deep oceans. Proceedings Royal Soc. 274: 3069-3077.

Bender, N.A., K. Crosbie and H.J. Lynch. 2016. Patterns of tourism in the Antarctic Peninsula region: A 20-year analysis. Antarctic Sci. 1: 1-10.

Bonugli-Santos, R.C., M.R. dos Santos Vasconcelos, M.R.Z. Passarini, G.A.L. Vieira, V.C.P. Lopes, P.H. Mainardi, et al. 2015. Marine-derived fungi: diversity of enzymes and biotechnological applications. Front Microbiol. 6: 269-284.

Bueno, J.L., P.A.D. Santos, R.R. Silva, I.S. Moguc, A. Pessoa Jr., M.V. Vianna, et al. 2019. Biosurfactant production by yeasts fromdifferent types of soil of the South Shetland Islands (Maritime Antarctica). J. Appl. Microbiol. 126: 1402-1413.

Cabrerizo, A., J. Dachs, D. Barceló and K.C. Jones. 2013. Climatic and biogeochemical controls on the remobilization and reservoirs of persistent organic pollutants in Antarctica. Environ. Sci. Technol. 47: 4299-4306.

Carrasco, M., J.M. Rozas, S. Barahona, J. Alcaino, V. Cifuentes and M. Baeza. 2012. Diversity and extracellular enzymatic activities of yeasts isolated from King George Island, the sub-Antarctic region. BMC Microbiol. 12: 251.

Clarke, A., E.J. Murphy, M.P. Meredith, J.C. King, L.S. Peck, D.K.A. Barnes, et al. 2006. Climate change and the marine ecosystem of the western Antarctic Peninsula. Phil. Trans. R. Soc. B. 362: 149-166.

Convey, P. and L.S. Peck. 2019. Antarctic environmental change and biological responses. Sci. Adv. 11(eaaz0888): 1-16.

Convey, P., K.A. Hughes and T. Tin. 2012. Continental governance and environmental Management mechanisms under the Antarctic Treaty System: Sufficient for the biodiversity challenges of this century? Biodiversity13: 1-15.

Crockett, A.B. and G.J. White. 2003. Mapping sediment contamination and toxicity in Winter Quarters Bay, Mc-Murdo Station, Antarctica. Environ. Monit. Assess. 85: 257-275.

Cui, X., G. Zhu, H. Liu, G. Jhang, J. Wang and W. Zhu. 2016. Diversity and function of the Antarctic krill microorganisms from *Euphausia superba*. Sci. Rep. 6: 36496.

Cury, J.C., H.E. de Jesus, H.D.M. Villela, R.S. Peixoto, C.E.G.R. Schaefer, M.C. Bícego, et al. 2011 Bioremediation of the diesel-contaminated soil of the Brazilian Antarctic Station. INCT-APA Annual Activity Report.

Damare, S., P. Singh and S. Raghukumar. 2012. Biotechnology of marine fungi. Prog. Mol. Subcell. Biol. 53: 277-297.

de Jesús, H.E., R.S. Peixoto and A.S. Rosado. 2015. Bioremediation in Antarctic soils. J. Pet. Environ. Biotechnol. 6(248): 2.

Deshmukh, R., A.A. Khardenavis and H.J. Purohit. 2016. Diverse metabolic capacities of fungi for bioremediation. Indian J. Microbiol. 56: 247-264.

Divya, L.M., G.K. Prasanth and C. Sadasivan. 2014. Potential of the salt-tolerant laccase-producing strain *Trichoderma viride* Pers. NFCCI-2745 from an estuary in the bioremediation of phenol-polluted environments. J. Basic Microbiol. 54: 542-547.

Donohue, K.A., K.L. Tracey, D.R. Watts, M.P. Chidichimo and T.K. Chereskin. 2016. Mean antarctic circumpolar current transport measured in drake passage. Geophysical Research Lett. 43(11): 760.

Duarte, A.W.F., I. Dayo-Owoyemi, F.S. Nobre, F.C. Pagnocca, L.S.C. Chaud, J.R. Pessoa, et al. 2013. Taxonomic assessment and enzymes production by yeasts isolated from marine and terrestrial Antarctic samples. Extremophiles 17: 1023-1035.

Duarte, A.W.F., M.R. Passarini, T.P. Delforno, F.M. Pellizzari, C.V.Z. Cipro, R.C. Montone, et al. 2016. Yeasts from macroalgae and lichens that inhabit the South Shetland Islands, Antarctica. Environ. Microbiol. Rep. 8: 874-886.

Duarte, A.W.F., D.O.S. Santos, M.V. Vianna, J.M.F. Vieira, V.H. Mallagutti, F.J. Inforsato, et al. 2018. Cold-adapted enzymes produced by fungi from terrestrial and marine Antarctic environments. Crit. Ver. Biotechnol. 38: 600-619.

Fernández, P.M., M.M. Martorell, M.G. Blaser, L.A.M. Ruberto, L.I.C. de Figueroa and W.P. Mac Cormack. 2017. Phenol degradation and heavy metal tolerance of Antarctic yeasts. Extremophiles 21(3): 445-457.

Flocco, C.G., W.P.M. Cormack and K. Smalla. 2019. Antarctic soil microbial communities in a changing environment: Their contributions to the sustainability of antarctic ecosystems and the bioremediation of anthropogenic pollution. pp. 133-161. *In*: A.G.S. Castro-Sowinski. (ed.). The Ecological Role of Micro-organisms in the Antarctic Environment. Springer Polar Sciences, Switzerland.

Fraser, C.I. and J.M. Waters. 2013. Algal parasite *Herpodiscus durvillaeae* (Phaeophyceae: Sphacelariales) inferred to have traversed the pacific ocean with its buoyant host. J. Phycol. 49: 202-206.

Frenot, Y., S.L. Chown, J. Whinam, P.M. Selkirk, P. Convey, M. Skotnicki, et al. 2005. Biological invasions in the Antarctic: Extent, impacts and implications. Biological Rev. 80: 45-72.

Furbino, L.E., V.M. Godinho, I.F. Santiago, F.M. Pellizzari, T.M. Alves, C.L. Zani, et al. 2014. Diversity patterns, ecology and biological activities of fungal communities associated with the endemic macroalgae across the Antarctic Peninsula. Microb. Ecol. 67: 775-787.

Furbino, L.E., F.M. Pellizzari, P.C. Neto, C.A. Rosa and L.H. Rosa. 2018. Isolation of fungi associated with macroalgae from maritime Antarctica and their production of agarolytic and carrageenolytic activities. Polar Biol. 41: 527-535.

Gachon, C.M.M., T. Sime-Ngando, M. Strittmatter, A. Chambouvet and G.H. Kimet. 2010. Algal diseases: Spotlight on a black box trends. Plant Sci. 15: 633-640.

Gao, G.R., Y.F. Yin, D.Y. Yang and D.F. Yang. 2013. Promoting behavior of fungal degradation Polychlorinated Biphenyl by Maifanite. Adv. Mater. Res. 662: 515-519.

Gerginova, M.G., N.M. Peneva, E.T. Krumova and Z.A. Alexieva. 2013. Biodegradation ability of fungal strains isolated from Antarctic towards pah. *In*: Proceedings of the 13th international conference of environmental science and technology Athens, Greece 5-7.

Glöckner, F.O., L.J. Stal, R.A. Sandaa, J.M. Gasol, F. O'Gara, F. Hernandez, et al. 2012. *In*: J.B. Calewaert, N. McDonough. (eds.). Marine Microbial Diversity and its role in Ecosystem Functioning and Environmental Change. Marine Board Position Paper 17. Marine Board-ESF, Ostend.

Godinho, V.M., L.E. Furbino, I.F. Santiago, F.M. Pellizzari, N. Yokoya, D. Pupo, et al. 2013. Diversity and bioprospecting of fungal communities associated with endemic and cold-adapted macroalgae in Antarctica. ISME J. 7: 1434-1451.

Godinho, V.M., M.T.R. de Paula, D.A.S. Silva, K. Paresque, A.P. Martins, P. Colepicolo, et al. 2019. Diversity and distribution of cryptic cultivable fungi associated with marine animals of Antarctica. Fungal Biol. 123(7): 507-516.

Gomes, J., I. Gomes and W. Steiner. 2000. Thermolabile xylanase of the Antarctic yeast Cryptococcus adeliae: production and properties. Extremophiles 4: 227-235.

Gonçalves, V.N., L.S. Campos, I.S. Melo, V.H. Pellizari, C.A. Rosa and L.H. Rosa. 2013. *Penicillium solitum*: A mesophilic, psychrotolerant fungus present in marine sediments from Antarctica. Polar Biol. 36:1823-1831.

Gonçalves, V.N., C.R. Carvalho, S. Johann, G. Mendes, T.M.A. Alves, C.L. Zani, et al. 2015. Antibacterial, antifungal and antiprotozoal activities of fungal communities present in different substrates from Antarctica. Polar Biol. 38: 1143-1152.

Gonçalves, V.N., F.S. Oliveira, C.R. Carvalho, C.E.G.R. Schaefer, C.A. Rosa and L.H. Rosa. 2017. Antarctic rocks from continental Antarctica as source of potential human opportunistic fungi. Extremophiles 21: 851-860.

Grasso, S., V. Bruni and G. Maio. 1997. Marine fungi in Terra Nova Bay (Ross Sea, Antarctica). New Microbiol. 20: 371-376.

Heesch, S., A.F. Peters, J.E. Broom and C.L. Hurd. 2008. Affiliation of the parasite *Herpodiscus durvillaeae* (Phaeophyceae) with the Sphacelariales based on DNA sequence comparisons and morphological observations. Eur. J. Phycol. 43: 283-295.

Henríquez, M., K. Vergara, J. Norambuena, A. Beiza, F. Maza, P. Ubilla, et al. 2014. Diversity of cultivable fungi associated with Antarctic marine sponges and screening for their antimicrobial, antitumoral and antioxidant potential. World J. Microbiol. Biotechnol. 30: 65-76.

Herrera, L.M., C.X. García-Laviña, J.J. Marizcurrena, O. Volonterio, R.P. de León and S. CastroSowinski. 2017. Hydrolytic enzyme-producing microbes in the Antarctic oligochaete *Grania* sp. (Annelida). Polar Biol. 40: 947-953.

Hickey, P. 2013. Toxicity of water soluble fractions of crude oil on some bacteria and fungi Isolated from marine water. Am. J. Anim. Res. 3: 24-29.

Hughes, K.A. and P.D. Bridge. 2010. Tolerance of Antarctic soil fungi to hydrocarbons and their potential role in soil bioremediation. pp. 277-300. *In*: A.K. Bej, J. Aislabie, R.M. Atlas. (eds.). Polar Microbiology: The Ecology, Biodiversity and Bioremediation Potential of Microorganisms in Extremely Cold Environments. Taylor and Francis, Boca Raton.

IAATO 2020. IAATO Data and Statistics. Acess: https://iaato.org/information-resources/data-statistics/

Jacob, U. 2007. Food web, marine. *In*: B. Riffenburgh. (ed.). Enciclopedia of the Antartic, 1: 409-410.

Kriwoken, L.K. and D. Rootes. 2000. Tourism on ice: Environmental impact assessment of Antarctic tourism. Impact Assessment Project Appraisal. 18: 138-150.

Lee, J.E. and K.A. Hughes. 2010. Focused tourism needs focused monitoring. Antarct. Sci. 22: 1.

López-García, P., F. Rodriguez-Valera, C. Pedros-Alio and D. Moreira. 2001. Unexpected diversity of small eukaryotes in deep-sea Antarctic plankton. Nature 409: 603-607.

Loque, C.P., A.O. Medeiros, F.M. Pellizzari, E.C. Oliveira, C.A. Rosa and L.H. Rosa. 2010. Fungal community associated with marine macroalgae from Antarctica. Polar Biol. 33: 641-648.

Maldonado, M., N. Cortadellas, I. Trillas and K. Rützler. 2005. Endosymbiotic yeast maternally transmitted in a marine sponge. Biol. Bull. 209: 94-106.

Margesin, R. 2014. Bioremediation and biodegradation of hydrocarbons by cold-adapted yeasts. pp. 465-480. *In*: P. Buzzini, R. Margesin. (eds.). Cold-adapted yeasts. Springer, Berlin, Heidelberg.

Martínez-Alvarez, L.M., A. Lo Balbo, W.P. Mac Cormack and L.A.M. Ruberto. 2015. Bioremediation of a hydrocarbon contaminated Antarctic soil: Optimization of a biostimulation strategy using response-surface methodology (RSM). Cold Regions Sci. Technol. 119: 61-67.

Martorell, M.M., L.A.M. Ruberto, P.M. Fernández, L.I.C. Figueroa and W.P. Mac Cormarck. 2017. Bioprospection of cold-adapted yeasts with biotechnological potential from Antarctica. J. Basic Microbiol. 57: 504-516.

Martorell, M.M., L.A.M. Ruberto, L.I.F. Castellanos and P.M. Cormack. 2019. Bioremediation abilities of antarctic fungi. pp. 517-534. *In*: S.M. Tiquia-Arashiro, M. Grube. (eds.). Fungi in Extreme Environments: Ecological Role and Biotechnological Significance. Springer Nature Switzerland.

Mays, H., D. Oehler, K. Morrison, A. Morales, A. Lycans, J. Perdue, et al. 2019. Phylogeography, population structure and species delimitation in rockhopper penguins (*Eudyptes chrysocome* and *Eudyptes moseleyi*). J. Hered. 110(7): 801-817.

Mishra, M. and I.S. Thakur. 2015. Biodegradation of lignocellulosic waste in the environment. pp. 169-194. *In*: R. Chandra. (ed.). Adv. Biodegradation and Bioremediation Ind. Waste. CRC Press.

Ogaki, M.B., L.C. Coelho, R. Vieria, A.A. Neto, C.L. Zani, T.M.A. Alves, et al. 2019. Cultivable fungi present in deep-sea sediments of Antarctica: Taxonomy, diversity and bioprospection of bioactive compounds. Extemophiles 24: 227-238.

Passarini, M.Z.R., M.V.N. Rodrigues, M. da Silva and L.D. Sette. 2011. Marine-derived filamentous fungi and their potential application for polycyclic aromatic hydrocarbon bioremediation. Mar. Poll. Bull. 62: 364-370.

Poveda, G., C. Gil-Durán, I. Vaca, G. Levicán and R. Chávez. 2018. Cold-active pectinolytic activity produced by filamentous fungi associated with Antarctic marine sponges. Biol. Res. 51: 28.

Pugh, G.J.F. and E.B.G. Jones. 1986 Antarctic marine fungi: A preliminary account. pp. 323-330 *In*: S.T. Moss. (ed.). The Biology of Marine Fungi. Cambridge: Cambridge University Press.

Raghukumar, C. 2006. Algal-fungal interactions in the marine ecosystem: Symbiosis to parasitism. Central Salt Research Institute, India

Raghukumar, S. 2008. Thraustochytrid marine protists: Production of PUFAs and other emerging technologies. Mar. Biotechnol. 10: 631-640.

Raghukumar, S. 2017. The marine environment and the role of fungi. pp. 17-38. *In*: Fungi in Coastal and Oceanic Marine Ecosystems. Springer, Cham.

Raikar, M.T., S. Raghukumar, V. Vani, J.J. David and D. Chandramohan. 2001. Thraustochytrid protists degrade hydrocarbons. Ind. J. Mar Sci. 30: 139-145.

Riddle, M.J. 2009. Human-mediated impacts on the health of Antarctic wildlife. pp. 241-262. *In*: K.R. Kerry, M.J. Riddle. (ed.). Health of Antarctic Wildlife: A Challenge for Science and Policy. Springer-Verlag, London, UK.

Rintoul, S.R. and A.C.N. Garabato. 2013. Dynamics of the southern ocean irculation. pp. 471-492. *In*: G. Siedler, S.N. Griffies, J. Gould, J.A. Church. (eds.). Ocean Circulation and Climate: A 21st Century Perspective. 2nd Ed. (International Geophysics, 103), Academic Press.

Rosa, L.H., F.M. Pellizzari, M.B. Ogaki, M.T.R. de Paula, A. Mansilla, J. Marambio, et al. 2019. Sub-antarctic and antarctic marine ecosystems: an unexplored ecosystem of fungal diversity. pp. 221-241. *In*: L.H. Rosa. (ed.). Fungi of Antarctica: Diversity, Ecology and Biotechnological Applications. Springer.

Rovati, J.I., H.F. Pajot, L. Ruberto, W. Mac Cormack and L.I. Figueroa. 2013. Polyphenolic substrates and dyes degradation by yeasts from 25 de Mayo/King George Island (Antarctica). Yeast 30: 459-470.

Ruisi, S., D. Barreca, L. Selbmann and L. Zucconi. 2007. Fungi in Antarctica. Ver. Environ. Sci. Biotechnol. 6(1-3): 127-141.

Schell, J. 1965. Introductory remarks. *In*: J. Van Mieghen, P. Van Oye. (ed.). Biogeography and ecology in Antartica. Springer, 15: p. ix-xxvii.

Schoenrock, K.M., C.B. Amsler, J.B. McClintock and B.J. Baker. 2015. A comprehensive study of Antarctic algal symbioses: Minimal impacts of endophyte presence in most species of macroalgal hosts. European J. Phycol. 50(3): 271-278,

Simões, J.C.O. 2011. ambiente antártico: domínio de extremos. *In*: L.C.O. Simões. (ed.). Antártica e as mudanças globais: um desafio para a humanidade. Blucher, São Paulo, Brazil, 9: p. 15-27.

Snape, I., M.J. Riddle, D.M. Filler and P.J. Williams. 2003. Contaminants in freezing ground and associated ecosystems: key issues at the beginning of the new millennium. Polar Rec. (Gr. Brit) 39: 291-300.

Song, C., G.L. Liu, J.L. Xu and Z.M. Chi. 2010. Purification and charac terization of extracelular beta-galactosidase from the psychrotolerant yeast *Guehomyces pullulans* 17-1 isolated from sea sediment in Antarctica. Process Biochem. 45: 954-960.

Tin, T., Z.L. Fleming, D.G. Hughes, D. Ainley, P. Convey, C. Moreno, et al. 2009. Impacts of local human activities on the Antarctic environment. Antarctic Sci. 21: 3-33.

Troncoso, E., S. Barahona, M. Carrasco, P. Villarreal, , J. Alcaíno, V. Cifuentes, et al. 2017. Identification and characterization of yeasts isolated from the South Shetland Islands and the Antarctic Peninsula. Polar Biol. 40: 649-658.

Turner, J., R. Bindschadler, P. Convey, G. di Prisco, E. Fahrbach, J. Gutt, et al. 2009. Antarctic Climate Change and the Environment. Scientific Committee on Antarctic Research 526 pp.

Tyagi, M., M.M.R. da Fonseca and C.C. de Carvalho. 2011. Bioaugmentation and biostimulation strategies to improve the effectiveness of bioremediation processes. Biodegradation 22(2): 231-241.

Vaca, I., C. Faúndez, F. Maza, B. Paillavil, V. Hernández, F. Acosta, et al. 2013. Cultivable psychrotolerant yeasts associated with Antarctic marine sponges. World J. Microbiol. Biotechnol. 29: 183-189.

Vacondio, B., W.G. Birolli, I.M. Ferreira, M.H. Seleghim, S. Gonçalves, S.P. Vasconcellos, et al. 2015. Biodegradation of pentachlorophenol by marine-derived fungus *Trichoderma harzianum* CBMAI 1677 isolated from ascidian *Didemnun ligulum*. Biocatal. Agric. Biotechnol. 4: 266-275.

Vaz, A.B.M., L.H. Rosa, M.L. Vieira, V. Garcia, L.R. Brandão, L.C.R.S. Teixeira, et al. 2011. The diversity, extracellular enzymatic activities and photoprotective compounds of yeasts isolated in Antarctica. Braz. J. Microbiol. 43: 937-947.

Velmurugan, N. and Y.S. Lee 2012. Enzymes from marine fungi: current research and future prospects. pp. 441-474. *In*: E.B.G. Jones. (ed.). Marine Fungi and Fungal-like Organisms (Marine and Freshwater Botany), Walter de Gruyter, Berlin.,

Vieira, G.A.L., M.J. Magrini, R.C. Bonugli-Santos, M.V.N. Rodrigues and L.D. Sette. 2018. Polycyclic aromatic hydrocarbons degradation by marine-derived basidiomycetes: optimization of the degradation process. Brazilian J. Microbiol. 49(4): 749-756.

Viswanath, B., B. Rajesh, A. Janardhan, A.P. Kumar and G. Narasimha. 2014. Fungal laccases and their applications in bioremediation. Enzyme Res. (163242): 1-21.

Wentzel, L.C.P., F.J. Inforsato, Q.V. Montoya, B.G. Rossin, N.R. Nascimento, A. Rodrigues, et al. 2019. Fungi from admiralty bay (King George Island, Antarctica) soils and marine sediments. Microb. Ecol. 77: 12-24.

WMO 2020. WMO: New record for Antarctic continent reported. Acess: https://public.wmo.int/en/media/news/new-record-antarctic-continent-reported.

Wynn-Williams, D.D. 1990. Ecological aspects of Antarctic microbiology. pp.71-146. *In*: K.C. Marshall. (ed.). Advances in Microbial Ecology. 11ed. Plenum Press, New York, USA.

Yang, S.Z., H.J. Jin, Z. Wei, R.X. HE, Y.J. Ji, X.M. Li, et al. 2009. Bioremediation of oil spills in cold environments: A review. Pedosphere 19: 371-381.

Yarden, O. 2014. Fungal association with sessile marine invertebrates. Front Microbiol 5: 1-6.

8

Bacterial Bioremediation of Petroleum Hydrocarbon in Ocean

Kamleshwar Singh[1], Sushma Kumari[1], Pratibha Kushwaha, K. Suresh Kumar[1,*] and Kyung-Hoon Shin[2,*]

INTRODUCTION

The transformation of organic remains (fossils of ancient marine organisms, such as plants, algae, and bacteria), coal, and natural gas into carbon-rich substances (over several million years) leads to the formation of crude oil or petroleum that has been an integral part of the ecosphere. Crude oils mainly comprise of hydrocarbons (Hazen et al. 2016, Varjani 2017); Colwell et al. (1977) categorized petroleum hydrocarbon into four classes: aromatics (cyclic hydrocarbon), saturates (aliphatics), asphaltenes (phenols, esters, ketones, porphyrins and fatty acids), and resins (amides, carbazoles, pyridines, sulfoxides and quinolines).

While certain amount of oil is released to the world's oceans through natural seeps (that comprise approximately half of the estimated 1.2 million tonnes of every year; Prince et al. 2017), the most catastrophic and concentrated release of oil occurs through accidental oil spills. Components of hydrocarbons are highly recalcitrant to degradation (due to their low water solubility), and therefore, they persist in nature for a long time causing extensive environmental and economic impact. Petroleum hydrocarbons being highly lipophilic are less soluble in water; their recalcitrant nature often leads to their biomagnification. Their carcinogenic, immunotoxic, mutagenic and teratogenic properties are of great concern (Mahmoud and Bagey 2018).

Marine oil spills are marked as greatest disasters for marine organisms (e. g. seabirds and marine animals). In one of the largest accidental marine oil spill in history, i.e. the Deep Water Horizon (DWH 2010) disaster, nearly 795 million litres of oil was released into the deep ocean; this disaster cost >US$61 billion (Mapelli et al. 2017). In the DWH spill, approximately 149,000 km^2 of the Gulf of Mexico was covered by oil slicks and closed for fishing (Berenshtein et al. 2020). Similarly, in the Exxon-Valdez oil spill in 1989, more than 250,000 seabirds were killed. A huge amount (11 million gallons) of oil from Exxon Valdez remained afloat harming the environment as compared to the 4.9 million barrels oil from DWH spill (Doshi et al. 2018). Although genotoxicity, immunotoxicity and endocrine toxicity of oil spills on exposed population are often well studied, one cannot deny the fact that an

[1] Department of Botany, University of Allahabad, Prayagraj 211002, India.
[2] Hanyang University ERICA Campus, 15588, Ansan, South Korea.
[*] Corresponding Author: ksuresh2779@gmail.com; shinkh@hanyang.ac.kr

awareness needs to be created regarding the toxicity of the PAH fraction of petroleum hydrocarbon as they are resistant to microbial degradation and persist in seawater sediment for ages. Incompletely processed PAHs along with alkyl radicals induce DNA impairment, growth anomalies, and malignant and non-malignant variations in tissues (Selvam and Thatheyus 2018). Among all PAHs known, benzopyrene is the most potent carcinogen among sixteen priority PAHs listed by U.S. EPA (Doshi et al. 2018). Benzene rings containing compounds, such as benzene, toluene, ethylbenzene and xylene (BTEX group), cause deformities in the foetus, infertility, and could lead to abortion, while toluene is associated with hepatic and gastrointestinal dysfunctions as well as teratogenic and a strong depressant of the central nervous system. Xylene at high concentrations, due to its impact on the central nervous system depression, could lead to unconsciousness and death (Claro et al. 2018).

Several clean-up and recovery approaches have been developed for oil spills; they are broadly classified into physical (controlled burning, skimming, and absorbing), chemical (dispersing and solidifying), and biological/bioremediation treatments. Although these physical and chemical methods are immediately used for rapidly controlling the diffusion and drift of oil, they are not suitable for ecological restoration. Selection of sorbent material depends on the nature of the oil spill. On the other hand, the biological methods involve use of natural biological material for remediation purpose, e.g. addition of microbes along with supplementation of nutrients and/or oxygen, to stimulate microbial growth, which would result in biodegradation of the oil (Doshi et al. 2018). Overall, bioremediation is considered eco-friendly and cost-effective for marine ecological restoration, as it leads to a complete decomposition of complex petroleum hydrocarbons (Xue et al. 2015). In fact, fertilizer-enhanced bacterial bioremediation has been reported as a microbial treatment process that was successfully used in the Exxon Valdez oil spill (Bragg et al. 1994).

Bacterial Bioremediation

Biodegradation is a type of weathering process that could result in complete mineralization of hydrocarbons; nevertheless, the rate and extent of degradation strongly depends on a variety of factors involving oil characteristics, environmental conditions, and the organisms prevailing in the contaminated region (Brakstad et al. 2018). Bioremediation is an environmental treatment technology that utilizes the natural metabolic activities of living organisms for removal of contaminants from soil, sediment, and water. When it comes to removal of petroleum hydrocarbons, bioremediation is a preferred strategy as it is cost-effective, eco-friendly and sustainable, as compared to the chemical and physical approaches (Gardner and Gunsch 2020). A few petroleum hydrocarbon-degrading bacteria include *Pseudomonas, Bacillus, Alcaligenes, Micrococcuse, Methylomonas, Achromobacter, Acinetobacter Arthrobacter, Nocardia, Flavobacterium, Corynebacterium, Methylobacterium,* and *Rhodococcus* (Parach et al. 2017).

According to Simpanen et al. (2016), microbes have a natural capacity to transform or mineralize naturally occurring organic molecules, which can be utilized for *in situ* remediation of oil-contaminated areas. Oil could be degraded by a variety of marine microbes, including bacteria, fungi, and algae. Hydrocarbon-degrading bacteria in temperate seawater usually include members of the classes Alphaproteobacteria and Gammaproteobacteria; oil-polluted Arctic seawater and marine ice often become enriched by Gammaproteobacteria (Brakstad et al. 2018). While biodegradation is the main mechanism of contaminant-removal (in strategies involving natural treatments of oil degradations), there are other physical and chemical processes such as evaporation, dispersion, sorption, and dilution that can also contribute to the contaminant reduction (Simpanen et al. 2016).

Overall, there are two treatment strategies that are applied to treat hydrocarbon-polluted environments when it comes to bioremediation; these include biostimulation and bioaugmentation. At times, the indigenous microbes adapt and degrade the contaminants; however, this could be a slow process. In order to enhance natural attenuation, microbial activity needs to be stimulated by providing optimal environmental conditions such as nutrient content, oxygen or other electron acceptor accessibility, pH, temperature, and redox conditions (Simpanen et al. 2016). *In situ* chemical oxidation (ISCO) is a strategy by which organic contaminants can be destroyed or converted into more biodegradable forms quickly (Sutton et al. 2011); this treatment involves distributing a chemical oxidant {e.g. hydrogen peroxide (H_2O_2), permanganate, persulfate, and ozone} into the subsurface. Biostimulation enhances the degrading activity of indigenous bacterial communities by improving carbon: nitrogen: phosphorus (C:N:P) ratios. In general, biostimulation involves either addition of electron acceptor (e.g. O_2, Fe^{3+}, or SO_4^{2-}) to promote oxidative reduction or addition of an electron donor (e.g. organic substrate) to reduce oxidized pollutants. Biostimulation efficiency is reported to have improved by nutrient microencapsulation for slow and sustained release of nutrients (Mapelli et al. 2017). Adams et al. (2015) carried out biostimulation of contaminated soil (containing more than 38,000 mg/kg TPH) by addition of organic nutrients and obtained 100% removal of 2-3 ringed PAHs within the first 3 months.

On the other hand, bioaugmentation comprises the addition of either a pure oil-degrading strain, or a consortium of microorganisms, having contaminant degrading capacity; it is generally employed in cases where the indigenous bacterial community is either not efficient or not sufficient for biodegradation (Gardner and Gunsch 2020). Further, genetic bioaugmentation involves transfer of phylogenetically similar biodegradation genes (responsible for hydrocarbon metabolism) from an exogenous donor to indigenous microbes by means of horizontal gene transfer (HGT) of catabolic plasmids (Gardner and Gunsch 2020, Miri et al. 2019). Mrozik and Piotrowska-Seget (2010) demonstrated efficient bioaugmentation of contaminated sites with a mixture of PAHs (naphthalene, phenanthrene, anthracene, pyrene, dibenzoanthracene, benzopyrene) using a reconstructed microbial consortium. However, in case of bioaugmentation treatment, it is necessary that the environmental conditions and the microorganisms should be well controlled and characterized.

Ruberto et al. (2006) evaluated biodegradation of PAHs in Antarctic soils, and determined the effect of biostimulation with a complex organic source of nutrients (fish meal) with surfactant (Brij 700), the effect of bioaugmentation with a psychrotolerant PAH-degrading bacterial consortium, and the combination of both strategies. They observed that a combination of biostimulation and bioaugmentation caused a significant removal of phenanthrene after 56 days under Antarctic environmental conditions; however, when individual strategies were applied no significant reduction in phenanthrene concentration was observed.

Many normal and extreme bacterial species have been isolated and utilized as biodegraders for dealing with petroleum hydrocarbons (Xu et al. 2018); although degradation pathways of a variety of petroleum hydrocarbons (e.g. aliphatics and polyaromatics) employ oxidizing reactions, these pathways differ due to the specific oxygenases found in different bacterial species. For example, certain bacteria metabolize specific alkanes, while others break down aromatic or resin fractions of hydrocarbons; thus, this phenomenon is related to the chemical structure of petroleum hydrocarbon components. Table 8.1 details a few petroleum hydrocarbon degraders. Xu et al. (2018) enlisted more than 79 bacterial genera that are capable of degrading petroleum hydrocarbons including *Achromobacter, Acinetobacter, Alkanindiges, Alteromonas, Arthrobacter, Burkholderia, Dietzia, Enterobacter, Kocuria, Marinobacter, Mycobacterium, Pandoraea, Pseudomonas, Staphylococcus, Streptobacillus, Streptococcus,* and *Rhodococcus*. They reported that some obligate hydrocarbonoclastic bacteria (OHCB), such as *Alcanivorax, Marinobacter, Thallassolituus, Cycloclasticus,* and *Oleispira*, that show a low abundance or undetectable status before pollution, could be dominant after petroleum oil contamination. This suggested that the fate of the environment is crucial to these bacteria and it influences degradation of petroleum hydrocarbons. Certain bacteria have abilities to degrade wide ranges of petroleum hydrocarbons, for example, *Dietzia* sp. DQ12-45-1b utilizes *n*-alkanes (C6-C40) and other compounds as the sole carbon sources (Wang et al. 2011). *Achromobacter xylosoxidans* DN002 degrades a variety of monoaromatic and polyaromatic hydrocarbons (Ma et al. 2015). The genus *Pseudomonas* has been considered accountable for biodegradation of the PAHs that contain three and four ring (Abdel-Shafy and Mansour 2018). Amenu (2014) reported that *Pseudomonas* sp. S3 and F3 efficiently degrade naphthalene under optimal temperature; they observed 61.11% of naphthalene biodegradation efficiency for F3 after 7 days. Pawar et al. (2013) reported that *Pseudomonas putida* strain had a gene for the degradation of naphthalene (NAH7). Bamforth and Singleton (2005) reported that *Bacillus* sp., *Pseudomonas* sp., *Acinetobacter calcoaceticus*, *Micrococcus* sp., *Nocardia erythropolis, Candida Antarctica, Ochrobactrum* sp., *Serratia marcescens, Acinetobacter* sp., *Alcaligenes odorans, Candida tropicalis*, and *Arthrobacter* sp. degraded alkane. Bacteria such as *Pseudomonas* sp., *Brevibacillus* sp., *B. stearothermophilus, Bacillus* sp., *Corynebacterium* sp., *Vibrio* sp., *Ochrobactrum* sp., and *Achromobacter* sp. degrade mono-aromatic hydrocarbons, while poly-aromatic hydrocarbons are degraded by *Alcaligenes odorans, Achromobacter* sp., *Mycobacterium* sp., *Sphingomonas paucimobilis, Mycobacterium flavescens, Pseudomonas* sp., *Arthrobacter* sp., *Bacillus* sp., *Rhodococcus* sp., *Xanthomonas* sp., *Alcaligenes,* and *Burkholderia cepacia* (Bamforth and Singleton 2005). They also mentioned that *Vibrionaceae, Pseudomonas* sp., *Moraxella* sp., and *Enterobacteriaceae* effectively degraded resin (Bamforth and Singleton 2005).

TABLE 8.1 Petroleum hydrocarbon degradation by various bacteria.

Chemical Name	Structure	Bacteria/ Bacterial Consortium	Phylum	Degradation Efficiency	Time	References
n-Hexane		*Ochrobactrum* sp.	α-Proteobacteria	92.3%	28 days	Zhong et al. 2020
		Bacillus sp.	Firmicutes	93.4%	28 days	Zhong et al. 2020
Hexadecane		*Ochrobactrum* sp.	α-Proteobacteria	85.4%	28 days	Zhong et al. 2020
		Bacillus sp.	Firmicutes	92.6%,	28 days	Zhong et al. 2020
		Alcanivorax borkumensis	γ-Proteobacteria	96.7%	7 days	Nekouei et al. 2017
		Bacillus flexus	Firmicutes	84%	20 days	Gomes et al. 2016
Benzene		*Alcanivorax borkumensis*	γ-Proteobacteria	93.9%	7 days	Nekouei and Nekouei 2017
		Bacillus subtilis	Firmicutes	93%	5 days	Mukherjee and Bordoloi 2012
		Pseudomonas aeruginosa	γ-Proteobacteria	75%	5 days	Mukherjee and Bordoloi 2012
		Marinobacter vinifirmus	γ-Proteobacteria	100%	3 days	Berlendis et al. 2010
Ethylbenzene	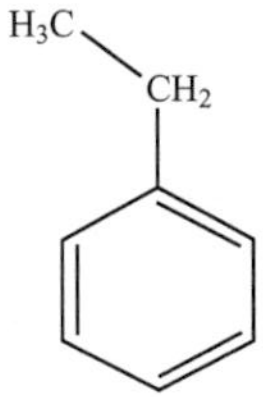	*Alcanivorax borkumensis*	γ-Proteobacteria	89.0%	7 days	Nekouei and Nekouei 2017
		Marinobacter vinifirmus	γ-Proteobacteria	65%	7 days	Berlendis et al. 2010
		Marinobacter hydrocarbonoclasticus	γ-Proteobacteria	60%	7 days	Berlendis et al. 2010
o-Xylene	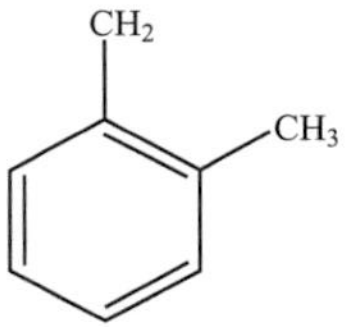	*Alcanivorax borkumensis*	γ-Proteobacteria	90.0%	7 days	Nekouei and Nekouei 2017

Contd.

TABLE 8.1 Contd.

Chemical Name	Structure	Bacteria/ Bacterial Consortium	Phylum	Degradation Efficiency	Time	References
m-Xylene	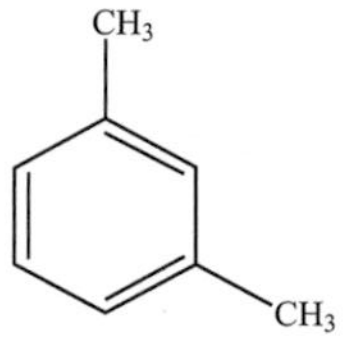	*Alcanivorax borkumensis*	γ-Proteobacteria	89.3%	7 days	Nekouei and Nekouei 2017
		Bacillus subtilis	Firmicutes	80%	5 days	Mukherjee and Bordoloi 2012
		Pseudomonas aeruginosa	γ-Proteobacteria	88%	5 days	Mukherjee and Bordoloi 2012
p-Xylene		*Alcanivorax borkumensis*	γ-Proteobacteria	89.0%	7 days	Nekouei and Nekouei 2017
		Marinobacter hydrocarbonoclasticus	γ-Proteobacteria	70%	7 days	Berlendis et al. 2010
Toluene	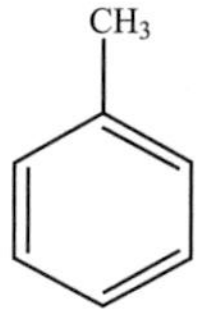	*Alcanivorax borkumensis*	γ-Proteobacteria	96.8%	7 days	Nekouei and Nekouei 2017
		Ochrobactrum sp.	α-Proteobacteria	53.7%	7 days	Zhong et al. 2020
		Bacillus subtilis	Firmicutes	72%	5 days	Mukherjee and Bordoloi 2012
		Marinobacter vinifirmus	γ-Proteobacteria	100%	3 days	Berlendis et al. 2010
Phenol	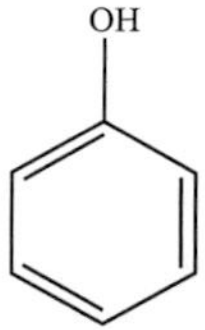	*Modicisalibacter tunisiensis*	γ-Proteobacteria	100%	20 days	Gomes et al. 2016
		M. tunisiensis	γ-Proteobacteria	100%	5 days	Bonfá et al. 2013
		Arhodomonas aquaeolei	γ-Proteobacteria	88%	8 days	Bonfá et al. 2013
		Halomonas organivorans	γ-Proteobacteria	75%	8 days	Bonfá et al. 2013

Contd.

TABLE 8.1 Contd.

Chemical Name	Structure	Bacteria/ Bacterial Consortium	Phylum	Degradation Efficiency	Time	References
Fluorene	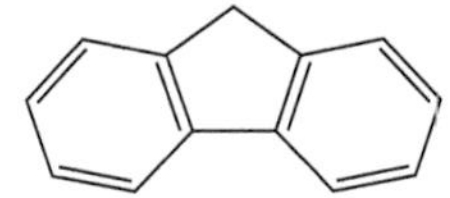	*Stenotrophomonas maltophilia*	γ-Proteobacteria	47.9%	45 days	Kumari et al. 2018
Benzo[a] pyrene	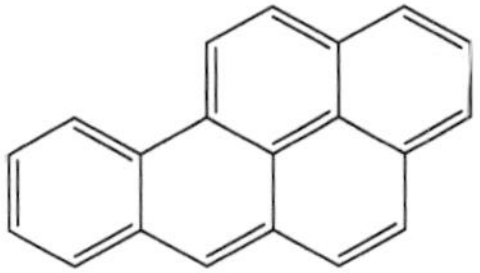	*Mesoflavibacter zeaxanthinifaciens*	Bacteroidetes	86%	42 days	Okai et al. 2015
		Olleya sp.	Bacteroidetes	79%	42 days	Okai et al. 2015
Benzo[b] fluoranthene	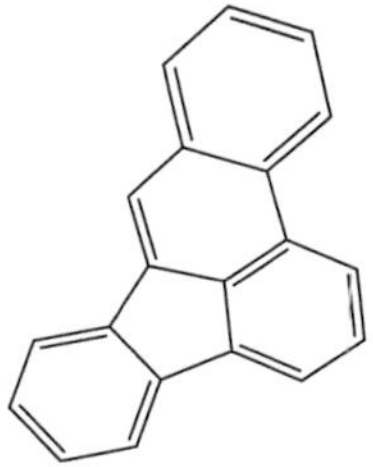	*Pseudomonas aeruginosa*	γ-Proteobacteria	61.2%	45 days	Kumari et al. 2018
Phenanthrene	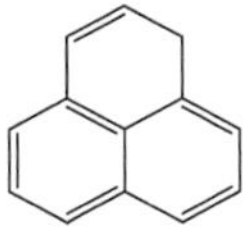	*Marinobacter flavimaris*	γ-Proteobacteria	87%	20 days	Gomes et al. 2016
		Pseudomonas aeruginosa	γ-Proteobacteria	67.1%	45 days	Kumari et al. 2018
Pyrene	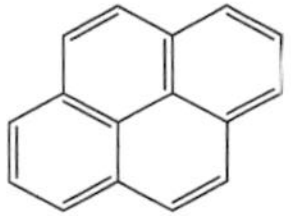	*Marinobacter flavimaris*	γ-Proteobacteria	78%	20 days	Gomes et al. 2016
		Mycobacterium flavescens	Actinobacteria	89.4%	14 days	Dean-Ross et al. 2002

Contd.

TABLE 8.1 Contd.

Chemical Name	Structure	Bacteria/ Bacterial Consortium	Phylum	Degradation Efficiency	Time	References
Naphthalene	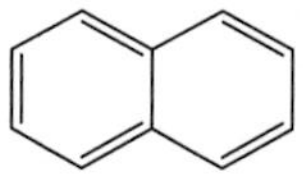	*Nitratireductor* sp. *Microbacterium esteraromaticum*	α-Proteobacteria Actinobacteria	87% 81.40%	20 days 45 days	Gomes et al. 2016 Kumari et al. 2018
Crude oil		*Rhodococcus ruber*	Actinobacteria	90%	7 days	Parach et al. 2017
Bacterial Consortium						
Naphthalene Fluorene Phenanthrene Benzo(b) fluoranthene		*Stenotrophomonas maltophilia, Ochrobactrum anthropi, P. mendocina, Microbacterium esteraromaticum and P. aeruginosa*	γ-Proteobacteria α-Proteobacteria γ-Proteobacteria Actinobacteria γ-Proteobacteria	89.1% 63.8% 81% 72.8%	45 days	Kumari et al. 2018
Naphthalene Phenanthrene Pyrene Crude oil		*Pseudomonas* sp., *Bacillus* sp., *Ochrobactrum* sp. *and Pseudomonas* sp.	γ-Proteobacteria Firmicutes α-Proteobacteria γ-Proteobacteria	96% 93% 61% 48%	5 days	Shen et al. 2015
Benzene Toluene m-Xylene		*B. subtilis* *and P. aeruginosa*	Firmicutes γ-Proteobacteria	99% 96% 98%	5 days	Mukherjee and Bordoloi 2012
Naphthalene Phenanthrene Pyrene		*Rhodococcus* sp. *Gordonia* sp. *Pseudomonas monteili* and *Pseudomonas* sp.	Actinobacteria Actinobacteria γ-Proteobacteria γ-Proteobacteria	100% 100% 42%	1 day 3 days 10 days	Isaac et al. 2015
Diesel oil		*Joostella* *and Pseudomonas* *Joostella* *and Alcanivorax*	Bacteroidetes, γ-Proteobacteria Bacteroidetes, γ-Proteobacteria	99.2% 99.4%	10 days 10 days	Rizzo et al. 2018 Rizzo et al. 2018

Factors Influencing Bacterial Degradation of Petroleum Hydrocarbons

Several factors such as the composition or type of crude hydrocarbon, concentration, microbial inoculum, temperature, mixing intensity, and nutrients, potentially influence rates of oil biodegradation (Prince et al. 2017). Petroleum hydrocarbon degradation in the marine environment is typically influenced by the bioavailability of nutrients (nitrogen and phosphorus), organisms and environmental factors (temperature, pH, oxygen accessibility, and salinity). An increasing number of studies demonstrating how bioremediation potential of pre-adapted microbes could be enhanced by changing environmental factors have been cited by Shen et al. (2015).

Type of Contaminant and its Concentration

Petroleum hydrocarbons are composed of mixture of complicated substances; the degradation rate of any petroleum hydrocarbons compound by microorganisms essentially depends on the chemical configuration of these compounds and their concentration (Abdel-Shafy and Mansour 2018). In other words, the microbial degradation of petroleum hydrocarbons depends on its composition, i.e. aliphatics, branched, cyclo-aliphatics, and aromatics (monoaromatic hydrocarbon and polyaromatic hydrocarbon-PAH). The composition of the petroleum hydrocarbon and its inherent biodegradability is of vital importance when suitability of a remediation approach is considered. Overall, petroleum hydrocarbons are categorized into saturated, aromatic, resin, as well as asphaltenes.

Bacterial degradation of HCs depends on the length of the carbon chain. Ivshina et al. (2017) stated that the most active bacterial growth is observed in case of C_{13}-C_{19}, while gaseous n-alkanes (up to C_4), short-chain liquid alkanes (C_5-C_{10}), and hard paraffins (C_{30}-C_{44}) are difficult to degrade; cycloalkanes, mono- and polycyclic aromatic HCs are suggestively resistant to microbial degradation. Abdel-Shafy and Mansour (2018) reported that n-alkanes (with average chain from C_{10} to C_{25}) are the most chosen compounds by the microbes that are easily biodegradable; contrarily, petroleum products with short chain are more toxic and less degradable. According to them, alkanes with long chain from C_{25} to C_{40} are water-repellent solids and are difficult of degrade due to their lack of dissolution and thereby lack of biological availability.

The rate of biodegradation of aliphatic compounds decreases progressively with increasing branching of the molecular structure: non-branched > less-branched > branched (Gros et. al 2014). Logeshwaran et al. (2018) state the order for susceptibility for microbial degradation of petroleum hydrocarbons as: n-alkanes > branched alkanes > low-molecular weight aromatics > cyclic alkanes. *Acinetobacter* sp., *Pseudomonas* sp., *Ralstonia* sp. and *Microbacterium* sp. are reported to degrade aromatic hydrocarbons (Simarro et al. 2013). Table 8.1 details the structure of various hydrocarbons and enlists their bacterial degradation. Abdel-Shafy and Mansour (2018) too reported that alkanes with branched chain as well as cyclic chain are considerably slowly biodegraded than the symmetric n-alkanes; further, highly condensed aromatic, tars, cycloparaffinic structures, asphaltic materials, and bitumen have higher boiling point, and are difficult to be degraded by microbes.

Brassington et al. (2007) suggested that the kind of materials remaining from the biodegradation of petroleum could be rather similar to humic materials.

Solubility has a major role to play in the degradation rate of petroleum hydrocarbons. Abdel-Shafy and Mansour (2018) described the biodegradation rate of the petroleum hydrocarbons that are extremely solubilized in the aqueous medium to be approximately proportionate to their concentration, while vice versa is true for the less aqueous hydrocarbons.

The microbial biodegradation capability tremendously depends on the hydrocarbon substrate concentration as it is used as carbon and energy source. Lower concentration of the hydrocarbon substrate would not provide adequate carbon for the growth of bacteria, while higher crude oil concentrations could cause accumulation of oil on the surface of aqueous phase results in decreasing the oxygen level that in turn would be inhibitory (Parach et al. 2017). A petroleum content >5% in a medium could cause a reduction in the action of microbes. A rise in concentration of oil leads to change in the C:N:P ratio, which could restrict oxygen consumption that could probably impede the process of biodegradation (Abdel-Shafy and Mansour 2018).

Individual vs. Consortium

Catalytic enzymes are known to vary with the bacteria (Xu et al. 2018). Therefore, with the variation in the catalytic enzyme composition, the role played by different indigenous bacteria at oil-contaminated sites would vary widely. According to Xu et al. (2018), most bacteria effectively degrade or utilize certain petroleum hydrocarbon components, while others are completely unavailable. Therefore, joint action of multiple bacteria at a petroleum hydrocarbon site would be more suitable to achieve the best environmental purification effect (Dombrowski et al. 2016).

While some reports emphasize on selective metabolization of certain petroleum hydrocarbons by single strains of bacteria, others emphasize on utilizing a microbial consortium of strains belonging to the same or dissimilar genera. For example, Al-Hawash et al. (2018) observed that a consortium showed more possibility of petroleum hydrocarbon degradation than the individual cultures. Different microbial species have distinct preferences for the degradation of hydrocarbons, for e.g. some prefer mono- or polynuclear aromatics, while others conjointly degrade both aromatics and alkanes as well as linear, branched, or cyclic alkanes. Xu et al. (2018) opined that bioremediation of complex hydrocarbon by individual strains of microorganism have narrow range of hydrocarbon metabolising enzymatic machinery that could lead to inefficient biodegradation; they suggest the use of microbial consortia, producing diverse source of enzymes and sharing complementary catabolic pathways leading to metabolism of range of hydrocarbon and improved biodegradation. Miri et al. (2019) also emphasized on the use of bacterial consortium as they are more efficient than the single strain; this is probably because of a wider catabolic potential and the enzymatic ability for the degradation of oil compounds. Wang et al. (2018) obtained an aboriginal bacterial consortium from Penglai oil spill accident (China) and reported that it had higher oil degradation efficiency as compared to the individual bacteria when dispersed in the marine ecosystem. Tao et al. (2017) utilized an indigenous bacterial consortium as well as exogenous bacteria (*Bacillus subtilis*) to effectively accelerate the degradation of crude oil. A halotolerant hydrocarbon utilizing bacterial consortium (HUBC) consisting of *Ochrobactrum* sp., *Stenotrophomonas maltophilia* and *Pseudomonas aeruginosa* has been reported to degrade crude oil (3% v/v with a degradation percentage as high as 83.49%; Varjani et al. 2015). Apart from indigenous population, addition of exogenous bacterial consortia often proves to be beneficial in oil degradation; for example, bioaugmentation with an artificial consortium containing *Aeromonas hydrophila, Alcaligenes xylosoxidans, Gordonia* sp., *Pseudomonas fluorescens, Pseudomonas putida, Rhodococcus equi, S. maltophilia,* and *Xanthomonas* sp. was reported to effectively biodegrade (89% efficiency; 365-day treatment) diesel oil-contaminated soil (Szulc et al. 2014). Thus, the use of bacterial consortia with multiple catabolic genes could be a rational strategy for hastening petroleum hydrocarbons degradation from contaminated environments (Xu et al. 2018).

Nonetheless, many common microbial consortia for petroleum hydrocarbon contamination are co-cultures of bacteria with microalgae and fungi.

Immobilization

Abdel-Shafy and Mansour (2018) explained that petroleum hydrocarbon degradation by microbial cells could be achieved by immobilizing cells. In cell immobilization, the targeted cell is immobilized in a limited space by variety of carriers such as alginate, mollusc shells and plant cells, facilitating the reuse of the organisms, making the process economic. In contrast, in enzyme immobilization, the enzyme is attached to a solid prop such as calcium alginate or activated poly (vinyl alcohol) (PVA) or activated polyethylene imine (PEI) (Abdel-Shafy and Mansour 2018). Immobilizing live cells could help avoid wash-out of cells, ensure a higher cell concentration in small volumes, embed the nutrients and resist external dispensable factors (Shen et al. 2015). Díaz et al. (2002) demonstrated the superiority of immobilized bacterial cells in promoting biodegradation rate of petroleum, as compared to free living cells within a broad domain of saltiness.

The process of immobilizing whole cells can be carried out in batch or continuous systems; in the continuous system, a packed bed reactor is generally employed for petroleum hydrocarbon biodegradation (Abdel-Shafy and Mansour 2018). Cunningham et al. (2004) demonstrated the PVA cryogelation technique using hydrogel and indigenous microorganisms for constructing laboratory bio-piles to compare bio-augmentation and bio-stimulation, reporting diesel degradation during 5 weeks. On the other hand, Rahman et al. (2006) studied the ability of alginate beads- immobilized bacteria for biodegradation of PH, observing no reduction in the degradation ability even after repeated reuse. Therefore, immobilization could be considered as an assured method for remediation of petroleum hydrocarbon contaminated sites.

Petroleum Surface Area

Most of the hydrocarbon degradation by microbes is reported to occur in or close to the interface of air and water at marine environment and the interface of air and soil in soil environment (Abdel-Shafy and Mansour 2018). The surface area of the petroleum hydrocarbon at the interface impacts biodegradation rate; the petroleum degradation increases with increase in surface area facing the air. With increase in availability of active sites on the surface layer, the tendency of the petroleum hydrocarbon to be degraded by microbes increases. Contrarily, a thick petroleum hydrocarbon blanket prevents the renewal of oxygen and nutrients for the microorganism (Abdel-Shafy and Mansour 2018). However, increased surface area concept for better degradation does not hold true for all HCs; for e.g. a tar ball being an insoluble substrate does not favour microbial growth (Atlas 1981).

According to Hassanshahian and Cappello (2013), dispersed oil particles are more vulnerable to biological attack (as compared to the undispersed ones) because they have a greater exposed surface area. They mention that addition of dispersants enhance the rate of natural biodegradation by forming water-in-oil emulsions (mousses) in seawater; the mousses get filled with the insoluble oil due to heavy wave action. Hassanshahian and Cappello (2013) further explain that these emulsions can form quickly in turbulent conditions and could contain 30 to 80% water. Atlas (1981) described that the degree of spreading of petroleum hydrocarbon and the surface area determines the oil available for microbial colonization by hydrocarbon-degrading microorganisms; further, in aquatic systems, the oil normally spreads, forming a thin slick. Availability of increased surface area should accelerate biodegradation (Atlas 1981).

A high surface area to volume ratio is desirable when it comes to bacterial degradation of HCs (Xu et al. 2018). Davis and Westlake (1978) reported that immobilization of cells onto inert surfaces increased available surface area to facilitate growth of biomass and also enhance degradation rate. Xua and Lu (2010) demonstrated that oil removal in a crude oil-contaminated soil was increased by application of hydrocarbon-degrading bacteria immobilized on peanut hull powder as biocarrier. This biocarrier provides large surface area and strong adsorption capability, in addition to improving oxygen diffusion and enhancing dehydrogenase activity in soil.

Surface Active Agents (Surfactants)

Surfactants are amphiphilic compounds that lower the interfacial tension (at the oil-water interface) and the surface tension of water, thereby favouring mass transport of hydrocarbons from the oil phase into the aqueous phase (Mohanty et al. 2013). Surfactants (even at low concentrations) reduce the repulsive forces between two dissimilar phases and help these two phases to mix and interact more easily; they change the surface tension and/or interfacial tension increasing solubility and bioavailability of hydrophobic compounds (Varjani and Upasani 2017). A few commonly used non-ionic surfactants include ethoxylates, ethylene and propylene oxide copolymers, and sorbitan esters (Cameotra and Makkar 2010). Commercially available ionic surfactants include fatty acids, ester sulfonates or sulfates (anionic), and quartenary ammonium salts (cationic).

According to Mohanty et al. (2013), surfactant-enhanced bioremediation (SEB) helps overcome bioavailability constraints encountered in biotransformation of non-aqueous phase liquid (NAPL) pollutants. Surfactants enhance the uptake of constituents through micellar solubilisation and emulsification; they also alter microbial cell surface characteristics. Even though there are abundant hydrocarbon degrading cultures in nature, due to the structural complexity of oil and culture specificities, complete biodegradation of oil is rarely achieved (even under favourable environmental conditions) (Mohanty et al. 2013). The success of a SEB assisted degradation process is dependent on the choice of appropriate surfactant type and dose. Nevertheless, hydrocarbons partitioned in micelles are not always readily bioavailable; here, the microorganism-surfactant interaction that is based on the type of PH is an essential factor to be considered (Mohanty et al. 2013). Apart from this, surfactant toxicity should also be considered.

Caemotra and Makkar (2010) reported improvement in biodegradation of naphthalene, phenanthrene, and anthracene by *E. coli* JM109 recombinant strains carrying naphthalene dioxygenase and regulatory genes cloned from *Pseudomonas fluorescens* N3 in the presence of surfactants such as Tween 60 and Triton X100. Kolomytseva et al. (2009) suggested the utilization of Tween 60 to enhance the fluorene biodegradation rates by *R. rhodochrous* VKM B-2469; Tween 60 works as an additional carbon source and decreases the fluorene toxicity to the bacterial cells.

Biosurfactants and Bioemulsifiers

Biosurfactants, a heterogeneous series of substances with chemically active surface, are generated by a broad diversity of microbes. These biological surfactants promote mobility, solubility, and remediation of pollutants (Abdel-Shafy and Mansour 2018). Biosurfactants, being amphipathic surface-active macromolecules (containing hydrophilic and hydrophobic regions), have good solubilisation, and increase the bioavailability of HCs (Miri et al. 2019, Doshi et al. 2018). Even more, biosurfactants help the microbes to use hydrocarbon as a source of carbon and energy by decreasing the surface tension of the medium (Parach et al. 2017). They increase the chance of direct contact between a bacteria and an oil droplet. Biosurfactants enhance degradation by facilitating microbial oil uptake and emulsifying the hydrocarbon. Cameotra and Singh (2008) achieved successful biodegradation of petroleum sludge using biosurfactants. In a study, Mapelli et al. (2017) observed that bacteria such as *Acinetobacter, Bacillus, Pseudomonas,* and *Alcanivorax* sp. predominated an oil-contaminated marine surface water site due to their glycolipid biosurfactant producing capacity. They stated that one of the main factors limiting the rate of HC degradation in seawater was low HC solubility; microorganisms increase and modulate HC bioavailability by producing biosurfactants and modifying cell membrane hydrophobicity, for e.g. Alcanivorax spp. (Mapelli et al. 2017). Biosurfactants overcome the toxicological disability of surfactants (Cameotra and Makkar 2010). Ornithine lipids and the subtilysin (produced by a gram-positive bacterium by *Bacillus subtilis*) are amongst the few most effective biosurfactants reported till date (Hua and Wang 2014).

Cameotra and Singh (2008) studied the effectiveness of biodegradation under varying nutritive element blends and a petroleum bio-surfactant synthesis; they reported that *Pseudomonas* possesses ability to utilize petroleum hydrocarbon substances (as a source of carbon and energy), and can produce biosurfactants. Fritsche and Hofrichter (2000) reported rhamnolipid bio-surfactant production by *Pseudomonas* sp., and its mechanism of forming micelles during the degradation of petroleum hydrocarbon substances. Abdel-Shafy and Mansour (2018) reported that *Pseudomonas* sp., *Aeromonas* sp., *Bacillus* sp., *Pseudomonas aeruginosa, Pseudomonas chlororaphis* and *Pseudomonas putida* are exceedingly studied for bio-surfactants of the glycolipid kind; further organisms such as *Pseudomonas aeruginosa* and *Rhodococcus erythropolis* have >90% hydrocarbon capability (6 weeks in aqueous medium; Abdel-Shafy and Mansour 2018).

Alternatively, addition of biosurfactant-producing bacteria could stimulate the growth of hydrocarbon-degrading bacterium strains, thereby enhancing their capability to biodegrade alkanes, alkenes, and aromatic compounds (Shi et al. 2019). Khan et al. (2017) reported that the bacterial strain *Pseudomonas rhizosphaerae* BP3 was incapable of producing biosurfactants; however, when a biosurfactant producing hydrocarbon degrading bacteria *Pseudomonas poae* BA1 was used, the degradation of hydrocarbons increased from 16 to 28%. Rhamnolipids from *Pseudomonas aeruginosa* 57SJ, *Renibacterium salmoninarum* 27BN, *P. putida* Z1 BN, and *P. aeruginosa* PA1 are used for bioremediation, while rhamnolipid of *P. aeruginosa* GL1 is a surface-active agent. Arthrofactin from *Arthrobacter* is an oil displacement agent (Caemeotra and Makkar 2010). Abdel-Shafy and Mansour (2018) observed a blend of 11 compounds (rhamnolipid) produced by microbes that could degrade > 91% of petroleum hydrocarbon compounds in a polluted soil by petroleum sludge on a 1% (v/v) basis within 5 weeks. Using this crude bio-surfactant led to more efficient hydrocarbon biodegradation (91-95% elimination of petroleum hydrocarbon in 4 weeks). *Pseudomonas* and *Pseudomonas aeruginosa* have been extensively studied for their hydrocarbon degrading ability and surfactant (glycolipid) production (Abdel-Shafy and Mansour 2018). Biosurfactant production has also been reported for other microbes, for e.g. glycolipid production by *Pseudomonas chlororaphis, Aeromonas* spp., *Bacillus* sp. and *Pseudomonas putida*, Sophorolipids by *Candida bombicola*, Rhamnolipids by *Pseudomonas aeruginosa* and *Pseudomonas fluorescens*, Lipomannan by *Candida tropicalis*, and Surfactin by *Bacillus subtilis* (Abdel-Shafy and Mansour 2018). Potential candidates for use in bioremediation applications or in biosurfactant production include: *Bacillus cereus* 28BN (a rhamnolipid-producing strain (Abdel-Shafy and Mansour 2018). *Rhodococcus wratislaviensis* BN38 (glycolipid-producing, grows on 2% n-hexadecane), *Rhodococcus ruber* (having enhanced crude oil desorption and mobilization ability with biosurfactant production), and *P. putida* (capable of utilizing two-, three- and four-ring PAHs).

Nazina et al. (2003) isolated 20 microbes with capacity to produce surface-active compounds in media with individual hydrocarbons, lower alcohols, and fatty acids from the Daqing oil field; according to them, aerobic saprotrophic bacteria belonging to the genera *Bacillus, Brevibacillus, Rhodococcus, Dietzia, Kocuria, Gordonia, Cellulomonas, Clavibacter, Pseudomonas*, and *Acinetobacter* could noticeably decrease the surface tension of cultivation media. Here, the biosurfactants were produced by *B. cereus, R. ruber,* and *Bacillus licheniformis*, while *Rhodococcus, Dietzia, Kocuria, and Gordonia* produced exopolysaccharides in media with hydrocarbons (Nazina et al. 2003). Urum et al. (2004, 2005, 2006) ascertained ability of aqueous biosurfactant solutions (aescin, lecithin, rhamnolipid, saponin, and tannin) in washing crude oil contaminated soil, stating that most of the action of the biosurfactant involved reduction of surface and inter facial tensions rather than the solubilisation and emulsification effects. According to them, chemical surfactants (e.g. SDS) were more effective for the aliphatics than aromatic hydrocarbons, whereas biosurfactants had more preference for the aromatic hydrocarbons than for the aliphatic hydrocarbons. Menezes et al. (2005) carried out surface tension and the E24 index results on bacteria from oil contaminated sites in USA and China, and isolated *B. cereus, B. sphaericus, B. fusiformis, Acinetobacter junii,* a non-cultured bacterium, *Pseudomonas* sp., and *B. pumilus* that displayed substantial potential for production of biosurfactants and could be applied in the bioremediation of soils contaminated with petroleum hydrocarbons.

In a unique study, Bodour et al. (2003) realized that the distribution of bacterium in contaminated sites was dependent on soil conditions, wherein the Gram-positive biosurfactant-producing isolates were located in the heavy metal- or uncontaminated soils, and the Gram-negative isolates were from hydrocarbon- or co-contaminated soils. Plaza et al. (2006) isolated bacteria from petroleum hydrocarbon-contaminated soils that produced biosurfactants/bioemulsifiers under thermophilic conditions.

Biosurfactants have more adaptable physicochemical properties making them suitable for applications in the oil industry, which explains why the large majority of the biosurfactants produced (estimated to be of the order of 400-500 tons yr^{-1}, including captive use for tertiary oil recovery or tank cleaning) are used in petroleum-related applications (Cameotra and Makkar 2010). They are used in tank oil recovery, oil spill management, microbial enhanced oil recovery, and as heavy oil dispersants and demulsifiers (Cameotra and Makkar 2010).

Xu et al. (2018) explained bacteria to have evolved counter-measures against petroleum contaminants, i.e. improved adhesion ability of cells by altering their surface components and secreting bioemulsifier to enhance their access to target hydrocarbon substrates. These bacteria have incredible prospective as environmental remediation agents, accelerating removal of petroleum hydrocarbon pollutants from the environment.

Environmental Factors

Effect of Pressure

Recent expansion in deep-sea drilling demands a better understanding of the effect of low temperature and increasing pressure on hydrocarbon degradation by microbes (Marietou et al. 2018). Pressure is a slightly limited factor as it is related mostly to the deep-sea environment. High pressure of deep-sea environment causes microbial growth deterrence, affecting the rate of petroleum hydrocarbon degradation. Pressure of 10 MPa (comparable to that in the DWH plume) is known to inhibit the growth of three *Alcanivorax* species on dodecane (as a sole carbon source); even 5 MPa pressure causes a significant reduction in cell replication. This happens due to synthesis of energy intensive osmolyte ectoine which is a pressure responsive compound (piezolyte) (Mapelli et al. 2017).

According to Schedler et al. (2014), *R. qingshengii* TUHH-12 degraded 0.035 mM/h n-hexadecane at ambient pressure (17 to 43 h), while at high pressure it could degrade 0.019 mM/h (16 to 44 h). They also reported utilization of a variety of aromatic compounds including biphenyl, anthracene, phenanthrene, naphthalene, toluene, cyclohexane and 1,3,5-trimethylbenzene, as carbon sources for growth by a model strain *S. yanoikuyae* B1; essentially, the growth of *S. yanoikuyae* B1 on naphthalene was strongly inhibited by high pressure. In their study after 66 h of incubation at elevated pressure, the initially colourless culture medium turned brown, while at ambient pressure the culture showed no change of colour. According to them, the first significant effects of high pressure on cellular components and processes of bacteria were found to commence at 20 MPa, affecting the RNA transcription. At pressures above 100 MPa, modifications in membrane fluidity were observed, while pressure above 400 MPa caused protein denaturation. In the experiment carried out by Schedler et al. (2014), at 0.1 MPa, *S. yanoikuyae* B1 was able to grow with naphthalene, whereas at 13.5 MPa there was no growth at all; in fact, after 66 h of incubation, cells were no longer viable. They concluded that *S. yanoikuyae* B1 was a piezosensitive strain that grew best and utilized naphthalene at an optimal rate at ambient pressure; however, it also metabolized naphthalene at 13.9 MPa, although this was much slower than that at 0.1 MPa.

Marietou et al. (2018) reported that pressure acts synergistically with low temperature to slow microbial growth; this in turn could slow down oil degradation in deep-sea environments.

Effect of Temperature

The ambient temperature influences the properties of hydrocarbons (oils) as well as occurrence of microorganisms and their activity (Shen et al. 2015); therefore, ambient temperature plays a critical role in case of oil remediation in the environment (especially spills). Temperature also influences the solubility of hydrocarbons. Although hydrocarbon biodegradation could occur at various temperatures, the rate of biodegradation generally decreases with the decreasing temperature (Das and Chandran 2011). Atlas (1975) observed that the viscosity of the oil increased at low temperatures; however, the volatility of the toxic low-molecular weight hydrocarbons was reduced (at lower temperatures), delaying the onset of biodegradation. Highest HC degradation rates generally occur at temperatures ranging from 30-40°C in the soil, which is much different from freshwater (20-30°C) and marine environments (15-20°C) (Das and Chandran 2011, Mahmoud and Bagey 2018).

Low temperatures change the properties of oil in a spill (increased oil viscosity), which in turn affects physiology of microorganisms, causing retardation in biodegradation. Nevertheless, significant biodegradation of HCs are also reported in psychrophilic environments in temperate regions; for example, Cowell et al. (1973) showed that temperature did not seem to be a limiting factor for petroleum degradation in the cold marine environment. They mentioned that there was an indigenous cold-adapted microbial community capable of utilizing hydrocarbons.

Effect of pH

Shen et al. (2015) stated that pH is an important factor that affects ability of bacteria in terms of nutrient utilization, interaction with the organic pollutants on bacterial surface, extracellular enzyme production capability, and stability of enzymatic compounds. The biodegradation of hydrocarbons could lead to production of organic acids and other metabolic products, which probably cause a reduction in pH. The optimum pH depends on type of microbe being used for HC degradation. Abdel-Shafy and Mansour (2018) reported that indigenous bacteria, fungi, and yeast could be sustained alive at pH 2; they also described alkaliphiles (that existed within the alkaline lakes at pH ranges of 7.5 to 10). These alkaliphiles included bacteria that grew well at pH > 9 and had slow growth at pH 6.5. Shen et al. (2015) reported alkaline pH ranging from 7.0 to 8.0 (like that of natural seawater) to be most suitable for petroleum hydrocarbon degradation. Similarly, Parach et al. (2017) observed that the optimum pH for maximum activity of the *Rhodococcus ruber* KE1 growth was observed at pH 8.5.

Pathak and Jaroli (2014) proposed that microbes present in polluted regions had efficiency for PAHs metabolism at high pH. Contrarily, Abdel-Shafy and Mansour (2018) recently reported the addition of calcium-containing inorganic mineral such as carbonates and oxides or hydroxides to increase the pH; they investigated phenanthrene degradation in an aquatic system with *Burkholderia cocovenenans* (isolated from the soil polluted by petroleum) at pH 5.5 to 7.5. According to them, the rate of phenanthrene degradation by bacteria was 40% after 16 days at pH 5.5; however, removal of phenanthrene was 80% at circumneutral pH values. Abdel-Shafy and Mansour (2018) further stated that the pH of the growth medium was a predominant factor influencing the efficacy of BA2 *Sphingomonas paucimobilis*; biodegradation of anthracene and phenanthrene was reduced at pH 5.2. Abdel-Shafy and Mansour (2018) also observed PAH biodegradation at a highly acidic soil (polluted with coal tar) of pH 2. As per their study, the local microbes in soil near the coal stack competently decreased the amount of naphthalene to a remarkable 50% (>28 days duration). Phenanthrene and anthracene removal rates ranged between 10-20%. They isolated several microorganisms that not only reduced the pH of the aqueous medium (from 9 to 6.5; 24h), but also degraded naphthalene from the environment (using it as a carbon source). Abdel-Shafy and Mansour (2018) reported that the efficacy of the microbes degrading naphthalene (*Pseudomonas fluorescens* and *Pseudomonas frederiksbergensis*) strongly reduced at high pH values.

Effect of Oxygen

Fungi, bacteria and yeast are considered as dynamic microbes involved in biodegradation of petroleum aliphatic and aromatic hydrocarbons; they can biodegrade petroleum hydrocarbons either aerobically or anaerobically. Oxygen has a significant influence on microbial oxidation of various hydrocarbons in a natural ecosystem. According to Mahmoud and Bagey (2018), the presence of a consortium, having broad and varied metabolic activity in presence or absence of oxygen, is rudimentary for degradation of complex mixtures of petroleum hydrocarbons (e.g.

oil degradation in soil, freshwater or marine environment). Thus, oxygen availability is an essential factor in determining the pathway of the HC breakdown (Mahmoud and Bagey 2018).

Generally, the upper/surface layer is exposed to air. Despite this, at times the hydrocarbon forms a thick layer on the surface (in both marine and freshwater ecosystems); this leads to a state of oxygen restriction (Mahmoud and Bagey 2018), which could influence the occurrence, growth and HC degrading efficiency of the existing microbes. Abdel-Shafy and Mansour (2018) clearly explained that the availability of dissolved oxygen is vital for microbial respiration. The dissolved molecular oxygen is important to the overall pathways of biodegradation. Due to the continuous air and water interface as well as the activity of wind and waves, the surface waters (seas, oceans, lakes, and, harbours) have unlimited supplies of oxygen, i.e. surface waters are saturated with dissolved oxygen (Abdel-Shafy et al. 1994). However, the dissolved oxygen decreases with depth, and thus, increased depth often leads to reduced biodegradation rate.

Furthermore, there is depletion in oxygen at the bottommost layers of water and sediment; thus in this case, the HC biodegradation takes place without oxygen, anaerobically. Abdel-Shafy and Mansour (2018) explained that petroleum hydrocarbon substances, dispersing and reaching to the deepest seas and oceans, are generally covered by sediment, and take very long time to be degraded by microorganisms. The heavy pool of petroleum on the top of water surface has a blanketing effect at interface of air and water, hindering the renewal of oxygen. Wave-motion and mechanical aeration of waters is also known to influence HC degradation; additionally, injecting hydrogen peroxide, soil venting, air sparging, and providing the necessary amounts of oxygen via injection could also change the degradation scenario (Abdel-Shafy and Mansour 2018).

The type of microbe chosen (aerobic or anaerobic) plays the most essential role along with the bacterial mass. Das and Chandran (2011) described bacterial degradation of a variety of HC (benzene, toluene, alkanes, etc.) in freshwater, marine environments, and soil.

Effect of Nutrients

Nutrients play a vital role in successful biodegradation of hydrocarbon pollutants, especially nitrogen, phosphorus, and at times iron; in fact, nutrient could become a limiting factor affecting the biodegradation processes (Das and Chandran 2011). Atlas (1975) studied oil spill in marine and freshwater environments and reported that the supply of carbon significantly increased, while the availability of nitrogen and phosphorus became the limiting factor for oil degradation. This is more obvious in marine environments due to low levels of nitrogen and phosphorous in seawater (Abdel-Shafy and Mansour 2018). An easy accessibility of a microbe to nutrient is profitable to degradation. The aerobic microbes derive benefits from nutritive elements such as phosphorus, nitrogen, as well as traces of potassium, calcium, sulphur, magnesium, iron, and manganese as micronutrients (but nitrogen and phosphorus are considered necessary; Abdel-Shafy and Mansour 2018). However, petroleum hydrocarbons do not contain significant amount of some nutrients (viz. such as nitrogen and phosphorous) required for microbial growth; therefore, supplementation of nutrients in the form of fertilizers is often recommended (Das and Chandran 2011). Atlas and Hazen (2011) reported that the addition of oleophilic fertiliser enhances the polycyclic aromatic hydrocarbons (PAH) and aliphatic hydrocarbon biodegradation rate by a factor of two and five, respectively, in Exxon Valdez. They observed that the rate of oil degradation was a function of the ratio of nitrogen (N): biodegradable oil and time. They obtained broad degradation of both polynuclear aromatic and aliphatic compounds in the oil. According to Atlas and Hazen (2011), from 1989 to 1991, about 107 000 lbs (48 600 kg) of N was applied in the form of fertilizers, at 2237 different shoreline areas of Prince William Sound; thereafter, most of the oil had been removed from shorelines by 1992.

On the other hand, freshwater wetlands are generally nutrient deficient (due to uptake of nutrients by the plants; thus, addition of nutrients seems requisite to enhance the biodegradation of oil pollutant in freshwater environments (Das and Chandran 2011).

Excessive nutrient concentrations are also reported to inhibit the biodegradation activity (Chaillan et al. 2006). For instance, high NPK levels could inhibit biodegradation of hydrocarbons (Chaîneau et al. 2005, Oudot et al. 1998), especially on aromatics (Carmichael and Pfaender 1997). Excessive nutrients disturb the C: N: P ratio which leads to oxygen limitations (Varjani and Upasani 2017). Das and Chandran (2011) emphasized on the effectiveness of fertilizers for the crude oil bioremediation in subarctic intertidal sediments, and the use of poultry manure as organic fertilizer in contaminated soil.

Effect of Salinity/Sodium Chloride (NaCl)

Salinity and pressure are considered as key factors for biodegradation of HCs in the estuaries and the deep seas; however, estuaries could vary in their salinity as compared to the sea and ocean (Abdel-Shafy and Mansour 2018). The salinity requirement of the indigenous microbes in each water body varies. Inorganic salts are requisite for microbial proliferation as they promote enzyme reaction, maintain cell membrane equilibrium, and regulate osmotic pressure. Shen et al. (2015) observed that in case of low salinity, increase of salt concentration could enhance biodegradation to a certain extent; however, on continued increase of NaCl concentration, higher degradation effciency would not be achieved. This can be explained by the fact that each organism has a specific optimum salinity for HC degradation; organisms belonging to only few genera have been found to be capable of degrading hydrocarbons under elevated salinity ($>10\%$ and even saturation). Members of the genera *Alcanivorax, Marinobacter, Halomonas, Halobacterium, Haloferax, Halococcus,* and *Haloarcula* are found in saline environments (Paniagua-Michel and Fathepure 2018).

Mechanism of Hydrocarbon Substance Degradation

Microbes use organic contaminants as a source of carbon and electrons (which the organisms can extract to obtain energy) (Miri et al. 2019, Rajendran et al. 2016). Generally, the microorganisms gain energy by catalysing energy-producing chemical reactions involving breaking of chemical bonds and transferring electrons away from the contaminant. This type of reaction is called *oxidation-reduction reaction*, wherein the organic contaminant is *oxidized* (loses electrons); further, the chemical that gains the electrons is *reduced*. The contaminant is called the *electron donor*, while the electron recipient is called the *electron acceptor* (Rajendran et al. 2016). Microbes can degrade pollutants under aerobic (in the presence of oxygen) and anaerobic (in absence of oxygen) conditions (Figure 8.1). Under both conditions (aerobic and anaerobic), oxidation involves electron transfer between donors and acceptors in the respiratory chain. The aerobic microbes use oxygen (O_2) as the final electron acceptor, while anaerobic microorganisms use sulphate, nitrate, manganese, iron and organic intermediates as electron acceptors (Miri et al. 2019). Under anaerobic conditions, microbes can carry out oxidation of substrates by electron transfer to a suitable acceptor. Oxidation is carried out using various respiratory pathways including nitrate respiration (denitrification), sulphate respiration in sulphate-reducing bacteria, iron and manganese respiration, acetogenesis and methanogenesis, as well as fermentation (Miri et al. 2019). Thus, under both conditions (aerobic and anaerobic), oxidation involves electron transfer between donors and acceptors in the respiratory chain; moreover, the proton transport across a membrane is used for the generation of adenosine triphosphate (ATP) and $NADPH_2$ as energy source for cell mass growth and other energy-requiring reactions (Miri et al. 2019).

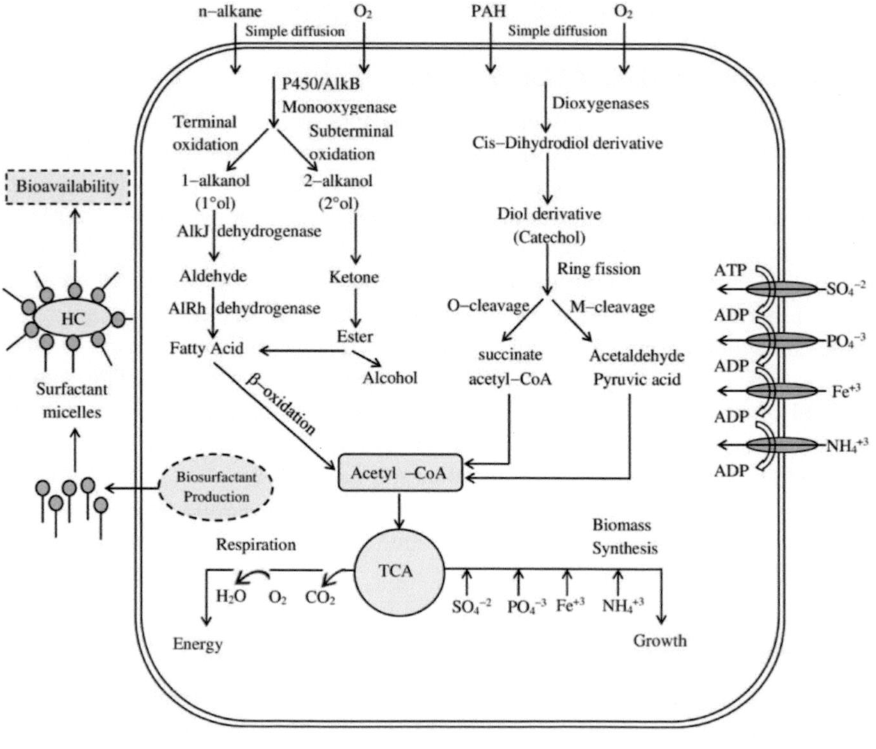

FIGURE 8.1 Petroleum hydrocarbon metabolism in a bacterial cell (based on Wang et al. 2020, Cruz et al. 2017, Logeshwaran et al. 2018, Varjani et al. 2017, Liu et al. 2019a,b).

Other mechanisms involved in biodegradation include: linking the microbial cells to the substrates and producing bio-surfactants (Das and Chandran 2011). Bacteria secrete surfactants to overcome the low bioavailability of substrate (reducing the surface tension of hydrophobic substrates) or by direct contact with hydrophobic droplet (by means of increasing the hydrophobicity of cell surface) (Fuentes et al. 2014).

In response to HC inputs, marine microbial communities often show ecological successions, wherein aliphatic HC degraders (e.g. *Alcanivorax* spp.) begin and subsequently enrichment of other microorganisms able to metabolize aromatic HCs (e.g. *Cycloclasticus* spp.) occurs (Mapelli et al. 2017). Overall, the three stages in the transformation of petroleum compounds by bacteria include: absorption of petroleum compounds on bacterial surface, followed by the transfer of these compounds to the cell membrane and finally their breakdown inside the bacterial cell (Mapelli et al. 2017). The HC degrading bacteria have hydrophobic surfaces which enable interactions with hydrocarbon molecules. Adsorption takes place on bacterial cell followed by a subsequent lateral diffusion transport mechanism that enables passage of hydrocarbon into the outer membrane. Solubilisation and dissolution are key factors in uptake and transport of hydrocarbon compounds. In case of high-molecular-weight (HMW) hydrocarbons in aquatic systems, there is low solubilisation or slow dissolution, but the uptake and rate of metabolism is slower as compared to low-molecular-weight (LMW) hydrocarbons. Both passive and active transportation systems for uptake of hydrocarbons have been found in microbes. Furthermore, in case of high oil concentration, uptake occurs through facilitated passive transport systems, while at lower concentrations active transport occurs (Abbasian et al. 2015). Although the degradation of petroleum components is an oxidizing process, the degradation mechanism differs due to the varying chemical structures and chemical constituents (saturated, aromatic, resin, and asphaltene fractions; Xue et al. 2015). The most easily biodegradable fraction of petroleum hydrocarbons is n-alkanes, and the least biodegradable are the PAHs and asphaltenes (Varjani and Upasani 2017).

Miri et al. (2019) described the concept of co-metabolism in petroleum hydrocarbon degradation; according to them, the petroleum hydrocarbon degradation rate is slow due to the complexity of this compound and its toxicity towards bacteria. In order to improve the degradation efficiency, an additional simple carbon source (chitosan, soluble starch, etc.) could be supplemented to the medium. For instance, *Pseudomonas zhaodongensis* degrades all the BTEX compounds from petroleum contaminated groundwater with xylene (Claro et al. 2018).

The entire complexity of the metabolic pathways involved in HC catabolism under *in situ* conditions is not yet understood, especially in relation to the environmental conditions. A few of the HC degradation pathways have been described below:

Aerobic Degradation Pathway

Biodegradation of petroleum hydrocarbon can occur as aerobic degradation pathway and anaerobic degradation pathway, but aerobic metabolism is the more common mode of biodegradation. Aerobic microorganisms use oxygen (O_2) as the final electron acceptor, while anaerobic microorganisms use substances, such as sulphate, nitrate, manganese, iron and organic intermediates, as electron acceptors. Biodegradation of organic contaminants occurring quickly and completely is generally achieved in the aerobic restraint (Abdel-Shafy and Mansour 2018). Various biochemical/enzymatic reactions viz. reduction, oxidation, hydroxylation and dehydrogenation are common for both aerobic and anaerobic pathways of degradation for petroleum hydrocarbon pollutants by microbes. In aerobic degradation, pollutants serve as a source of carbon and energy for microbes that are capable of producing degrading enzymes such as monooxygenases and dioxygenases (that catalyze reactions by incorporating O_2 into the structures; Miri et al. 2019). Abdel-Shafy and Mansour (2018) described that the chief requirement for aerobic bacterial degradation of PAH rings is dissolved oxygen; in the first step, both monooxygenase (aliphatic) and dioxygenase (aromatic) act as catalyst in the oxidation processes. Initially, the enzyme dioxygenase breaks the aromatic ring and then forms the first intermediate (cis-dihydrodiols). They described the involvement of a multicomponent enzymatic process to contain several proteins (indispensable for the functioning of the enzyme) and metal ions (that work as a cofactor). Fritsche and Hofrichter (2000) described aerobic biodegradation of hydrocarbon substances to involve intracellular crack of organic contaminants in the environment via oxidation; thereafter, the environmental paths of HC biodegradation gradually transform such organic contaminants to fragments through a centrally mediated metabolism (tricarboxylic acid cycle). They describe that the biological synthesis of the cell mass mainly occurs from metabolism of the centric precursors, for e.g. acetyl-CoA, pyruvate, and succinate. In addition, sugars that are required in different biological synthesis are mainly synthesized by gluconeogenesis process (Fritsche and Hofrichter 2000). Abdel-Shafy and Mansour (2018) stated that it is essential to consider particular enzymatic involvement of oxygenases (mono- and dioxygenases) as the first crack on the HC.

In a water column or surface water, biodegradation of hydrocarbon is primarily due to presence of aerobic hydrocarbon degrading bacteria that belongs to α and γ-Proteobacteria taxa. In order to overcome the problem of low bioavailibility of hydrocarbon substrate for uptake, bacterial cell secretes biosurfactants that emulsifies the oil droplets and facilitates uptake (Cruz et al. 2017). The n-alkanes are oxidized to primary alcohols, which are oxidized to aldehydes (that are further oxidized to carboxylic acids and by β-oxidation leading to the formation of carbon dioxide via Krebs cycle). In the aerobic pathway, the involvement of oxygen and enzymes such as mono- and dioxygenases (that add oxygen to the hydrocarbons) is remarkable. The n-alkanes are oxidised by an electron carrier dependent monooxygenase system, i.e. the alkane hydroxylases, to their corresponding alkane-1-ol. The initial intracellular attack on the organic hydrocarbon is an oxidative process; the activation and incorporation of oxygen is the enzymatic key reaction catalysed by oxygenases and peroxidases (Varjani and Upasani 2017). Bacteria degrading hydrocarbons need oxygen at two metabolic sites, i.e. during the initial attack on the HC substrate, and at the end of the respiratory chain (Varjani and Upasani 2017, Das and Chandran 2011).

Mahmoud and Bagey (2018) reported that the chief pathways and enzymes involved in petroleum hydrocarbon degradation are cytochrome P450 family and alkane monooxygenase. Pathways such as terminal oxidation, sub-terminal oxidation, ω-oxidation and β-oxidation are involved in the degradation of the diversity of compounds found in crude oil. Peripheral degradation pathways (the tricarboxylic acid (TCA) cycle) convert organic compounds systematically step-wise into intermediates of central intermediary metabolism (Varjani and Upasani 2017). According to Varjani and Upasani (2017), the biosynthesis of cell biomass occurs from central precursor metabolites viz. acetyl-CoA, succinate, pyruvate, etc.

Petroleum hydrocarbons have the following order of decreasing susceptibility to degradation: n-alkanes > branched alkanes > low-molecular-weight aromatics > cyclic alkanes (Chandra et al. 2013). Cyclohexane are reported to be degraded by *Brachymonas petroleovorans* from oil refinery wastewater sludge, while cyclic alkanes represent a component of crude oil which is resistant to microbial attack (Varjani and Upasani 2017). The terminal methyl group of cyclohexane makes primary attack complicated; however, the alkyl side chains of cycloalkanes facilitate the degradation (Varjani and Upasani 2017).

Varjani and Upasani (2017) proposed a pathway for biodegradation of cyclohexane, wherein monooxygenase initiates the oxidation of cyclohexane and converts it to cyclohexanol. Cyclohexanol is subsequently processed by dehydrogenases and hydrolases enzymes resulting in formation of adipic acid. Adipic acid then enters in an intermediary metabolism through β-oxidation. Varjani and Upasani (2017) reported that the aromatic HCs are less biodegradable than saturated HCs. Further, in case of aromatic HCs, after the oxygenolytic cleavage step, the -diol intermediate ring structure is cleaved by intradiol- or extradiol oxygenases. If there is a cleavage between the hydroxyl groups, then it is referred to as meta-cleavage and if the cleavage is before the hydroxyl groups, then it is an ortho-cleavage yielding protocatechuate and catechol that can be further converted to tricarboxylic acid cycle (TCA) intermediates (Varjani and Upasani 2017).

According to Varjani and Upasani (2017), the benzene ring is cleaved by microbes through different pathways by the enzymes viz. monooxygenases or dioxygenases. The metacleavage genes are located on plasmids. Genes for ortho-cleavage are located on chromosome; however, genetically modified orthogenes are located on catabolic plasmids (Varjani 2017). Varjani and Upasani (2017) described that the first enzymatic attack is initiated by phenol monooxygenase which results in catechol formation. Catechol then can be degraded through either ortho- or meta-cleavage through appropriate enzymes, and sequential reactions result in formation of intermediates of the TCA cycle.

Anaerobic Degradation Pathway

Anaerobic degradation pathways are significant because of the richness of anaerobic electron acceptors in seawater; however, they are relatively slower than the aerobic degradation pathways (Selvam and Thatheyus 2018). According to Varjani and Upasani (2017), anaerobic degradation of HCs is widespread; it occurs under nitrate-, iron-, manganese- and sulphate-reducing conditions, as well as methanogenic conditions in presence of suitable anaerobic bacteria. In other words, under anaerobic conditions, oxidation is carried out using different respiratory pathways as mentioned below:

- Nitrate respiration (denitrification) wherein numerous aerobic as well as many anaerobic bacteria can reduce nitrate to molecular nitrogen.
- Sulphate respiration in sulphate-reducing bacteria can reduce sulphate to hydrogen sulphide and mineralized BTEX.
- Iron and manganese respiration, where Fe^{+3} and Mn^{+4} are reduced to Mn^{+2} or Fe^{+2} and some compounds such as toluene can be mineralized.
- Acetogenesis and methanogenesis, where CO_2 is reduced to acetic acid or methane that can lead to reduction of some chemicals.
- Fermentation, wherein an organic compound is oxidized and its intermediates function as an electron acceptor (Miri et al. 2019).

The sulphate-reducing bacterium was the first anaerobe capable of utilizing n-alkanes under strictly anoxic conditions. Under anaerobic conditions, microorganisms utilize different electron acceptors in respiration (e.g. manganese, nitrate, sulphate and iron, instead of oxygen). Nitrate as electron acceptor is required in anaerobic degradation of alkanes and aromatics and changed over to N_2 or N_2O, and carbon is changed over to CO_2 (Mahmoud and Bagey 2018).

Yeung et al. (2013) studied bacterial degradation of petroleum hydrocarbon under anaerobic condition in contaminated fractured rock environments of cold regions (approx. 5°C). Sulphate reducers (*Desulfosarcina-Desulfococcus* group) from marine oil and gas cold seeps of Hydrate Ridge in the Gulf of Mexico are reported to degrade propane and butane (Jaekel et al. 2013).

Contrarily, Yoshikawa et al. (2017) observed that an integrated anaerobic-aerobic biodegradation was more suitable for complete degradation of seven volatile organic compounds (tetrachloroethylene, trichloroethylene, cis-dichloroethylene, vinyl chloride, benzene, toluene, and dichloromethane).

Alkanes Hydrocarbon Degradation

Petroleum hydrocarbon compounds differ in their susceptibility to microbial attack. As mentioned earlier, the n-alkanes are considered as the most readily degradable fraction amongst petroleum hydrocarbon constituents; the degradative pathway starts with oxidative attack at the terminal methyl group of alkanes forming alcohol, aldehyde, and fatty acid. Carboxylic acid (through β-oxidation) produces acetyl-CoA which passes into Krebs cycle (Xue et al. 2015, Selvam and Thatheyus 2018). The degradation of alkanes in the presence of oxygen by bacteria is as follows: n-alkanes > branched alkanes > low-molecular-weight aromatics > cyclic alkanes. Cyclic alkanes without terminal methyl group are resistant to microbial attack. The methyl group is the site of primary attack facilitating degradation (Varjani and Upasani 2017).

According to current concepts, there are three possible alkane oxidation pathways: (1) Monoterminal oxidation of the alkane methyl group resulting in a primary alcohol further converted to an aldehyde and monocarboxylic acid, (2) Subterminal oxidation to a corresponding methyl ketone via a secondary alcohol, and (3) Diterminal oxidation when terminal methyl groups of the alkane are oxidized simultaneously or consecutively to form dicarboxylic acids similar to the parent HC molecule in carbon chain length. Further, monoterminal oxidation is the most widespread HC assimilation pathway (Ivshina et al. 2017). Enzymes such as alkane monooxygenase and dehydrogenases (fatty alcohol and fatty aldehyde) are vital to biodegradation efficiency (Selvam and Thatheyus 2018).

In case of cycloalkane, an initial oxidation reaction involving monooxygenase occurs and the cycloalkane is converted to cyclohexanol; this cyclohexanol is subsequently processed by dehydrogenases and hydrolases enzymes resulting in formation of adipic acid. The adipic acid thereafter enters intermediary metabolism through β-oxidation.

Under anaerobic condition, the n-alkanes are degraded by sulphur-reducing bacterium strain Hxd3 in the following manner. First, carboxylation of alkyl chain occurs at the C3 position, followed by removal of two carbon atoms from carbon 1 and carbon 2 to produce fatty acid (which is either mineralized to carbon dioxide by β-oxidation, or gets assimilated into the microbial cell). However, the β-proteobacterium strain HxN1, which is a denitrifier, converts n-alkane and fumarate to succinate which then changes to CoA-thioester. β-oxidation and further degradation leads to regeneration of fumarate (recycled) and activation of n-alkane (Selvam and Thatheyus 2018).

Aromatic Hydrocarbon Degradation

Mahmoud and Bagey (2018) suggested aerobic degradation as the speedy method for complete elimination of aromatic hydrocarbons. Essentially, aromatic hydrocarbons, such as biphenyl, naphthalene, phenanthrene, toluene, xylene, and pyrene are less biodegradable than saturated hydrocarbons. In phenol degradation, the enzymatic attack is performed by phenol monooxygenase, which results in catechol formation. The catechol can then be degraded through either ortho- or meta-cleavage through appropriate enzymes and sequential reactions resulting in formation of intermediates of TCA cycle (Varjani and Upasani 2017).

Typically, PAHs are considered highly carcinogenic, mutagenic, and teratogenic substances that are resistant to bacterial degradation and have high affinity for deep sea sediments. Microbial enzymes such as oxygenase, dehydrogenase and peroxidases are reported to be capable of degrading aromatic compounds such as PAHs (Logeshwaran et al. 2018). These PAHs could be degraded under both aerobic and anaerobic conditions. Nevertheless, the rate of anaerobic PAH degradation is relatively slower as compared to aerobic degradation; moreover, comparatively little is known about the genes, plasmids and enzymes involved in anaerobic PAH degradation (Xue et al. 2015, Varjani et al. 2017, Logeshwaran et al. 2018). The degrading process of aromatic hydrocarbon in aerobic pathway involves the following steps: first, the aromatic hydrocarbon undergoes dioxygenation by ring hydroxylating dioxygenase (RHD) to dihydrodiol. In contrast, in the presence of toluene and the non-substrate o-xylene, *P. putida* mt-2 induces efflux pumps and several stress response related proteins are up-regulated, and antioxidant enzymes alkyl hydroperoxide reductase and catalase alleviate cellular oxidative damage caused by reactive oxygen species (Fuentes et al. 2014). Fuentes et al. (2014) explained that the xylene/toluene pathway from *P. putida* mt-2 starts with monooxygenation step. Dihydrodiol is degraded to a dihydroxydiol derivative like catechol, in the presence of dehydrogenases. Further, the diol-derivative undergoes a ring fission reaction via an ortho- and meta-cleavage. The orthocleavage produces Cis cis-muconic acid which further metabolizes to succinate and acetyl co-A, while metacleavage produces 2-hydroxymuconic semialdehyde that further metabolizes to acetaldehyde and pyruvic acid. These ortho and metaclevage products enter intermediary metabolic pool via β-oxidation pathway (Xue et al. 2015, Logeshwaran et al. 2018, Fuentes et al. 2014).

Contrarily, in anaerobic degradation, the aromatic compounds are oxidized to phenols/organic acids and transformed to fatty acids, which ultimately metabolize to methane and carbon dioxide by anaerobic bacteria such as sulphate-reducing bacteria (Varjani and Upasani 2017). Moreover, addition of fumarate, methylation of aromatic compound, followed by hydroxylation and carboxylation also occurs. These reactions yield ring saturation, β-oxidation and/or ring cleavage reaction, and produce central metabolites such as benzoyl-coA that are eventually incorporated in biomass or completely oxidized (Varjani and Upasani 2017).

Various microbes such as bacteria, fungi, and algae have the ability to degrade aromatic compounds such as PAHs. An aromatic hydrocarbon could serve as carbon source and a toxic signal for bacteria. Bacterial species belonging to the genera *Pseudomonas, Agrobacterium, Acenitobacter, Bacillus, Burkholderia, Rhodococcus, Rhodotorula, Mycobacterium, Paenibacillus* and *Sphingomonas* degrade PAHs (Logeshwaran et al. 2018). Investigations on marine and non-marine hypersaline environments reveal that *Halomonas* spp. degrade mainly phenolics and aromatic acids, while *Marinobacter* spp. degrade mainly aliphatic and PAHs. *Alcanivorax,* having relatively low salt tolerance, are n-alkane-utilizing specialists (Paniagua-Michel and Fathepure 2018).

Enzymes and Genes Involved in Petroleum Hydrocarbon Degradation Metabolism

Gkorezis et al. (2016) reported that the expression of genes involved in petroleum hydrocarbon degradation is tightly controlled; further, specific gene regulation mechanisms ensure that the genes involved in alkane degradation are expressed only under certain conditions, i.e. in the presence of the appropriate alkanes, when other preferred substrates are not available. Based on the chain length and type of petroleum hydrocarbon pollutant, various enzymes are required to introduce oxygen in substrate to start biodegradation. The differential regulation of multiple alkane hydroxylases has been described for *P. aeruginosa* RR1 and *P. aeruginosa* PAO1; these bacteria contain the alkane hydroxylases alkB1 (which oxidizes C16-C24 n-alkanes) and alkB2 (which oxidizes C12-C20 n-alkanes) (Gkorezis et al. 2016). When C10-C22 alkanes are present, both genes are expressed; however, the expression of alkB1 is twice as that of alk B2. Furthermore, alkB2 is preferentially induced in the beginning of the exponential phase, and alkB1 is preferentially induced during the late exponential phase, with expression of both genes decreasing during the stationary phase (Gkorezis et al. 2016).

Varjani and Upasani (2017) suggested some of the important genes (alkA, alkM, alkB, LadA, assA1, assA2, nahA-M, napA, amoA, dsrAB and mcrA) and enzymes (oxygenases, peroxidases, reductases, hydroxylases and dehydrogenases) for both aerobic and anaerobic pathways of microbial degradation petroleum hydrocarbon pollutants. The major genes involved in microbial aerobic degradation of hydrocarbon pollutants include: alkB gene for alkane monooxygenase, xylE gene for catechol dioxygenase and nahAc gene for naphthalene dioxygenase. The most studied aerobic enzyme includes alkane hydroxylases, encoded by the alkB, p450 and alma genes. Under anaerobic conditions (typical in sediments), HC catabolism is activated by the addition of fumarate to the secondary carbon and catalysed by alkyl-succinate synthase (Mapelli et al. 2017). Historically, the critical point for the analysis of PAH degradation by aerobic bacteria started with the discovery, in *P. putida* strain G7, of naphthalene catabolic genes (nah) located on the plasmid NAH7 (Gkorezis et al. 2016).

The major classes of monooxygenases isolated in prokaryotes are encoded by the gene alkB in most of bacteria, and alkM in *Acinetobacter* sp.; furthermore, a flavoprotein-dependent monooxygenase LadA isolated from a thermophilic bacteria (*Geobacillus thermodenitrificans* NG80-2) is known to activate the long-chain alkanes (C15 to C36) for degradation (Abbasian et al. 2015). The cytochrome P450 enzymes containing di-iron alkane hydroxylases (e.g. alkB), membrane-bound copper-containing methane monooxygenases, and soluble di-iron methane monooxygenases (associated with hydrocarbon degradation in aerobic condition) are reported to be found in both eukaryotes and prokaryotes (Al-Hawash et al. 2018). Methane monooxygenases (MMO) are the first line of enzymes in the degradation of shorter (one to four carbons) chain length n-alkanes (Selvam and Thatheyus 2018).

Genetically Engineered Microbes

Bioremediation, being the most intensive area of research, still needs more biological advances to develop practically efficient bacterial bioremediation processes. Bacteria utilising the hydrocarbon as substrate for carbon and energy source are still not proficient in their successful elimination from ocean since bacterial populations have the tendency to evolve themselves in the direction of ecological fitness rather than biotechnological efficiency (Ghoshal et al. 2017). The catabolic efficiency of bacteria for petroleum hydrocarbon utilization can be enhanced by employing bioengineering. Genetically engineered microorganisms (GEMs) with petroleum hydrocarbon remediation potential can be developed with the help of modern molecular biology techniques such as gene conversion, gene duplication, and transposon or plasmids mediated gene delivery, which might play vital roles to boost up the biodegrading potential of naturally occurring microorganisms (Ghoshal et al. 2017). GEMs are prepared in laboratory conditions by transfer of plasmids containing desired genetic constructions from exogenous to indigenous hydrocarbon degrading microorganisms. Desired exogenous genetic constructions may have genes responsible for desirable biodegradation pathways, improved catabolic enzymes affinity and specificity, enhanced bioavailability of substrate and its utilization, and stable genetic system.

With the selection of suitable catabolic gene(s), the recombinant DNA fragment construct is inserted into a suitable vector and introduced into suitable bacterial host cells (Miri et al. 2019). The degradation of some petroleum components is controlled by an extra-chromosomal plasmid; therefore, superbugs can be constructed by introducing plasmids with capabilities for degrading different components in a single cell (Liu et al. 2019b). Bacterial strains with proven suitability as hydrocarbon degraders (such as *Pseudomonas* sp., Acinetobacter sp., Alcanivorax sp. and *Bacillus* sp.; Jiang et al. 2020) could be selected for GEM.

Many successful gene manipulations in natural bacterial strains have been carried out till date. For example, Liu et al. (2019b) constructed a recombinant *Acinetobacter baumannii* S30 pJES by inserting the lux gene into the chromosome of the *A. baumannii* S30 to enhance the biodegradation efficiency for total petroleum hydrocarbon (TPH) of crude oil. On the other hand, a gene (alkB) coding for alkane monooxygenase was inserted in non-alkane degrading strain of *Streptomyces coelicolor* M145 strain. Moreover, xylE gene coding for catechol 2, 3 dioxygenase was cloned from plasmid DNA of *Pseudomonas putida* BNF1 and inserted to the alkanes degrading strain *Acinetobacter* sp. BS3; these gene insertions enhance the degradation of alkane n-hexadecane and aromatic hydrocarbons, respectively (Mishra et al. 2019, Liu et al. 2019b). The Tou cluster gene coding for enzyme toluene o-xylene monooxygenase (ToMO) from the mesophile *Pseudomonas stutzeri* has been cloned and expressed in the Antarctic *Pseudoalteromonas haloplanktis*, which improved its capacity for degradation of aromatic hydrocarbons at low temperature (Miri et al. 2019).

However, modification of bacteria to increase their hydrocarbon degrading efficiency and the release of these genetically engineered microorganisms (GEMs) in the environment is an environmental risk, posing biosafety issues which has to be regulated (Ivshina et al. 2017, Miri et al. 2019). In most countries, the field trials of GEM are restricted. In USA, GEM are only approved for bioremediation purpose; the genetically modified *P. fluorescens* HK44 possessing a naphthalene catabolic plasmid (pUTK21) with inserted lux genes is approved here. However, bioaugmentation with GEMs or non-indigenous microorganisms is banned in Norway, Sweden and Iceland; this could be due to the fact that little information is available regarding the total impact of modified organisms on ecology and vulnerability of ecosystems to the potentially invasive microorganism (Miri et al. 2019).

CONCLUSIONS

The oceans comprise Earth's most valuable natural resources. Escalating economic growth has led to unparalleled increase in global petroleum product consumption and hiked industrialization, which in turn has increased the pollution load. Vast quantities of petroleum products are transported across the oceans, and often tankers accidently spill their cargo in oceans or seas. The oil spill often finds its way to shore damaging both marine and coastline organisms. It therefore becomes essential that significant attention is paid to remediation of the petroleum-contaminated sites. Biodegradation of hydrocarbons is a lucrative eco-friendly technique for eliminating the toxicity of these compounds. Several microorganisms from the sediment, water, and soil have been investigated for bioremediation of hydrocarbons. This review highlights bioremediation as the most appropriate tool for enhancement of cleansing and treating petroleum hydrocarbon compounds. It substantiates the applicability and discusses the factors involved in petroleum hydrocarbon degradation.

Although investigations over the past decades demonstrating utility of both aerobic and anaerobic processes are known, extensive and targeted research on biostimulation and bioaugmentation for enhancing biodegradation of petroleum contaminations needs to be undertaken along with applicability of biosurfactant and genetically engineered bacteria. This would provide a clear understanding regarding the exact rate of biodegradation and could prove potentially useful.

ACKNOWLEDGEMENTS

The authors are sincerely thankful to the Head, Department of Botany, University of Allahabad for his support. The author Kamleshwar Singh is especially thankful to CSIR-JRF Fellowship for the financial support provided.

REFERENCES

Abbasian, F., R. Lockington, M. Mallavarapu and R. Naidu. 2015. A comprehensive review of aliphatic hydrocarbon biodegradation by bacteria. Appl. Biochem. Biotechnol. 176: 670-699.

Abdel-Shafy, H.I., M.F. Abdel-Sabour and M.R. Farid. 1994. Distribution pattern of metals in the environment of the little Lake. J. Environ. Protect. Eng. 20: 5-16.

Abdel Shafy, H.I. and M.S.M. Mansour. 2018. Microbial degradation of hydrocarbons in the environment: An overview. pp. 353-386. *In*: V. Kumar, M. Kumar, R. Prasad (eds.). Microbial Action on Hydrocarbons. Springer, Singapore.

Adams, G.O., P.T. Fufeyin, S.E. Okoro and I. Ehinomen. 2015. Bioremediation, biostimulation and bioaugmention: A review. Int. J. Environ. Bioremed. Biodegrad. 3: 28-39.

Agnello, A.C., M. Bagard, E.D. van Hullebusch, G. Esposito and D. Huguenot. 2016. Comparative bioremediation of heavy metals and petroleum hydrocarbons co-contaminated soil by natural attenuation, phytoremediation, bioaugmentation and bioaugmentation-assisted phytoremediation. Sci. Total Environ. 563: 693-703.

Al-Hawash, A.B., M.A. Dragh, S. Li, A. Alhujaily, H.A. Abbood, X. Zhang, et al. 2018. Principles of microbial degradation of petroleum hydrocarbons in the environment. Egyptian J. Aquatic Res. 44: 71-76.

Amenu, D. 2014. Isolation of poly aromatic hydrocarbons (PAHs) degrading bacteria's. Landmark Res. J. Med. Med. Sci. 1: 1-3.

Atlas, R.M. 1975. Effects of temperature and crude oil composition on petroleum biodegradation. J. Appl. Microbiol. 30(3): 396-403.

Atlas, R.M. 1981. Microbial degradation of petroleum hydrocarbons: An environmental perspective. Microbiol. Rev. 45(1): 180-209.

Atlas, R.M. and T.C. Hazen. 2011. Oil biodegradation and bioremediation: A tale of the two worst spills in U.S. history. Environ. Sci. Technol. 45: 6709-6715.

Bamforth, S.M. and I. Singleton. 2005. Bioremediation of polycyclic aromatic hydrocarbons: Current knowledge and future directions. J. Chem. Technol. Biotechnol. 80: 723-736.

Berenshtein, I., C.B. Paris, N. Perlin, M.M. Alloy, S.B. Joye and S. Murawski. 2020. Invisible oil beyond the Deepwater Horizon satellite footprint. Sci. Adv. 6: 8863.

Bodour, A.A, K.P. Drees and R.M. Maier. 2003. Distribution of biosurfactant-producing bacteria in undisturbed and contaminated arid Southwestern soils. Appl. Environ. Microbiol. 69: 3280-3287.

Bragg, J.R., R.C. Prince, E.J. Harner and M.A. Ronald. 1994. Effectiveness of bioremediation for the Exxon Valdez oil spill. Nature 368: 413-418.

Brakstad, O.G., E.J. Davies, D. Ribicic, A. Winkler, U. Brönner and R. Netzer. 2018. Biodegradation of dispersed oil in natural seawaters from Western Greenland and a Norwegian fjord. Polar Biol. 41: 2435-2450.

Brassington, K.J., R.L. Hough, G.I. Paton, K.T. Semple, G.C. Risdon, J. Crossley, et al. 2007. Weathered hydrocarbon wastes: A risk management primer. Crit. Rev. Environ. Sci. Technol. 37: 199-232.

Cameotra, S.S. and P. Singh. 2008. Bioremediation of oil sludge using crude biosurfactants. Int. Biodeter. Degrad. 62: 274-280.

Cameotra, S.S. and R.S. Makkar. 2010. Biosurfactant-enhanced bioremediation of hydrophobic pollutants. Pure Appl. Chem. 82: 97-116.

Carmichael, L.M. and F.K. Pfaender. 1997. The effect of inorganic and organic supplements on the microbial degradation of phenanthrene and pyrene in soils. Biodegradation. 8: 1-13.

Chaillan, F., C.H. Chaîneau, V. Point, A. Saliot and J. Oudot. 2006. Factors inhibiting bioremediation of soil contaminated with weathered oils and drill cuttings. Environ. Pollut. 144: 255-265.

Chaîneau, C.H., G. Rougeux, C. Yéprémian and J. Oudot. 2005. Effects of nutrient concentration on the biodegradation of crude oil and associated microbial populations in the soil. Soil Biol. Biochem. 37: 1490-1497.

Claro, E.M.T., J.M. Cruz, R.N. Montagnolli, P.R.M. Lopes, J.R.M. Júnior and E.D. Bidoia. 2018. Microbial degradation of petroleum hydrocarbons: Technology and mechanism. pp. 125-141. *In*: V. Kumar, M. Kumar, R. Prasad. (eds.). Microbial Action on Hydrocarbons. Springer, Singapore.

Cowell, E.B. 1973. The ecological effects of oil pollution on littoral communities. Essex England Applied Science Publishers. 88-98.

Colwell, R.R., J.D. Walker and J.J. Cooney. 1977. Ecological aspects of microbial degradation of petroleum in the marine environment. Crit. Rev. Microbiol. 5(4): 423-445.

Cruz, A., A.J. Cavaleiro, A.M. Paulo, A. Louvado, M.M. Alves, A. Almeida, et al. 2017. Microbial remediation of organometals and oil hydrocarbons in the marine environment. pp. 41-66. *In*: M. Naik, S. Dubey. (eds.). Marine Pollution and Microbial Remediation. Springer, Singapore.

Cui, C.Z., C. Zeng, X. Wan, D. Chen, J.Y. Zhang and P. Shen. 2008. Effect of rhamnolipids on degradation of anthracene by two newly isolated strains, *Sphingromonas* sp 12A and *Pseudomonas* sp 12B. J. Microbiol. Biotechnol. 18: 63-66.

Das, N. and P. Chandran. 2011. Microbial degradation of petroleum hydrocarbon contaminants: An overview. Biotechnol. Res. Int. 2011: 941810.

Davis, J.S. and W.D.S. Westlake. 1978. Crude oil utilization by fungi. Can. J. Microbiol. 25: 146-56.

Devi, R.S., V.R. Kannan, K. Natarajan, D. Nivas, K. Kannan, S. Chandru, et al. 2016. The role of microbes in plastic degradation. pp. 341-370. *In*: R. Chandra. (ed.). Environmental Waste Management. CRC Press, United States.

Díaz, M.P., K.G. Boyd, S.J.W. Grigson and J.G. Burgess. 2002. Biodegradation of crude oil across a wide range of salinities by an extremely halotolerant bacterial consortium MPD-M, immobilized onto polypropylene fibers. Biotechnol. Bioeng. 79: 145-153.

Dombrowski, N., J.A. Donaho, T. Gutierrez, K.W. Seitz, A.P. Teske and B.J. Baker. 2016. Reconstructing metabolic pathways of hydrocarbon-degrading bacteria from the Deepwater Horizon oil spill. Nat. Microbiol. 1: 16057.

Doshi, B., M. Sillanpää and S. Kalliola. 2018. A review of bio-based materials for oil spill treatment. Water Res. 135: 262-277.

Ferguson, A., H. Solo-Gabriele and K. Mena. 2020. Assessment for oil spill chemicals: Current knowledge, data gaps and uncertainties addressing human physical health risk. Mari. Pollut. Bulletin. 150: 110746.

Fritsche, W. and M. Hofrichter. 2000. Aerobic degradation by microorganisms. pp. 146-155. *In*: J. Klein. (ed.). Environmental Processes-Soil Decontamination. Wiley-VCH, Weinheim.

Fuentes, S., V. Méndez, P. Aguila and M. Seeger. 2014. Bioremediation of petroleum hydrocarbons: Catabolic genes, microbial communities and applications. Appl. Microbiol. Biotechnol. 98: 4781-4794.

Gardner, C.M. and C.K. Gunsch. 2020. Environmental microbiome analysis and manipulation. pp. 113-133. *In*: D. O'Bannon. (ed.). Women in Water Quality. Women in Engineering and Science. Springer, Cham.

Ghosal, D., S. Ghosh, T.K. Dutta and Y. Ahn. 2016. Current state of knowledge in microbial degradation of polycyclic aromatic hydrocarbons (PAHs): A review. Front. Microbiol. 7: 1369.

Gkorezis, P., M. Daghio, A. Franzetti, J.D. Van Hamme, W. Sillen and J. Vangronsveld. 2016. The interaction between plants and bacteria in the remediation of petroleum hydrocarbons: An environmental perspective. Front. Microbiol. 7: 1836.

Gros, J., C.M. Reddy, C. Aeppli, R.K. Nelson, C.A. Carmichael and J.S. Arey. 2014. Resolving biodegradation patterns of persistent saturated hydrocarbons in weathered oil samples from the Deepwater Horizon disaster. Environ. Sci. Technol. 48: 1628-1637.

Hassanshahian, M. and S. Cappello. 2013. Crude oil biodegradation in the marine environments. INTECH 101-135.

Hazen, T.C., R.C. Prince and N. Mahmoudi. 2016. Marine oil biodegradation. Environ. Sci. Technol. 50: 2121-2129.

Hua, F. and H.Q. Wang. 2014. Uptake and trans-membrane transport of petroleum hydrocarbons by microorganisms. Biotechnol. Biotechnol. Equip. 28: 165-175.

Ivshina, I.B., M.S. Kuyukina and A.V. Krivoruchko. 2017. Hydrocarbon-oxidizing bacteria and their potential in eco-biotechnology and bioremediation. pp. 121-148. *In*: I. Kurtböke. (ed.). Microbial Resources: From Functional Existence in Nature to Industrial Applications. Elsevier, Amsterdam.

Jaekel, U., N. Musat, B. Adam, M. Kuypers, O. Grundmann and F. Musat. 2013. Anaerobic degradation of propane and butane by sulfate-reducing bacteria enriched from marine hydrocarbon cold seeps. ISME J. 7: 885-895.

Jiang, B., Y. Song, Z. Liu, W.E. Huang, G. Li, S. Deng, et al. 2020. Whole-cell bioreporters for evaluating petroleum hydrocarbon contamination. Crit. Revi. Environ. Sci. Technol. 1-51.

Jurelevicius, D., V.M. Alvarez, J.M. Marques, L.R.F. de Sousa Lima, F. de Almeida Dias and L. Seldin. 2013. Bacterial community response to petroleum hydrocarbon amendments in freshwater, marine and hypersaline water-containing microcosms. Appl. Environ. Microbiol. 79: 5927-5935.

Khan, A.H.A., S. Tanveer, S. Alia, M. Anees, A. Sultan, M. Iqbal, et al. 2017. Role of nutrients in bacterial biosurfactant production and effect of biosurfactant production on petroleum hydrocarbon biodegradation. Ecol. Eng. 104: 158-164.

Kolomytseva, M.P., D. Randazzo, B.P. Baskunov, A. Scozzafava, F. Briganti and L.A. Golovleva. 2009. Role of surfactants in optimizing fluorene assimilation and intermediate formation by *Rhodococcus rhodochrous* VKM B-2469. Bioresour. Technol. 100: 839-844.

Kostka, J.E., S.B. Joye, W. Overholt, P. Bubenheim, S. Hackbusch, S.R. Larter, et al. 2020. Biodegradation of petroleum hydrocarbons in the deep sea. pp. 107-124. *In*: S. Murawski, et al. (eds.). Deep Oil Spills. Springer, Cham.

Liu, J., Y. Zheng, H. Lin, X. Wang, M. Li, Y. Liu, et al. 2019a. Proliferation of hydrocarbon-degrading microbes at the bottom of the Mariana Trench. Microbiome 7: 47.

Liu, L., M. Bilal, X. Duan and H.M. Iqbal. 2019b. Mitigation of environmental pollution by genetically engineered bacteria-current challenges and future perspectives. Sci. Total Environ. 444-454.

Logeshwaran, P., M. Megharaj, S. Chadalavada, M. Bowman and R. Naidu. 2018. Petroleum hydrocarbons (PH) in groundwater aquifers: An overview of environmental fate, toxicity, microbial degradation and risk-based remediation approaches. Environ. Technol. Inno. 10: 175-193.

Mahmoud, G.A.E. and M.M.K. Bagy. 2018. Microbial degradation of petroleum hydrocarbons. pp. 125-143. *In*: V. Kumar, M. Kumar, R. Prasad. (eds.). Microbial Action on Hydrocarbons. Springer, Singapore.

Mapelli, F., A. Scoma, G. Michoud, F. Aulenta, N. Boon, S. Borin, et al. 2017. Biotechnologies for marine oil spill cleanup: Indissoluble ties with microorganisms. Trends Biotechnol. 35: 860-870.

Menezes Bento F., F.A. de Oliveira Camargo, B.C. Okeke and J.W.T. Jr Frankenberger. 2005. Microbiol. Res. 160: 249- 255

Miri, S., M. Naghdi, T. Rouissi, S. Kaur Brar and R. Martel. 2019. Recent biotechnological advances in petroleum hydrocarbons degradation under cold climate conditions: A review. Crit. Rev. Environ. Sci. Technol. 49: 553-586.

Mishra, B., S. Varjani, G.P. Iragavarapu, H.H. Ngo, W. Guo and B. Vishal. 2019. Microbial fingerprinting of potential biodegrading organisms. Curr. Pollution Rep. 5: 181-197.

Mohanty, S., J. Jasmine and S. Mukherji. 2013. Practical considerations and challenges involved in surfactant enhanced bioremediation of oil. Biomed. Res. Int. 2013: 328608.

Mrozik, A. and Z. Piotrowska-Seget. 2010. Bioaugmentation as a strategy for cleaning up of soils contaminated with aromatic compounds. Microbiol. Res. 165: 363-375.

Nazina, T.N., D.S. Sokolova, A.A. Grigor'ian, Y.F. Xue, S.S. Beliaev and M.V. Ivanov. 2003. Production of oil-processing compounds by microorganisms from the Daqing oil field, China. Mikrobiologiia. 72: 206-211.

Oudot, J., F.X. Merlin and P. Pinvidic. 1998. Weathering rates of oil components in a bioremediation experiment in estuarine sediments. Mar. Environ. Res. 45: 113-125.

Paniagua-Michel, J. and B.Z. Fathepure. 2018. Microbial consortia and biodegradation of petroleum hydrocarbons in marine environments. pp. 1-20. *In*: V. Kumar, M. Kumar, R. Prasad. (eds.). Microbial Action on Hydrocarbons. Springer, Singapore.

Parach, A., A. Rezvani, M.M. Assadi and B. Akbari-Adergani. 2017. Biodegradation of heavy crude oil using Persian Gulf autochthonous bacterium. Int. J. Environ. Res. 11: 667-675.

Pathak, S.S.H and D.P. Jaroli. 2014. Factors affecting the rate of biodegradation of polyaromatic hydrocarbons. Int. J. Pure App. Biosci. 2: 185-202.

Pawar, R.M. 2012. The effect of soil pH on degradation of polycyclic aromatic hydrocarbons. PhD thesis, University of Hertfordshire.

Plaza, G., I. Zjawiony and I. Banat. 2006. Use of different methods for detection of thermophilic biosurfactant-producing bacteria from hydrocarbon-contaminated bioremediated soils. J. Petro. Science. Eng. 50: 71-77.

Pornsunthorntawee, O., P. Wongpanit, S. Chavadej, M. Abe and R. Rujiravanit. 2008. Structural and physicochemical characterization of crude biosurfactant produced by *Pseudomonas aeruginosa* SP4 isolated from petroleum contaminated soil. Bioresour. Technol. 99: 1589-1595.

Prince, R.C., J.D. Butler and A.D. Redman. 2017. The rate of crude oil biodegradation in the sea. Environ. Sci. Technol. 51: 1278-1284.

Schedler, M., R. Hiessl, A.G. Valladares Juárez, G. Gust and R. Müller. 2014. Effect of high pressure on hydrocarbon-degrading bacteria. AMB Exp. 4: 7.

Selvam, A.D.G. and A.J. Thatheyus. 2018. Microbial degradation of petroleum hydrocarbons: An overview. pp. 485-504. *In*: V. Kumar, M. Kumar, R. Prasad. (eds.). Microbial Action on Hydrocarbons. Springer, Singapore.

Shen, T., Y. Pi, M. Bao, N. Xu, Y. Li and J. Lu. 2015. Biodegradation of different petroleum hydrocarbons by free and immobilized microbial consortia. Environ. Sci. Processes Impacts. 17: 2022-2033.

Shi, K., J. Xue, X. Xiao, Y. Qiao, Y. Wu and Y. Gao. 2019. Mechanism of degrading petroleum hydrocarbons by compound marine petroleum-degrading bacteria: surface adsorption, cell uptake and biodegradation. Energy Fuels 33: 11373-11379.

Simarro, R., N. Gonzalez, L.F. Bautista and M.C. Molina. 2013. Assessment of the efficiency of *in situ* bioremediation techniques in a creosote polluted soil: Change in bacterial community. J. Hazard. Mater. 262: 158-167.

Simpanen, S., M. Dahl, M. Gerlach, A. Mikkonen, V. Malk, J. Mikola, et al. 2016. Biostimulation proved to be the most efficient method in the comparison of *in situ* soil remediation treatments after a simulated oil spill accident. Environ. Sci. Pollut. Res. 23: 25024-25038.

Sutton, N.B., J.T.C. Grotenhuis, A.A.M. Langenhoff and H.M.M. Rijnaarts. 2011. Efforts to improve coupled *in situ* chemical oxidation with bioremediation: A review of optimization strategies, J. Soils Sediments. 11: 129-140.

Szulc, A., D. Ambrozewicz, M. Sydow, Ł. Ławniczak, A. Piotrowska-Cyplik, R. Marecik, et al. 2014. The influence of bioaugmentation and biosurfactant addition on bioremediation efficiency of diesel-oil contaminated soil: Feasibility during field studies. J. Environ. Manage. 132: 121-128.

Tao, K., X. Liu, X. Chen, X. Hu, L. Cao and X. Yuan. 2017. Biodegradation of crude oil by a defined co-culture of indigenous bacterial consortium and exogenous Bacillus subtilis. Bioresour. Technol. 224: 327-332.

Urum, K. and T. Pekdemir. 2004. Evaluation of biosurfactants for crude oil contaminated soil washing. Chemosphere. 57: 1139-1150.

Urum, K., T. Pekdemir, D. Ross and S. Grigson. 2005. Crude oil contaminated soil washing in air sparging assisted stirred tank reactor using biosurfactants. Chemosphere. 60: 334-343.

Urum, K., S. Grigson, T. Pekdemir and S. McMenamy. 2006. A comparison of the efficiency of different surfactants for removal of crude oil from contaminated soils. Chemosphere. 62: 1403-1410.

Van Dorst, J., S.D. Siciliano, T. Winsley, I. Snape and B.C. Ferrari. 2014. Bacterial targets as potential indicators of diesel fuel toxicity in subantarctic soils. Appl. Environ. Microbiol. 80: 4021-4033.

Varjani, S.J. 2017. Microbial degradation of petroleum hydrocarbons. Bioresour. Technol. 223: 277-286.

Varjani, S.J. and V.N. Upasani. 2017. Review on A new look on factors affecting microbial degradation of petroleum hydrocarbon pollutants. Int. Biodeterior. Biodegrad. 120: 71-83.

Varjani, S.J., D.P. Rana, A.K. Jain, S. Bateja and V.N. Upasani. 2015. Synergistic ex-situ biodegradation of crude oil by halotolerant bacterial consortium of indigenous strains isolated from on shore sites of Gujarat. India. Int. Biodeterior. Biodegrad. 103: 116-124.

Varjani, S.J., E. Gnansounou and A. Pandey. 2017. Comprehensive review on toxicity of persistent organic pollutants from petroleum refinery waste and their degradation by microorganisms. Chemosphere. 188: 280-291.

Wang, X.B., C.Q. Chi, Y. Nie, Y.Q. Tang, Y. Tan, G. Wu and X.L. Wu. 2011. Degradation of petroleum hydrocarbons (C_6-C_{40}) and crude oil by a novel *Dietzia* strain. Bioresour. Technol. 102: 7755-7761.

Wang, C., X. Liu, J. Guo, Y. Lv and Y. Li 2018. Biodegradation of marine oil spill residues using aboriginal bacterial consortium based on Penglai 19-3 oil spill accident, China. Ecotoxicol. Environ. Saf. 159: 20-27.

Wang, H., R. Jiang, D. Kong, Z. Liu, X. Wu, J. Xu, et al. 2020. Transmembrane transport of polycyclic aromatic hydrocarbons by bacteria and functional regulation of membrane proteins. Front. Environ. Sci. Eng. 14: 9.

Xu, X., W. Liu, S. Tian, W. Wang, Q. Qi, P. Jiang, et al. 2018. Petroleum hydrocarbon-degrading bacteria for the remediation of oil pollution under aerobic conditions: A perspective analysis. Front. Microbiol. 9: 2885.

Xua, Y. and M. Lu. 2010. Bioremediation of crude oil-contaminated soil: Comparison of different biostimulation and bioaugmentation treatments. J. Hazard. Mater. 183: 395-401.

Xue, J., Y. Yu, Y. Bai, L. Wang and Y. Wu. 2015. Marine oil-degrading microorganisms and biodegradation process of petroleum hydrocarbon in marine environments: A review. Curr. Microbiol. 71: 220-228.

Yeung, C.W., D.R. Van Stempvoort, J. Spoelstra, G. Bickerton, J. Voralek and C.W. Greer. 2013. Bacterial community evidence for anaerobic degradation of petroleum hydrocarbons in cold climate groundwater. Cold Reg. Sci. Technol. 86: 55-68.

Yoshikawa, M., M. Zhang and K. Toyota. 2017. Integrated anaerobic-aerobic biodegradation of multiple contaminants including chlorinated ethylenes, benzene, toluene and dichloromethane. Water, Air, Soil Pollut. 228: 25.

9

Heavy Metals: Applications, Hazards and Biosalvage

Bhagirath M. Baraiya, Shailesh C. Dabhi, Apoorva Bhayani, Anjana K. Vala* and Bharti P. Dave*

INTRODUCTION

Heavy metals are the natural elements that can be found throughout the whole of Earth's upper layer. The term "Heavy Metals" attributes to any metal ion, or metalloid element which is distinguished for its high atomic mass and high density. Broadly, elements that have atomic mass higher than 23 amu and atomic number exceeding 20 are considered as heavy metals (Figure 9.1, Table 9.1). Several researchers have defined them as metals having higher density (>5 g cm^{-3}-except Selenium) as compared to light metals (Gautam et al. 2014, Nieboer and Richardson 1980).

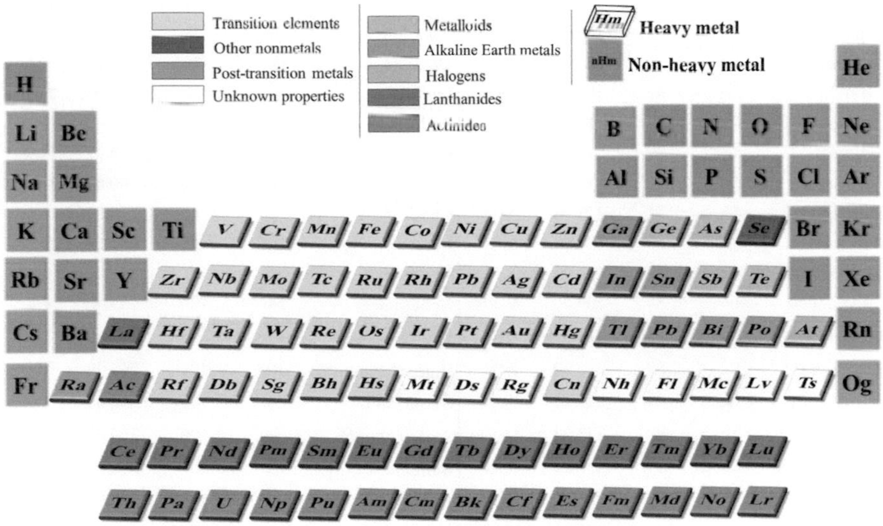

FIGURE 9.1 Heavy metal position in periodic table.

Department of Life Sciences, Maharaja K. rishna kumar sinhji Bhavnagar University, Bhavnagar, India.
* Corresponding Author: akv@mkbhavunai.edu.in; anjana_vala@yahoo.co.in

TABLE 9.1 Chemical properties of heavy metals.

Heavy Metal	Symbol	Atomic Number	Atomic Mass (amu)	Density (g cm^{-3})	Heavy Metal	Symbol	Atomic Number	Atomic Mass (amu)	Density (g cm^{-3})
Vanadium	V	23	50.94	6.00	**Cerium**	Ce	58	140.12	6.77
Chromium	Cr	24	51.99	7.15	**Praseodymium**	Pr	59	140.91	6.77
Manganese	Mn	25	54.94	7.30	**Neodymium**	Nd	60	144.24	7.01
Iron	Fe	26	55.84	7.87	**Promethium☢¶**	Pm	61	145.00	7.26
Cobalt	Co	27	58.93	8.86	**Samarium**	Sm	62	150.36	7.52
Nickel	Ni	28	58.69	8.90	**Europium**	Eu	63	151.96	5.24
Copper	Cu	29	63.55	8.96	**Gadolinium**	Gd	64	157.25	7.90
Zinc	Zn	30	65.38	7.13	**Terbium**	Tb	65	158.93	8.23
Gallium	Ga	31	69.72	5.91	**Dysprosium**	Dy	66	162.50	8.55
Germanium†	Ge	32	72.63	5.32	**Holmium**	Ho	67	164.93	8.80
Arsenic†	As	33	74.92	5.75	**Erbium**	Er	68	167.26	9.07
Selenium†	Se	34	78.97	4.81	**Thulium**	Tm	69	168.93	9.32
Zirconium	Zr	40	91.22	6.52	**Ytterbium**	Yb	70	173.04	6.90
Niobium	Nb	41	92.91	8.57	**Lutetium**	Lu	71	174.97	9.84
Molybdenum	Mo	42	95.95	10.20	**Hafnium**	Hf	72	178.49	13.30
Technetium☢¶	Tc	43	98.00	11.00	**Tantalum**	Ta	73	180.95	16.40
Ruthenium	Ru	44	101.07	12.10	**Tungsten**	W	74	183.84	19.30
Rhodium	Rh	45	102.91	12.40	**Rhenium**	Re	75	186.21	20.80
Palladium	Pd	46	106.42	12.00	**Osmium**	Os	76	190.23	22.59
Silver	Ag	47	107.87	10.50	**Iridium**	Ir	77	192.21	22.56
Cadmium	Cd	48	112.41	8.69	**Platinum**	Pt	78	195.08	21.50
Indium	In	49	114.82	7.31	**Gold**	Au	79	196.97	19.30
Tin	Sn	50	118.71	7.29	**Mercury**	Hg	80	200.60	13.53
Antimony†	Sb	51	121.76	6.68	**Thallium**	Tl	81	204.38	11.80
Tellurium†	Te	52	127.60	6.32	**Lead**	Pb	82	207.20	11.30
Lanthanum	La	57	138.90	6.15	**Bismuth**	Bi	83	208.98	9.79
Polonium☢¶	Po	84	209.00	9.20	**Nobelium☢**	No	102	259.00	UK
Astatine☢‡¶	At	85	210.00	UK	**Lawrencium☢**	Lr	103	262.00	UK
Radium☢¶	Ra	88	226.00	5.00	**Rutherfordium☢**	Rf	104	267.00	UK
Actinium☢¶	Ac	89	227.00	10.00	**Dubnium☢**	Db	105	268.00	UK
Thorium☢	Th	90	232.04	11.70	**Seaborgium☢**	Sg	106	269.00	UK
Protactinium☢	Pa	91	231.04	15.40	**Bohrium☢**	Bh	107	270.00	UK
Uranium☢	U	92	238.03	19.10	**Hassium☢**	Hs	108	269.00	UK
Neptunium☢¶	Np	93	237.00	20.20	**Meitnerium☢**	Mt	109	278.00	UK
Plutonium☢¶	Pu	94	244.00	19.70	**Darmstadtium☢**	Ds	110	281.00	UK
Americium☢	Am	95	243.00	12.00	**Roentgenium☢**	Rg	111	280.00	UK
Curium☢	Cm	96	247.00	13.51	**Copernicium☢**	Cn	112	285.00	UK
Berkelium☢	Bk	97	247.00	14.78	**Nihonium☢**	Nh	113	286.00	UK
Californium☢	Cf	98	251.00	15.10	**Flerovium☢**	Fl	114	289.00	UK
Einsteinium☢	Es	99	252.00	UK	**Moscovium☢**	Mc	115	289.00	UK
Fermium☢	Fm	100	257.00	UK	**Livermorium☢**	Lv	116	293.00	UK
Mendelevium☢	Md	101	258.00	UK	**Tennessine☢**	Ts	117	294.00	UK

(† = Metalloids; ¶ = Predicted to be metal; ‡ = too less in concentration for viable extraction; ☢ = Radioactive elements UK = Unknown)

Heavy metals occur in rather low concentrations between the low ppb ranges (noble metals) and up to 5% (iron); heavy metals are mainly found chemically bound in oxide, carbonate, sulfate or silicate rocks or also occur in their metallic, elemental form. These metals are found widely in the Earth's upper layer and are non-biodegradable in nature. They enter into the human body via air, water and food. Earth's crust consists of approximately 5% heavy metals by weight, with iron comprising 95% in that aggregate. Remaining 95% comprise of nonmetals (~75%) and light metals (~20 Kommel et al. 2013).

Heavy metals vary in their chemical properties, and are used enormously in electronics, machines and the artifacts of everyday life, as well as in high-tech applications. Some heavy metals like copper, selenium, and zinc are essential trace elements, playing important role in various biological processes (Shaw et al. 2004). Many heavy metals have technological importance, e.g. copper, lead, iron, zinc, tin, tungsten, etc. Recently, various heavy metals have been observed to act as the central atom of synthetically designed "bioinorganic" catalysts for special chemical transformations. Precious elements like gold, silver, iridium, rhodium, and platinum are used for various applications due to their low reactivity (Rao and Reddi 2000). Several heavy metals including mercury, cadmium, arsenic, chromium, thallium and lead show toxic effects even at very low concentration (Duruibe et al. 2007). As a result, they are competent to invade into not only the aquatic but also the food chains of humans and animals from diverse anthropogenic sources as well as from the natural geochemical weathering of soil and rocks. The main sources of contamination include industrial wastewaters, landfill leaches mining wastes, municipal wastewater and urban runoff, particularly from the electroplating, electronic and metal-finishing industries (Gautam et al. 2014). An attribute of heavy metal physiology is that even though many of them are essential for growth, they are also reported to have comprehensively toxic effects on cells, mainly according to their ability to denature protein molecules.

APPLICATIONS OF HEAVY METALS

Heavy metals such as iron (Fe), copper (Cu), zinc (Zn), manganese (Mn), cobalt (Co), magnesium (Mg), chromium (Cr), molybdenum (Mo), nickel (Ni) and selenium (Se) are essential nutrients required for various biochemical and physiological functions (World Health Organization 1996). Scanty supply of these micro nutrients could lead to various syndromes or inadequacy diseases (Wang and Shi 2001). For obligate aerobic beings as we are, it would not be possible to survive without having cytochromes which contain heme consisting of an iron ion that acts as a cofactor to an enzyme (Groves et al. 1981). Iron also plays a crucial role in our respiration system as an atom of the blood pigment heme. 70% of our body's iron is found in the red blood cells' hemoglobin and muscle cells' myoglobin. Copper acts in the transport of oxygen and electrons for activation of some enzymes, which is vital in photosynthesis reaction and other enzymes including catalase, superoxide dismutase and cytochrome C oxidase (Montgomery 1930, Stern 2010, Harvey and McArdle 2008, Dorsey and Ingerman 2004). Zinc plays a crucial role in the constitution of zinc finger enzymes (Maret 2013). Selenium is an antioxidant and is also involved in hormone biosynthesis (Stadtman 1990). Cobalt was perceived to be remarkable in the biosynthesis of complex compounds and cellular metabolism, especially in vitamin B12 as the median atom, which is required for cell division, in propionic acid biosynthesis, blood formation and the nervous system (Hudson et al. 1984). Additionally, vanadium and manganese are salient for regulation and reaction of several enzymes (Egami 1975), and some toxic elements like chromium, arsenic and nickel are also involved in enzymatic reactions. Arsenic is disclosed as a natural constituent in herring caviar (Viczek et al. 2016). Molybdenum also acts in some redox reactions (Mendel 2005), and cadmium plays a key role in the metabolism of some microalgae from the Diatomophyceae class (Lane et al. 2005). Furthermore, the role of tungsten in the metabolism of prokaryotes has been substantiated by scientists (Andreesen and Makdessi 2008). According to Iyengar (1998), 0.01% of our total body mass consists of heavy metals (65-75 kg) approximately. (Iyengar 1998). Zinc also serves as a co-factor in carbonic anhydrase activity and dehydrogenase (Holum 1979); it is a necessary element that balances copper in our body and also indispensable for male reproductive activity (Nolan 1983). Due to inadequacy of zinc, anemia can occur and hamper growth as well (Duruibe et al. 2007). Nickel is the element which occurs in volcanic rocks mainly, used enormously on an industrial scale for manufacturing of nickel-cadmium batteries. It is also utilized for manufacturing of automobile and aircraft parts, stainless steel, spark plugs, coins, cosmetic and in electroplating techniques (Dojlido and Best 1993). Arsenic is obtained from the upper layer of Earth. 'Arsenic' word came from comparable word 'Zarnikh' of Persian literature (Mudhoo et al. 2011). Arsenic is used in various fields like veterinary medicines for elimination of tapeworm in some cattle and sheep, and is also often utilized for the manufacture of agricultural application variants such as fungicides, insecticides, algicides, herbicides, and wood preservatives (Tchounwou 1999). It has been used for treating amoebic dysentery, trypanosomiasis, yaws and syphilis for more than a century, and is also presently used for the treatment of filariasis, amoebic dysentery and African sleeping sickness (Tchounwou 1999, Centeno et al. 2006).

PERILOUS EFFECTS OF HEAVY METALS

In biosystems, heavy metals have been described to affect cell organelles and cell constituents such as nuclei, cell membrane, endoplasmic reticulum, mitochondria, lysosome and some enzymes involved in damage repair metabolism and detoxification (Wang and Shi 2001). Metal ions have been observed to interface with cell constituents such as DNA and nuclear proteins, causing DNA damage and conformational changes that may lead to cell-cycle modulation, carcinogenesis, or apoptosis (Chang et al. 1996, Wang and Shi 2001, Beyersmann and Hartwig 2008). Some case studies have illustrated that oxidative stress and reactive oxygen species (ROS) simulate a significant role in carcinogenicity and toxicity of metals such as arsenic (Yedjou and Tchounwou 2006, 2007a, Tchounwou et al. 2004), lead (Yedjou and Tchounwou 2007b, Tchounwou et al. 2004a) and mercury (Sutton and Tchounwou 2007, Sutton et al. 2002) and other heavy metals like cadmium and chromium (Tchounwou et al. 2001, Patlolla et al. 2009a, b). These five elements are ranked prior among other metals due to their intense toxicity that are of great public health concern. These metals are categorized as either known or probable human carcinogen, as affirmed by the US Environmental Protection Agency (US-EPA) and the International Agency for Research on Cancer (IARC) (Tchounwou et al. 2012). Arsenic is an omnipresent element which is noticed in almost all environment systems. Environment adulteration by arsenic mainly occurs by anthropogenic activity and some natural events like soil erosion and volcanic eruptions (Abadin et al. 2007).

CHEMICAL REMEDIATION OF HEAVY METALS

Industries follow strict rules and regulations, yet heavy metal pollution has become now a major environmental issue in recent years. As mentioned above, heavy metals can exert toxic effects on ecosystems and their components. Hence, it is necessary to get rid of such contaminants. Conventionally, various physico-chemical processes are employed for heavy metal removal such as chemical precipitation, ion exchange, adsorption, membrane filtration, electro chemical treatment technologies, etc. (Fu and Wang 2011). Some techniques are described here (Figure 9.2).

Chemical Precipitation

Chemical precipitation is inexpensive, reasonably simple and effective method used most extensively in industries.

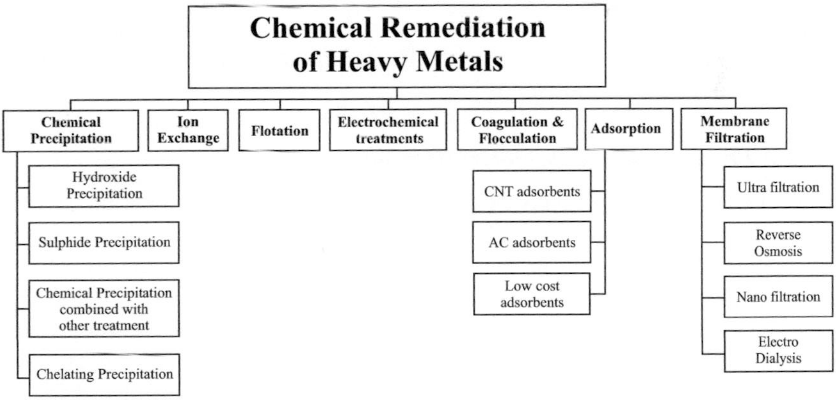

FIGURE 9.2 Types of chemical heavy metal remediation.

In this process, heavy metal ions will be converted to insoluble precipitates that can be extracted from water by filtration or sedimentation. The customary chemical precipitation involves both sulfide and hydroxide precipitation processes (Ku and Jung 2001).

Sulfide Precipitation

Sulfide precipitation is an effective process with certain advantages. Solubility of metal sulfide precipitation is lower than hydroxide precipitation and sulphides are not amphoteric, so they achieve high degree of metal removal above a broad pH spectrum in comparison to other techniques. It has been investigated that pyrite and synthetic iron sulphide guide the reaction to remove Cu^{2+}, Cd^{2+} and Pb^{2+} as chemical precipitation at low pH (<3) owing to H_2S production (Özverdi and Erdem 2006). Sulphate Reducing Bacteria (SRB) is emerging as the new sulphide precipitation method that anaerobically transfers the sulphate into hydrogen sulphate by oxidizing simple organic compound. Heavy metals could be found toxic (H_2S formation) in acidic conditions. It is required that this process be performed in neutral or basic medium. The separation problem could also be there in either settling or filtration led to form colloidal precipitates due to metal sulphide.

Hydroxide Precipitation

The hydroxide precipitation is widely used because of its low cost, relative simplicity and pH control (Huisman et al. 2006). Metal hydroxides can be eliminated by flocculation and sedimentation. It has been used to precipitate metals; lime is the desired base as per industrial settings to precipitate metals (Baltpurvins et al. 1997). Cu(II) and Cr(VI) ions are removed by hydroxide precipitation process using Ca(OH) and NaO (Mirbagheri and Hosseini 2005). Cr(VI) was converted to Cr(III) using $FeSO_4$. The size of the precipitate and efficiency of heavy metal removal has been increased by the use of the fly ash carbonation treatment. For enhancement of the removal of heavy metals from sewage, addition of coagulants such as alum, organic polymers and iron salts will assist; in addition, if coagulant was added the remaining heavy metal can be reduced more (Charerntanyarak 1999). However, there are constraints such as generation of large volume of low-density sludge, mixed metal can create problem in ideal hydroxide precipitation and other complexing agents can inhibit precipitation (Kongsricharoern and Polpraser 1995).

Chemical Precipitation with Other Methods

Successful affiliation has been observed with chemical precipitation by other methods (Gonzalez-Munoz et al. 2006). Upon nanofiltration of sulfide precipitated heavy metal ions, reduction in metal content resulted, further, the nanofiltered solution was reusable indicating successful sulfide precipitation combined with other method. Electro-Fenton method has also been used for the treatment of some fabric and rayon industry effluent treatment (Ghosh et al. 2011). Some reports showed that chemical precipitation was applied for the nickel removal from wastewater combined with ion-exchange (Papadopoulos et al. 2004, Feng et al. 2000).

Heavy Metal Chelating Precipitation

To overcome limitations of traditional precipitation and to meet environmental regulations, some companies started using chelating precipitation procedure to remediate heavy metal pollution. There are three heavy metal precipitants often used commercially: sodium dimethyledithiocarbamate (SSDC), tri-mercaptotriazine (TMT), and potassium/sodium thiocarbonate (Matlock et al. 2002a). 1,3-benzene di-amidoethanethiol ($BDET^{-2}$) dianion can effectively precipitate heavy metals in acid mine drainage, especially mercury in leachate (Matlock et al. 2002b, c) and N, N^0-bis-(dithiocarboxy)piperazine (BDP), 1,3,5-hexahydro triazinedithiocarbamate (HTDC). Both BDP and HTDC proved effective reducers of heavy metals (Fu et al. 2006). Later on, one more organic chelator of heavy metals was designed, namely dipropyldithiphosphate, which can effectively act on mercury, cadmium, lead and copper (Ying and Fang 2006).

Ion Exchange

Ion exchange is used extensively due to their advantages like high removal efficiency, high treatment capacity and swift kinetics (Kang et al. 2004). The processes carried out using synthetic resins or natural solid resins, which have the ability to exchange their cations, among both synthetic resins are effective to remove heavy metals. The main cation exchangers are sulfonic acid group and carboxylic acid group, which are strongly acidic and weakly acidic, respectively. From these functional groups, hydrogen ions can serve as interchangeable with metal cations; thus, heavy metal ions exchanged6 as effluent passes through the column by hydrogen ions.

Apart from resins, natural zeolites are broadly utilized to remove heavy metals owing to their high abundance and low cost. Many cases are reported that show that zeolites evince great cation exchange capacities for heavy metals (Motsi et al. 2009, Ostroski et al. 2009, Taffarel and Rubio 2009).

Adsorption

The adsorption method is a more effective and economical process to treat heavy metal pollution due to its efficacy to produce high quality treated effluent offering resilience in design and operations.

Activated Carbon Adsorbents

In these methods, mainly activated carbon (AC) as adsorbents are frequently used to pull out heavy metal contamination. Granular AC and AC fibers have been observed to remove Pb^{2+}, Cd^{2+} and Cu^{2+} (Jusoh et al. 2007, Kang et al. 2008). Consortium of AC and other supplements like tannic acid, surfactants, magnesium and alginate could be the most fruitful adsorbent (Park et al. 2007, Üçer et al. 2006, Yanagisawa et al. 2010, Ahn et al. 2009).

Carbon Nanotubes Adsorbents

Since carbon nanotubes (CNTs) were discovered in 1991, they have been widely used because of their splendid properties to adsorb various heavy metals like cadmium, chromium, lead, nickel and copper by their functional groups present on wall surface (Iijima 1991, Wang et al. 2007, Kabbashi et al. 2009, Kuo and Lin 2009, Pillay et al. 2009, Li et al. 2010, Kandah and Meunier 2007, Rao et al. 2007). The raw CNTs have low sorption capacity which can be increased after oxidized using $KMnO_4$, NaClO and HNO_3 solutions (Wang et al. 2007).

Low Cost Adsorbents

It is really important to look after cost effectiveness during the removal of heavy metal waste by industrialist prospective. Activated Carbon is a very good way to do so but it is expensive as well, so researchers are looking for alternative of AC. Till date, many reports have been published, which mainly suggest to use natural substances like zeolites, clay, natural kaolinite, peat, montmorilloinate, aragonite and agricultural waste (Bhattacharyya and Gupta 2008, Sud et al. 2008, Köhler et al. 2007, Apiratikul and Pavasant 2008, Al-Jlil and Alsewailem 2009, Gu and Evans 2008, Liu et al. 2008) and some modified compounds like modified plant waste and industrial byproducts, modified kaolinite and waste such as lignin, iatomite, and clinopyrrhotite (Betancur et al. 2009, Reyes et al. 2009, Sheng et al. 2009, Lu et al. 2006, Mohan and Chander 2006, Wan Ngah and Hanafiah 2008).

Membrane Filtration

Membrane technologies are useful for heavy metal remediation because of their easy operation, high efficacy and space saving approach.

Ultra-filtration

Ultra-filtration (UF) works at very low transmembrane pressure to remove colloidal and deliquesce material. Generally, the pore size of UF membrane is larger than hydrated ions of heavy metals; for getting excessive removal potency, the micellar enhanced ultrafiltration (MEUF) and polymer enhanced ultrafiltration (PEUF) was recommended.

MEUF proved to be an effective separation technique due to its ability to remove dissolved organic compound and multivalent ions of the heavy metal from aqueous system. The surfactants accumulate into micelle and can react and bind with metal ions to form large metal surfactant complex. Sodium dodecyl sulphate (SDS) is frequently used as an anionic surfactant in MEUF and linear alkyl benzene sulfonate (LAS) in a small-scale membrane.

PEUF is also a reliable method to isolate the metal ions. There are complexing agents that are used to form complexes with metal ions such as diethylamino ethyl cellulose, polyacrylic acid (PAA), humic acid and polyethylenimine (PEI). PEUF have high binding efficiency, high removal capacity and can also be reused (Kim et al. 2005, Molinari et al. 2008, Aroua et al. 2007, Labanda et al. 2009).

Reverse Osmosis

Reverse osmosis (RO) is a semipermeable membrane which allows fluid to pass by avoiding contamination. Copper and nickel were successfully removed by RO. In this technology, there is one limitation that it requires high power consumption and restoration of membranes (Shahalam et al. 2002).

Nano Filtration

Nano filtration (NF) is the interpose between UF and RO, and is a propitious technique for the purging of heavy metal ions like nickel copper, nickel, arsenic and cadmium. This technique ensures the ease of operation, high efficiency and low energy expenditure (Murthy and Chaudhari 2008, Muthukrishnan and Guha 2008, Cséfalvay et al. 2009, Ahmad and Ooi 2010, Nguyen et al. 2009, Figoli et al. 2010, Eriksson 1988). Pentavalent arsenic removal study was done by using two NF membrane (NF90 and N30F) (Figoli et al. 2010).

Electro Dialysis

Electro dialysis is another method using membrane to remove heavy metal contamination from brackish water and sea water. Here the separation of ions through charged membrane used electric field as incentive using two types of ion-exchange membrane: cation ion-exchange and anion ion-exchange membrane. The method is used to separate ions of lead, chromium, iron and copper.

The physico-chemical methods available so far are many a times costly, not so effective and produce lots of sludge. In this regard, there is a need to develop newer technologies which are not only effective and cost-effective but also meet the environmental standards. Bioremediation is one such potential alternative, and is employed in various fields.

BIOREMEDIATION OF HEAVY METALS

Bioremediation can be described as an approach to transform/degrade hazardous contaminants to non-hazardous ones by using specific biota. Algae, plants as well as microorganisms including bacteria and fungi are used for bioremediation (Figure 9.3).

Some important strategies for bioremediation of heavy metals are as follows, among which biosorption is one of the outstanding techniques.

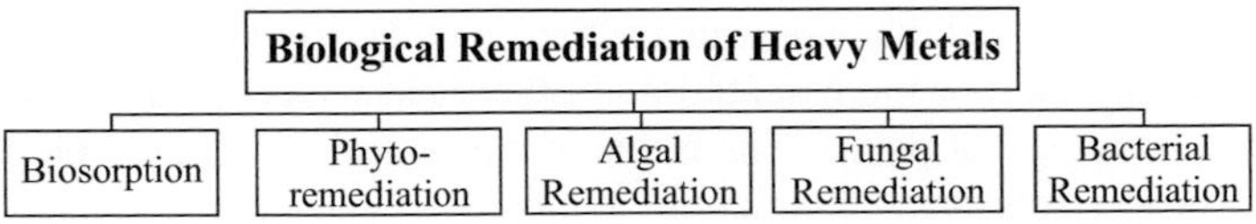

FIGURE 9.3 Classification of biological heavy metal remediation.

Biosorption

With increasing requirement for remediation of heavy metal contamination and awareness among scientific communities and policy makers, world is looking for more reliable ways over the conventional ones. In this scenario, biosorption can be pivotal as a technique offering several advantages, viz. cost effectiveness, reduced system size due to decrease in biological/chemical sludge, efficiency, lesser requirements and possibility of metal recovery. Biosorption is "a non-directed physiochemical interaction that may occur between metal/radionuclide species and microbial cells" (Strandberg et al. 1980, Alluri et al. 2007).

Biosorption can be classified as:

(i) On the basis of the cell metabolism, it can be (a) metabolism dependent and (b) metabolism independent.
(ii) Based on site, it can be categorized as (a) extracellular, (b) cell surface sorption and (c) intracellular.

After biosorption, the adsorbed heavy metal ions are delivered to the cytosol by various means including:

(i) Ion exchange: copper biosorbed by using fungi *Ganoderma lucidium* and *Asperigillus niger* (Muraleedharan and Venkobachar 1990).
(ii) Physical adsorption: electrostatic interaction responsible for copper biosorption by bacterium *Zooglea ramigera* and alga *Chorella vulgaris* (Aksu et al. 1992).
(iii) Complexation: both adsorption and formation of co-ordinate bonds between metals and amino or carboxyl groups of cell walls lead to biosorption of copper by *Chlorella vulgaris* and *Zoogloearamigera* (Aksu et al. 1992).

It is crucial to select the preferred biomass on the basis of its origin and system nature. The fast-growing living beings can be used for rapid remediation, for e.g. seaweeds and crab shells (Vieira and Volesky 2000). Sometimes agricultural side products like rice straw, exhausted coffee, walnut skin, cork biomass, wheat bran, mustard seed cakes, wool, coconut husks, waste tea, seed of *Ocimumbasilicum*, coconut fiber, rice hulls, pea pod and peat moss etc. can also be utilized as biosorbent (Dakiky et al. 2002, Ahluwalia and Goyal 2005, Chubar et al. 2003, Melo and D'souza 2004, Saeed et al. 2002). Besides this, living organisms including several seaweeds, bacteria, yeast and molds have also shown remarkable results in studies related to metal biosorption (Vieira and Volesky 2000).

These are few examples of the bio sorbent with their respective heavy metal ion species (Table 9.2).

Phytoremediation

Plants require certain metallic elements for their metabolism and growth. However, excessive concentration of the essential and other heavy metals can prove toxic to growth and physiology of plants. The elevated concentration of these elements can produce adverse effects due to interactions at molecular and cellular level. Malignity can lead to the inhibition or disruption of protein structure (Van Assche and

TABLE 9.2 Examples of biosorbents with their respective heavy metal ion species.

Microorganism	Example of Microorganisms	Heavy Metals	References
Algae	Brown Seaweed	Cr	(Figueira et al. 2000)
	Chlorella vulgaris	Cd	(Hosea et al. 1986)
	Eckloniamaxima	Cu	(Figueira et al. 2000)
	Sargassum sp.	Cu, Cd	(Volesky et al. 1999)
Fungi	*Aspergillus niger*	Pb, Cd	(Kapoor and Viraraghavan 1997)
	Mucor rouxii	Cd, Cu, Lt, Ag	(Mullen et al. 1992)
	Phanerochetechrysosporium	Ni, Pb	(Mullen et al. 1992)
	Penicillium chrysogenum	Au	(Niu and Volesky 1999)
Yeast	*Saccharomycese cerevisiae*	Cd, Ur, Hg	(Strandberg et al. 1981)
	Kluyveromyces fragilis	Cd	(Collins and Stotzky 1992)
Bacteria	*Escherichia coli*	Hg, Cu, Cr, Ni	(Churchill et al. 1995)
	Bacillus coagulans	Cr	(Brierley 1990)
	Pseudomonasauroginosa	Cr, Cu, Cd, Ni	(Hu et al. 1996)
	Bacillus polymyxa	Cu	(Brierley 1990)
	Bacillus subtilis	Cd	(Collins and Stotzky 1992)

Clijsters 1990). Moreover, heavy metals may prompt formation of reactive oxygen species (ROS) and free radicles possibly responsible for oxidative stress in plants (Dietz et al. 1999). Some plant species have adapted to these heavy metals stresses by using various strategies. Mycorrhizas, wall, extracellular exudates, plasma membrane and efflux pumping participate in tolerating the heavy metals. Metallothioneins or heat shock proteins are used to repair stress damaged proteins. Amino acids, organic acids and vacuolar compartmentalization are involved to reduce heavy metals toxicity.

Mycorrhizas

In mycorrhizae, which are partially regarded as plant's heavy metal tolerance mechanism, the type ectomycorrhizas found in plants and shrubs are effective to enhance metal resistance (Marschner 1995, Prasad and Strzałka 1999, Jentschke and Godbold 2000).

Some of the mechanisms which succor to resist and/or reduce heavy metal toxicity are: i) reduced access to apoplast caused by hydrophobicity of fungal sheath, ii) binding to extracellular materials or segregation in vacuolar compartment, iii) binding to extra hyphal slime, iv) chelation by fungal exudates, v) adsorption onto external mycelium and vi) binding metal to cell wall and accumulation in the vacuole. [137-143]

Aarbuscular mycorrhizal and ectomycorrhizal fungi including *Paxillus involutus* and *Thelephora terrestris* have been observed to mitigate heavy metal stress

(Jentschke and Godbold 2000, Van Assche and Clijsters 1990, Prasad and Strzałka 1999). *Suillus bovines* is another example having resistance against Zn, Cd and Cu (Van Assche and Clijsters 1990). *Pisolithustinetorius* shows resistance towards Cu, Zn and Pb. *Pisolithusarrhizus* is also resistant against Cd (Prasad and Strzałka 1999).

The Cell Wall and Root Exudates

The cell wall of root cells is in direct contact with the heavy metals which are present in groundwater. The role of cell wall in the tolerance and removal of heavy metal is difficult to explain due to limited capacity by cell wall and plasma membrane. Some reports describe that some plants are able to tolerate heavy metals in a limited way, such as *Silene vulgaris* and *Humilis* sp. (epidermal cell walls). Root exudates secrete chemicals that can chelate metals; oxalic acid helps tolerate metals.

Plasma Membrane

It is considered as the heavy metals' first living structure to target in order from outer side to inner side which can be effective in very less time due to the heavy metal exposure. The effect can be exerted in the form of increased leakage from the cell, particularly due to Cu. Also, K^+ efflux, cell membrane damage and ion leakage increased in the presence of copper in *Agrostis capillaries, Silene vulgaris*, wheat and *Mimulusguttatus*. Oxidation, cross linking of protein, change to the composition and fluidity of membrane lipids or inhibition of membrane proteins also play role in tackling metal stress. Aluminum activated anion channels in plasma membrane is the main mechanical process to tolerate the metals (Ernst et al. 1992, Bringezu et al. 1999, Clarkson and Hanson 1980, Salt et al. 2000, Ma et al. 2001). Oxidation can directly be affected in the present of Cu and Cd, affecting membrane permeability. In *Nitella,* Cu induces modification in the cell permeability, which can cause hindered light stimulated H^+-ATPase pump. Presence of Cd reduced ATPase activity of plasma membrane found in sunflower roots and wheat. In general, Cu and Zn are essential elements to plants but the tolerance also helps plants in plasma membrane protection. In heavy metal stressed conditions, free radicles or relative oxygen species count increased but mainly depends on the metal homeostasis (Wainwright and Woolhouse 1977, De Vos et al. 1991, Strange and Macnair 1991, Dietz et al. 1999).

Heat Shock Proteins

Heat Shock Proteins (HSP) are known to express reaction in the stress conditions, incorporating heavy metals as well, besides elevated temperature (Vierling 1991, Lewis et al. 1999). There are many reports on role of HSPs in combating metal stress (Tseng et al. 1993). The roots of *Armeria maritime* plants show very good growth on Cu rich soil expressing HSP17 (Neumann et al. 1995). The small heat shock proteins like HSP17 also exhibit increased expression against heavy metals in *Lycopersiconperuvianum* and *Silene* vulgaris (Wollgiehn and Neumann 1999). Expression of larger HSPs like HSP70 have been observed in presence of Cd

(Neumann et al. 1994). *Enteromorpha intestinalis,* a seaweed species, tolerates Cu with the expression of HSP70 (Lewis et al. 1999).

The other important mechanical arrangements which are involved in phytoremediation to tolerate heavy metals are vacuolar compartmentalization, metallothioneins, phytochelatins, organic acids and amino acids (Hall 2002).

Algal Remediation

Adsorption is the way to take up heavy metals by algae. There are two modes for adsorption of pollutants like heavy metals: (a) physical adsorption proceeds for quick adsorption of metal ions on to cell surface. Thereafter, (b) chemisorption leads to bringing the metal ions into cytoplasm.

Many reports have been published that prove the heavy metal remediation capacity of algae. *Scenedesmus obliquus* is able to accumulate elevated Zn and Cd with higher phosphorus content. Moreover, the species is able grow in the presence of Cu, Ni, Cd, Zn and Pb in higher heavy metal concentration (Shehata et al. 1980). The polyphosphate bodies in green algae provide storage lagoon for metal ions and lead to detoxification of heavy metals including Pb, Cd, Zn, Co, Hg, Mg, Ti, Cu, Ni and Sr. Algae play the major role after bacteria in controlling the heavy metal content in oceans and lakes (Goncalves et al. 1985, Sigg et al. 1987). The ability of algae to tolerate and accumulate metals has been seen for many years (Megharaj et al. 2003).

Many species of algae such as *Anabaena inaequalis, Stigeoclonium tenue, Westiellopsisprolifica*, *Synechococcus* sp. and *Chlorella* sp. tolerate heavy metals; among them, some marine algal species of *Anabaena* and *Chlorella* are able to remove heavy metals (Rai et al. 1998).

The mechanism of action to remove heavy metal in microalgae is mediated by phytochelatins and class III metallothioneins which adequately bind to metal ions. *Spirulina* has the capability to eliminate aqueous lead (Chen and Johns 1991). Various reports suggest the role of some important microalgae in the bioremediation of heavy metals: *Oedogoniumrivulare* and *Cladophora glomerata* candidates remove Cd, Cu, Co and Pb, filamentous algae *Spirogyra hatillensis* also contribute in the remedy of Fe, Ni, Mn and Cr. Some other marine microalgae play their role in removing Cd, Pb and Hg from marine environment (Dwivedi 2012).

Fungal Remediation

Some fungi and yeasts can amass various heavy metals such as Zn, Cu, Mn (essential micronutrients) and Cd, Hg, U, Sn, Ni (non-essential nutrient metals) in plenty or sometimes in excess than nutritionally required (Gadd and Mowll 1985). The fungal biomass has proven to be an important biosorbent for the removal of heavy metals (Lee et al. 2000, Merrin et al. 1998, Vala and Patel 2011).

Living and dead both types of cells are involved in metal uptake. One mode of uptake is binding of metal ions to cell wall and auxiliary cellular material. The intracellular uptake is metabolism dependent bioaccumulation or active metal uptake beyond the cell membrane (Sen and Dastidar 2010). The first mode is simple adsorption of metal carried out by both type of cells, dead and live. The second mode, being metabolism dependent, can only be seen in the living cells. Dead cells are the more preferred mode for the bio removal of metal ions due to absence of requirement of growth media and nutrients and absence of toxicity limitation. After the completion of process, the biosorbed metal can be easily recovered and biomass can be reused. The metal uptake reactors can be easily designed mathematically. Since the dead fungal cells show a greater capability to bind metals than live cells, the pretreated live cells (killing of live cells) methods are used (Kapoor and Viraraghavan 1996, Barros Júnior et al. 2003, Kogej and Pavko 2001).

Dead cells prove to be more beneficial over living cells in some industries. Dead cells are less sensitive towards high metal concentration, dreadful operation conditions and of course does not require continuous nutrients and recovery like living cells do. Non-living cells are used as waste product from fermentation processes. Dead cells of various fungi which can be used in heavy metal remediation are used in various industrials applications, like yeast (*Saccharomyces cerevisiae*) in food and beverages industries. *Aspergillus niger, Penicillium chrysogenum, Mucor miehei* and *Rhizopus arrhizus* are used in fermentation industries and *Aspergillus niger, Rhizopus arrhizus, Trichoderma reesei* and *Rhizopus nigricans* are used in pharmaceutical and chemical industries which has been found effective in biosorption of metal ions (Guibal et al. 1992, Luef et al. 1991, Mattuschka and Straube 1993, Niu et al. 1993, Rao et al. 1993, Tobin et al. 1993).

Cells can be killed by various physical methods such as heat treatment, mechanical disruption, autoclaving and vacuum drying (Siegel et al. 1986, Siegel and Siegel 1987, Huang et al. 1988a, Merrin et al. 1998, Huang et al. 1988b, Akthar et al. 1996), chemical treatment using acids, alkalis, formaldehyde and detergents (Barros Júnior et al. 2003, Tan and Cheng 2003, Akthar et al. 1996, Vasudevan et al. 2003, Fernandes and Nazareth 1999, Huang et al. 1988a).

Metal Uptake by Fungal Cells

A broad spectrum of fungi have been reported to biosorb metal ions, especially living cells of *Rhizopus, Penicillium, Mucor, Fusarium, Aspergillus* and *Sachharomyces* (Merrin et al. 1998, Barros Júnior et al. 2003, Kogej and Pavko 2001, Galun et al. 1983a, b, Volesky et al. 1993). Biomass of *Penicillium* has been reported to absorb heavy metals like Zn, As, Cr, Pb and Ni (Tan and Cheng 2003) and radionuclides such as Th, U and Sr can be biosorbed by *Saccharomyces, Penicillium* and *Rizopus* (Galun et al. 1983c, Tsezos and Volesky 1982, Tsezos et al. 1997). Marine-derived fungi have been observed to play important role in heavy metal removal (Vala et al. 2004, Vala 2010, Vala and Patel 2011, Vala and Sutariya 2012). Heavy metal uptake by fungal cells depends on the pH, temperature, composition of growth media, cell age, contact time with metal and equilibrium time. The accumulation of metals like Pb, Zn, Cu, U and Cd outreach equilibrium by *Saccharomyces, Penicillium, Aspergillus, Mucor* and *Rhizopus* non-growing cells into 1-4 hours (Huang et al. 1990, Gadd 1988, Mullen et al. 1992). *Aspergillus niger, Penicillium spinulosum* and *Trichoderma viride* also follow the same pattern (Townsley et al. 1986). The baker's yeast is also able to biosorb uranium and it depends on growth media composition and culture conditions; some reports suggest that 2.6 times more uranium is absorbed in 12-hours culture than 24 hours (Volesky and May-Phillips 1995). pH is also an important factor when things come to biosorption of metal ions. *Penicillium chrysogenum* has been reported to absorb ions of Ni, Pb, Cr and Zn from pH 2.0 to 7.0 (Volesky et al. 1993, Tan and Cheng 2003). Some pH sensitive fungi have been reported in Cd biosorption like *Aspergillus oryzae, Fusarium solani, Apsergillusniger* and *Candida utilis* in the acidic range (Barros Júnior et al. 2003).

Biomass concentration also leaves an effect on the biosorption of heavy metal ions. Some reports are available that show that the absorption activity decreased due to the high concentration of live cells; this situation is caused due to electrostatic interaction of the functional group on the cell surface, cells tend to attach to each other in the suspension and this leads to lowering of cell surface area in the proximity of solution (Merrin et al. 1998). Cd uptake is lower when *Aspergillusniger* concentration is high (Barros Júnior et al. 2003). Cr biosorption is also decreased when the concentrations of *Rhizopus arrhizus, Aspergillus oryzae, Rhizopus nigricans* and *Aspergillus niger* are increased (Niyogi et al. 1998, Lee et al. 2000, Abraham 2001).

The phenomenon of metal uptake can be explained in two ways: passive and active mode. There are many events which follow the unassertive ways like ion exchange, precipitation, complexation and precipitation (Galun et al. 1983a, Tsezos and Volesky 1982, Tsezos et al. 1997, Remade 1990, Kuyucak 1990, Gee and Dudcney 1988). On the other hand, assertive ways rely on metabolic processes of cell after ions were extradited to cell material. This way had been remarked for Mg, Cd, Co, Cu, Ca, Ni, Mn, Sr and Zn (Volesky 1994, White and Gadd 1987, Parkin and Ross 1986a, b). In the case of filamentous fungi, active mode might not be significant towards high metal concentration (Gadd 1990).

Bacterial Remediation

It is better known to us that bacteria are the most prominent contestant in terms of the remedial application compared to the chemical remediation and other biological means due to higher efficiency and occurrence in almost every compartment of environment. More specifically, marine bacteria are the most robust for the purpose. In context of heavy metal resistance/tolerance/remediation, they are potential candidates for the enhanced bioremediation of metal contaminated sites.

Due to the presence of metal resistance genes on bacterial plasmids, they exhibit resistance towards many heavy metals like zinc, copper, mercury, arsenic, gold, cadmium, nickel, tin, chromium, cobalt and many more. Besides the plasmid, some organisms also contain chromosomal gene for heavy metal resistance such as *Escherichia coli* for arsenic, *Bacillus* sp. for mercury and gene for P type ATPase found in *Bacillus* sp. which shows resistance towards cadmium. *Haemophilus influenzae* has *arsC, merP* and *merT* for arsenic and mercury resistance with the sequenced 1.8 Mbpgenome (Fleischmann et al. 1995). A cyanobacterium *Synechocystis*

PCC6808 that bears 1 Mbp and shows resistance towards cadmium, zinc and cobalt (Kanamaru et al. 1994). Novel operons are reported for bacterial mercury resistance. Enzyme arsenic reductase has been detected for arsenic and energy coupling has been observed for arsenic efflux. Some consortium also led to potential remediation like bacterial metallothioneins establishment with cyanobacteria.

Bacteria show some specification for heavy metal resistance:

(i) Most of the bacteria contain metal ion resistance system, but presence of unfamiliar resistance is an antecedent to evolve tolerance.
(ii) Mostly, bacteria have highly specific plasmid determined metal resistance instead of general mechanism.
(iii) In the mechanism, there are mostly two ways: enzyme detoxification and efflux pumping.

There are many studies that report bacterial metal remediation approaches and mechanisms. Here, resistance against Hg, As and Pb are discussed.

Resistance Towards Mercury (Hg)

Bacteria show prominent response against Hg^{2+}, mainly using mercury resistance determinants and *mer* is a well-known example (Silver 1996, Silver and Phung 1996).

Bacteria show resistance against mercury and these cells are competent to reduce Hg^{2+} to the metal (Figure 9.4). After that, mercury ions are transported outside the cell by passive diffusion (Silver 1996, Silver and Phung 1996), where it may be oxidized further by another bacteria (Smith et al. 1998).

In gram-negative bacteria, at first step Hg^{2+} is bounded to MerP (periplasmic binding protein). Hg^{2+} will be transported inside the cell using defined uptake system to avoid toxic effect of mercury ions on periplasmic proteins (Qian et al. 1998). MerP clubs the cation to MerT (mercury transporter) to deliver unto cytoplasm (Hobman and Brown 1996). The other or additional way to transport Hg^{2+} is provided by MerC protein (Hamlett et al. 1992, Sahlman et al. 1997).

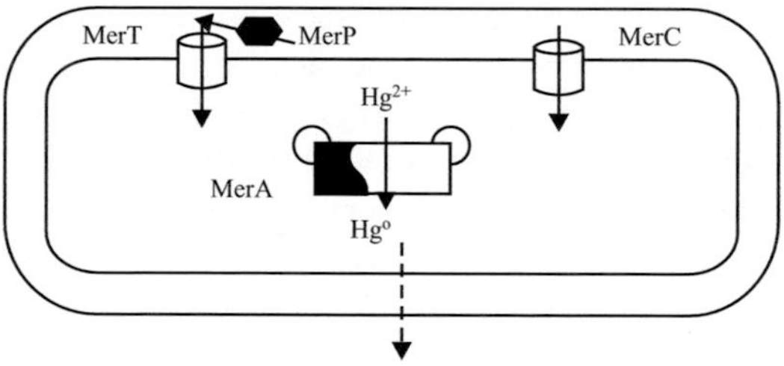

FIGURE 9.4 Mercury uptake and resistance mechanism in Bacteria.

After going inside, Hg^{2+} will be reduced to $Hg^{(0)}$ with NADPH by MerA protein supported by glutathione and other proteins (Schiering et al. 1991).

On the other hand, in addition to this Mer proteins organic mercurial remediation depends on the *mer* resistance determinant which encodes for Mer B organic mercurial lyase diffusion (Silver 1996, Silver and Phung 1996). After cleavage of organomercurials, the formed Hg^{2+} will be reduced by MerA, thus the mechanism decreases toxicity of organo mercurial (methylated and alkylated heavy metal compounds) resulting by chemical modification. Expressivity of MerA reductase was first detected in yeast (Rensing et al. 1992), then observed in plant as well (Rugh et al. 1998a, b, 1996). The *mer* – operator-based *lux* biosensor leads to the unique way of remediation and sensing approach (Selifonova et al. 1993).

Resistance for Lead (Pb)

Generally, lead is less toxic than other heavy metals but at a certain concentration, it would also affect living beings in various ways. Bacteria have been found that precipitate lead phosphate (Trajanovska et al. 1997, Levinson and Mahler 1998).

Lead resistance moderated by P-type ATPase in *Ralstonia* sp. CH34 and CadA P-type ATPase can also deliver Pb^{2+} (Rensing et al. 1998). Lead resistance is mainly carried out due to metal ion efflux.

Resistance to Arsenic (As)

Bacteria employ arsenic resistance efflux system that transports As(III) either involving two component ATPase (ArsA and ArsB) or involving only ArsB as a chemiosmotic transporter. Intracellular arsenate is converted to arsenite, the substrate of the efflux system by ArsC (arsenate reductase) (Nies and Silver 1995; Ji et al. 1994; Ji et al. 1992; Silver and Phung 1996).

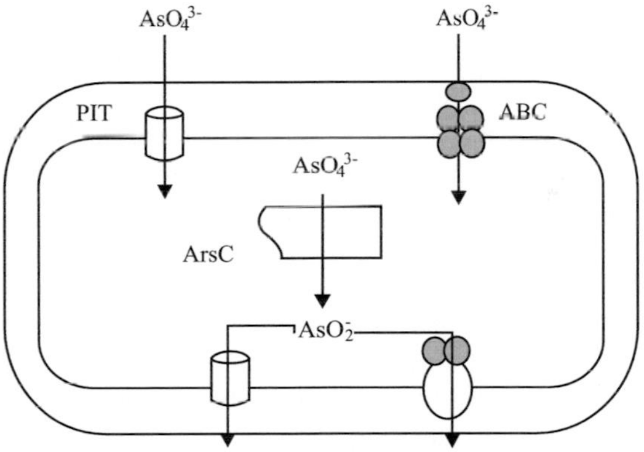

FIGURE 9.5 Arsenic uptake and resistance mechanism in Bacteria.

In *Escherichia coli,* arsenate reduction is done by ArsC protein which is allied to glutathione through glutaredoxin (Oden et al. 1994, Gladysheva et al. 1994, Liu and Rosen 1997). In *Staphylococcus aureus,* the electron donor for ArsC is thioredoxin (Ji et al. 1994). After that, arsenite leaves the cell which is mostly driven by chemiosmotic gradient and ArsB product which is an arsenite efflux system made up of just one efflux protein (Wu et al. 1992). In some cases, transporters are made up of ArsB protein assisted by ArsA ATPase. The ArsB protein is able to work alone in these systems (Kuroda et al. 1997), but when the mechanism goes by using $ArsA_2B$ complex, it is driven by chemiosmotic pressure and ATP (Dey and Rosen 1995). ArsA works as dimer along with four ATP – binding site and associated proteins in animals, plants, fungi, bacteria and archaea (Li et al. 1996, Li and Rosen 1998, Zhou and Rosen 1997). ArsB have been also reported in *Saccharomyces cerevisiae* and man (Rosenstein et al. 1992, Wysocki et al. 1997, Kurdi-Haidar et al. 1998a).

In some bacteria, plasmid system is found, like in *Staphylococcus xylosus*, while *E. coli* possesses chromosomally encoded system (Rosenstein et al. 1992, Diorio et al. 1995). The best explored example for arsenical resistance is plasmid encoded system in *E. coli* (Chen et al. 1986).

CASE STUDY: BIOLOGICALLY INDUCED REMOVAL OF SELECTED HEAVY METALS USING MARINE BACTERIA ISOLATED FROM ASSBRY SEA COAST

Introduction

In the context of environmental concern, the pollution of various hazardous materials like heavy metals, total petroleum hydrocarbons, and PAHs affect the biodiversity of the surrounding including human beings. There are so many sources which are brutally modifying nature. Some of them are various industries, transportations and other ways to contaminate natural habitats which leads to air, soil, and water pollution. Among them, marine pollution has eminent impact. It results in deleterious effects to marine diversity and living resources. The main sources of marine pollution are industrial and agricultural run-off, shipping transportation, ship breaking and recycling and tourism. Ship breaking practice leads to release of hazardous pollutants like PAHs, petroleum hydrocarbons, heavy metals and other materials. Heavy metals exert devastating impact on marine environment. In this study, As, Hg and Pb tolerating bacteria isolated from Alang – Sosia Ship Breaking and Recycling Yard (ASSBRY), Gujarat, India have been screened and their metal removal potential has been examined.

ASSBRY costal area has huge economic importance regarding shipping industries from past few decades. Ship breaking activity generates hazardous wastes like heavy metals, aromatic compounds, petroleum derivatives, various non-degradable and toxic compounds. Metal remediation using conventional physio-chemical techniques is quite expensive and not eco-friendly. Biological approach is one of the best alternative tool in recent years (Kurdi-Haidar et al. 1998b). Microbes (bacteria/microalgae/fungi) are useful as an alternative for metal removal; such a practice could be technically and economically viable. Some of the native microbial flora are capable of withstanding and coping elevated concentrations of toxic pollutants and are also found to transform/accumulate/degrade such recalcitrant compounds.

Materials and Methods

Sampling and Isolation

ASSBRY coast, which is situated on the Gulf of Kambhat, Gujarat, India, was selected as sampling site as it is Asia's largest ship breaking and recycling yard. Two sites situated in the north (Sitel) and south (Site2) part of ASSBRY coast line were selected for the study (Figure 9.6a, b). Water samples were examined for physicochemical parameters and were also investigated for the heavy metal content using Inductively Coupled Plasma – Mass Spectrometry (ICP-MS). Water samples were collected in sterile BOD bottles and brought to the laboratory in ice box. The samples were analyzed within 6h of collection.

FIGURE 9.6a Geographical Map of ASSBRY Sea Coast (Courtesy: Google Maps).

FIGURE 9.6b Sampling Site (Site 1: 21.41°N 72.19°E); (Site 1: 21.39°N 72.17°E) (Courtesy: Google Maps).

TABLE 9.3 Physico-chemical parameters of water samples collected from Site-1 and Site-2.

Parameters	Site-1	Site-2
Temperature (°C)	29.5	31.8
pH	8.5	8.13
Conductivity (mS/cm)	41.8	68.2
Total Dissolved Solid (TDS) (g/L)	28.2	34.5
Salinity (ppt)	41.7	43.2
DO (mg/L)	4.85	5.16
BOD (mg/L)	35.01	29.04

TABLE 9.4 Heavy metal quantification of sea water samples by using ICP-MS.

Heavy Metals	Pre-Monsoon (ppm)	Monsoon (ppm)	Post-Monsoon (ppm)
Arsenic (As)	0.2	0.16	0.2
Chromium (Cr)	9.6	227.6	135.9
Copper (Cu)	109.4	196.7	56.8
Lead (Pb)	52	36	48
Mercury (Hg)	181	389.7	156.7
Nickel (Ni)	38.3	47.5	32.9
Zinc (Zn)	294	266.3	185.9

For the isolation of metal (Hg, Pb and As) tolerating/removing bacteria, diluted (10^{-3}/ mL and 10^{-4}/mL) water samples were spread (in triplicates) on Zobell Marine Agar plates having different metal concentrations, i.e. 25 ppm, 50 ppm and 100 ppm. One set of media without added metal was also inoculated. The metals were selected on the basis of their distribution at study site and toxic nature.

The isolates, which exhibited tolerance to 100 ppm metal were exposed to higher concentrations (500 ppm, 1000 ppm, 1500 ppm and 2000 ppm) of respective metals on Zobell marine agar plates. For further confirming the organisms' ability to withstand such a high metal concentration, the organisms were also grown on Bromfield medium (an oligotrophic medium) with respective metal concentration (Azad et al. 2013).

Heavy Metal Removal Assay in Bacteria

In order to assess the ability of the test organism to remove metal from the medium, the screened isolates were grown in the liquid medium with highest tolerated concentration of the respective heavy metal. Based on screening results for agar plates, Hg concentration was kept 500 ppm and Pb and As test concentration was kept 1500 ppm. After 48 hours incubation, cell pellets and remaining medium were separated by centrifugation and then metal content was quantified from both samples using ICP-MS, following standard protocol and comparison with control in two ways: i) metal accumulated in bacterial cells and ii) metal remaining in the medium. The calculations were done by the resulting quantification of metal concentration in both systems.

Results and Discussion

In the first steps of the experiment, the sea water and sediment sample were collected from the heavy metal contaminated site (ASSBRY coast line). The physico-chemical parameters were measured as showed in Table 9.3. Along with them, the quantification of heavy metal pollution at the site was performed (Table 9.4). The reconnaissance indicates the presence of the heavy metal contamination and enough biotic community to do study.

Bacterial isolation on Zobell marine agar yielded a total of 116 isolates (CFUs) at the lowest heavy metal concentration, i.e. 25 ppm, as shown in Table 9.5. The highest mercury concentration where one isolate could grow was 500 ppm; in case of lead and arsenic, at 1500 ppm, 2 and 3 isolates could be obtained. There was no growth on plates amended with 2000 ppm metal. Microorganisms have several mechanisms to cope up with the heavy metals; in simple words, they can do three things: transform intracellularly/extracellularly or/and tolerate extracellularly or/and avoid metal content and grow utilizing other nutrients from the media, sometimes metal ions may get trapped in solid complex media. Hence, in the later stage, screening using the oligotrophic medium, Bromfield medium, was also carried out to further confirm occurrence of true metal tolerant bacteria.

TABLE 9.5 Number of isolates from screening respective to specific heavy metals.

Heavy Metals	25 ppm	50 ppm	100 ppm	500 ppm	1000 ppm	1500 ppm
Mercury (Hg)	12	8	5	1	–	–
Lead (Pb)	51	29	26	7	5	2
Arsenic (As)	53	46	16	6	5	3

Using the same oligotrophic medium, the metal removal efficiency analysis was carried out using ICP-MS technique. The results (Fig. 9.7) were quite interesting: the organisms were taking the respective metal inside the cell (or adsorbing) (Fig. 9.7a, b, c). For this study, the organisms selected were those which tolerated the highest respective heavy metal concentration. Bacterium no. 6 showed 40% accumulation/removal of mercury, whereas in the case of lead isolate no. 13 and 21 showed 30% and 20% accumulation/removal, respectively. In the case of arsenic, isolate no. 32, 36 and 35 accumulated/removed 20%, 30% and 20% arsenic (Fig. 9.7d). The results suggest that the organisms could uptake and remove the metals from the medium efficiently; hence, they have potential to possibly remove them from the contaminated sites. Further detailed analysis may help understand the mechanism of heavy metal removal by marine bacteria.

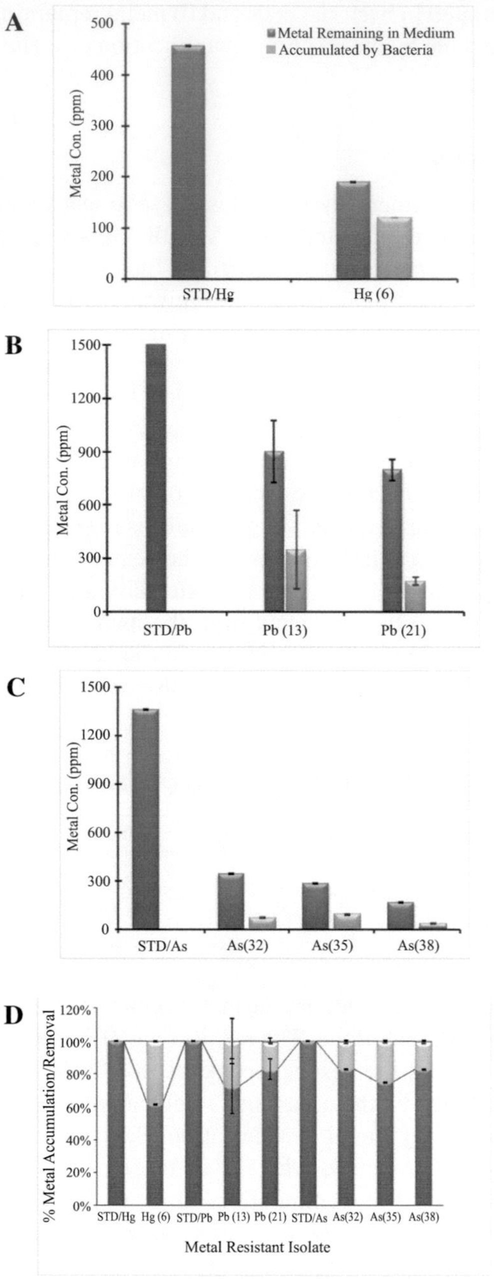

FIGURE 9.7 Resulted accumulation of metal by bacteria (Heavy Metal tolerance/removal assay).

A: Mercury Accumulation
B: Lead Accumulation
C: Arsenic Accumulation
D: Percentage Accumulation of all above three

ACKNOWLEDGEMENTS

Kind support of Dr Bharatsinh M Gohil, I/cHead, Department of Life Sciences, MK Bhavnagar University is thankfully acknowledged.

REFERENCES

Abadin, H., A. Ashizawa, Y.W. Stevens, F. Llados, G. Diamond, G. Sage, et al. 2007. Agency for toxic substances and disease registry (ATSDR) toxicological profiles. Toxicological Profile for Lead. Atlanta (GA): Agency for Toxic Substances and Disease Registry (US).

Abraham, T. Emilia. 2001. Biosorption of Cr (VI) from aqueous solution by *Rhizopus nigricans*. Bioresour. Technol. 79(1): 73-81.

Ahluwalia, S.S. and D. Goyal. 2005. Removal of heavy metals by waste tea leaves from aqueous solution. Eng. Life Sci. 5(2): 158-162.

Ahmad, A.L. and B. Seng Ooi. 2010. A study on acid reclamation and copper recovery using low pressure nanofiltration membrane. Chem. Eng. J. 156(2): 257-263.

Ahn, C.K., D. Park, S.H. Woo and J.M. Park. 2009. Removal of cationic heavy metal from aqueous solution by activated carbon impregnated with anionic surfactants. J. Hazard. Mater. 164(2-3): 1130-1136.

Aksu, Z., Y. Sag and T. Kutsal. 1992. The biosorpnon of copperod by *C. vulgaris* and *Z. ramigera*. Environ. Technol. 13(6): 579-586.

Akhtar, M.N., K.S. Sastry and P. Maruthi Mohan. 1996. Mechanism of metal ion biosorption by fungal biomass. Biometals 9(1): 21-28.

Al-Jlil, S.A. and F.D. Alsewailem. 2009. Saudi Arabian clays for lead removal in wastewater. Appl. Clay Sci. 42(3-4): 671-674.

Alluri, H.K., S.R. Ronda, V.S. Settalluri, J.S. Bondili, V. Suryanarayana and P. Venkateshwar. 2007. Biosorption: An eco-friendly alternative for heavy metal removal. Afr. J. Biotechnol. 6(25): 2924-2931.

Andreesen, Jan R. and Kathrin Makdessi. 2008. Tungsten, the surprisingly positively acting heavy metal element for prokaryotes. Ann. N.Y. Acad. Sci. 1125(1): 215-229.

Apiratikul R. and P. Pavasant. 2008. Sorption of Cu^{2+}, Cd^{2+} and Pb^{2+} using modified zeolite from coal fly ash. Chem. Eng. J. 144(2): 245-258.

Aroua M.K., F.M. Zuki and N.M. Sulaiman. 2007. Removal of chromium ions from aqueous solutions by polymer-enhanced ultrafiltration. J. Hazard. Mater. 147(3): 752-758. doi: 10.1016/j.jhazmat.2007.01.120.

Azad, A.K., A. Nahar, M.M. Hasan, K. Islam, Md Azim, Md S. Hossain, et al. 2013. Fermentation of municipal solid wastes by bacterial isolates for production of raw protein degrading proteases. Asian J. Microbiol. Biotechnol. Environ. Sci. 15: 365-374.

Baltpurvins, K.A., R.C. Burns, G.A. Lawrance and A.D. Stuart. 1997. Effect of electrolyte composition on zinc hydroxide precipitation by lime. Water Res. 31(5): 973-980.

Barros Júnior, L.M., G.R. Macedo, M.M.L. Duarte, E.P. Silva and A.K.C.L. Lobato. 2003. Biosorption of cadmium using the fungus *Aspergillus niger*. Braz. J. Chem. Eng. 20(3): 229-239.

Betancur, M., P.R. Bonelli, J.A. Velásquez and A.L. Cukierman. 2009. Potentiality of lignin from the Kraft pulping process for removal of trace nickel from wastewater: Effect of demineralisation, Bioresour. Technol. 100(3): 1130-1137.

Beyersmann, D. and A. Hartwig. 2008. Carcinogenic metal compounds: Recent insight into molecular and cellular mechanisms. Arch. Toxicol. 82(8): 493.

Bhattacharya, K.G. and S. SenGupta. 2008. Adsorption of a few heavy metals on natural and modified kaolinite and montmorillonite: A review. Adv. Colloid Interface Sci. 140(2): 114-131.

Blaudez, D., B. Botton and M. Chalot. 2000. Cadmium uptake and subcellular compartmentation in the ectomycorrhizal fungus *Paxillus involutus*. Microbiology 146(5): 1109-1117.

Brierley, J.A. 1990. Production and application of a *Bacillus*-based product for use in metals biosorption. pp. 305-312. *In*: B. Volesky. (ed.). Biosorption of Heavy Metals. CRC Press, Boca Raton, Florida.

Bringezu, K., O. Lichtenberger, I. Leopold and D. Neumann. 1999. Heavy metal tolerance of Silene vulgaris. J. Plant Physiol. 154(4): 536-546.

Bromfield, C.S. and A.F. Shride. 1956. Mineral resources of the San Carlos Indian Reservation, Arizona. US Government Printing Office.

Centeno, J.A., P.B. Tchounwou, A.K. Patlolla, F.G. Mullick, L. Murakata, E. Meza, et al. 2006. Environmental pathology and health effects of arsenic poisoning: A Critical Review. pp. 311-327. *In*: R. Naidu, E. Smith, G. Owens, P. Bhattacharya, Nadebaum. (eds.). Managing Arsenic in the Environmennt: From Soil to Human Health. CSIRO Publishing Melbourne, Austrlia.

Chang, L.W., L. Magos and T. Suzuki. (eds.). 1996. Toxicology of Metals. Taylor & Francis US.

Charerntanyarak, Lertchai. 1999. Heavy metals removal by chemical coagulation and precipitation. Water Sci. Technol. 39(10-11): 135-138.

Chen, C.M., T.K. Misra, S. Silver and B.P. Rosen. 1986. Nucleotide sequence of the structural genes for an anion pump. The plasmid-encoded arsenical resistance operon. J. Biol. Chem. 261(32): 15030-15038.

Chen, F. and M.R. Johns. 1991. Effect of C/N ratio and aeration on the fatty acid composition of heterotrophic *Chlorella sorokiniana*. J. Appl. Phycol. 3(3): 203-209.

Chubar, N., M. J.R. Carvalho and J.N. Correia. 2003. Cork biomass as biosorbent for Cu (II), Zn (II) and Ni (II). Colloids Surf. A Physicochem. Eng. Asp. 230(1-3): 57-65.

Churchill, S.A., J.V. Walters and P.F. Churchill. 1995. Sorption of heavy metals by prepared bacterial cell surfaces. J. Environ. Eng. 121(10): 706-711.

Ciszewski, D. and I. Malik. 2004. The use of heavy metal concentrations and dendrochronology in the reconstruction of sediment accumulation, MałaPanew River Valley, southern Poland. Geomorphology 58(1-4): 161-174.

Clarkson, D.T. and J.B. Hanson. 1980. The mineral nutrition of higher plants. Annu. Rev. Plant Physiol. 31(1): 239-298.

Collins, Y.E. and G. Stotzky. 1992. Heavy metals alter the electrokinetic properties of bacteria, yeasts and clay minerals. Appl. Environ. Microbiol. 58(5): 1592-1600.

Colpaert, J.V. and J.A. Van Assche. 1992. The effects of cadmium and the cadmium-zinc interaction on the axenic growth of ectomycorrhizal fungi. Plant Soil 145(2): 237-243.

Cséfalvay, E., V. Pauer and P. Mizsey. 2009. Recovery of copper from process waters by nanofiltration and reverse osmosis. Desalination 240(1-3): 132-142.

Dakiky, M., M. Khamis, A. Manassra and M. Mer'Eb. 2002. Selective adsorption of chromium (VI) in industrial wastewater using low-cost abundantly available adsorbents. Adv. Environ. Res. 6(4): 533-540.

De Vos, C.H.R., H. Schat, M.A.M. De Waal, R. Vooijs and W.H.O. Ernst. 1991. Increased resistance to copper-induced damage of the root cell plasmalemma in copper tolerant *Silene cucubalus*. Physiol. Plant. 82(4): 523-528.

Dey, S. and B.P. Rosen. 1995. Dual mode of energy coupling by the oxyanion-translocating ArsB protein. J. Bacteriol. 177(2): 385-389.

Dietz, K-J., M. Baier and U. Krämer. 1999. Free radicals and reactive oxygen species as mediators of heavy metal toxicity in plants. Heavy metal stress in plants. Springer, Berlin, Heidelberg, pp. 73-97.

Diorio, C., J. Cai, J. Marmor, R. Shinder and M.S. DuBow. 1995. An Escherichia coli chromosomal ars operon homolog is functional in arsenic detoxification and is conserved in gram-negative bacteria. J. Bacteriol. 177(8): 2050-2056.

Dojlido, J. and G.A. Best. 1993. Chemistry of Water and Water Pollution. Ellis Horwood Limited.

Dorsey, A., L. Ingerman and S. Swart. 2004. Toxicological profile for copperhttps://www.atsdr.cdc.gov/toxprofiles/tp132.pdf.

Duruibe, J. Ogwuegbu, M.O.C. Ogwuegbu and J.N. Egwurugwu. 2007. Heavy metal pollution and human biotoxic effects. Int. J. Phys. Sci. 2(5): 112-118.

Dwivedi, S. 2012. Bioremediation of heavy metal by algae: Current and future perspective. J. Adv. Lab. 3(3): 195-199.

Egami, F. 1975. Origin and early evolution of transition element enzymes. J. Biochem. 77(6): 1165-1169.

Eriksson, P. 1988. Nanofiltration extends the range of membrane filtration. Environ. Prog. 7(1): 58-62.

Ernst, W.H.O., J.A.C. Verkleij and H. Schat. 1992. Metal tolerance in plants. Acta Bot. Neerl. 41(3): 229-248.

Feng, D., C. Aldrich and H. Tan. 2000. Treatment of acid mine water by use of heavy metal precipitation and ion exchange. Miner. Eng. 13(6): 623-642.

Fernandes, O.S. and S.W. Nazareth. 1999. Studies on lead sorption by *Fusarium solani*. Poll. Res. 18(3): 211-216.

Figoli, A., A. Cassano, A. Criscuoli, M.S.I. Mozumder, M.T. Uddin, M.A. Islam, et al. 2010. Influence of operating parameters on the arsenic removal by nanofiltration. Water Res. 44(1): 97-104.

Figueira, M.M., B. Volesky, V.S.T. Ciminelli and F.A. Roddick. 2000. Biosorption of metals in brown seaweed biomass. Water Res. 34(1): 196-204.

Fleischmann, R.D., M.D. Adams, O. White, R.A. Clayton, E.F. Kirkness, A.R. Kerlavage, et al. 1995. Whole-genome random sequencing and assembly of *Haemophilus influenzae* Rd. Science 269(5223): 496-512.

Fu, Fenglian and Qi Wang. 2011. Removal of heavy metal ions from wastewaters: A review. J Environ. Manage. 92(3): 407-418.

Fu, Fenglian, Runming Chen and Ya Xiong. 2006. Application of a novel strategy—coordination polymerization precipitation to the treatment of Cu^{2+}-containing wastewaters. Sep. Purif. Technol. 52(2): 388-393.

Gadd, G.M. 1990. Fungi and yeasts for metal accumulation. pp. 249-276. *In*: H.L. Ehrlich, C.L. Brierley. (eds). Microbial mineral recovery. McGraw-Hill, New York.

Gadd, G.M. and J.L. Mowll. 1985. Copper uptake by yeast-like cells, hyphae and chlamydospores of *Aureobasidium pullulans*. Exp. Mycol. 9(3): 0-40.

Gadd, G.M., C. White, L. de. Rome. 1988. Heavy metal and radionuclide uptake by fungi and yeast. pp. 421-436. *In*: P.R. Norris, D.P. Kelly. (eds.). Biohydrometallurgy: Proceedings of an International Symposium. Science Technology Letters, Kew Surrey, UK.

Galun, M., P. Keller, H. Feldstein, E. Galun, S. Siegel and B. Siegel. 1983a. Recovery of uranium (VI) from solution using fungi II. Release from uranium-loaded *Penicillium* biomass. Water, Air and Soil Pollution 20(3): 277-285.

Galun, M., P. Keller, D. Malki, H. Feldstein, E. Galun, S. Siegel, et al. 1983b. Recovery of uranium (VI) from solution using precultured *Penicillium* biomass. Water, Air and Soil Pollution 20(2): 221-232.

Galun, M., P. Keller, D. Malki, H. Feldstein, E. Galun, S.M Siegel, et al. 1983c. Removal of uranium (VI) from solution by fungal biomass and fungal wall-related biopolymers. Science 219(4582): 285-286.

Gautam R.K., S.K. Sharma, S. Mahiya and M.C. Chattopadhyaya. 2014. Contamination of heavy metals in aquatic media: transport, toxicity and technologies for remediation. pp. 1-24. *In*: S. Sharma. (ed.). Heavy Metals in Water: Presence, Removal and Safety. Royal Society of Chemistry, London.

Gee, A.R. and A.W.L. Dudeney. 1988. Adsorption and crystallization of gold at biological surfaces. pp. 437-451. *In*: P.R. Norris, D.P. Kelly. (eds.). Biohydrometallurgy. Science & Technology Letters: London, UK.

Ghosh, P., A.N. Samanta and S. Ray. 2011. Reduction of COD and removal of Zn^{2+} from rayon industry wastewater by combined electro-Fenton treatment and chemical precipitation. Desalination 266(1-3): 213-217.

Gladysheva, T.B., Oden K.L.P. Rosen. 1994. Properties of the arsenate reductase of plasmid R773. Biochemistry 33(23): 7288-7293.

Goncalves, M.L.S., L. Sigg and W. Stumm. 1985. Voltammetric methods for distinguishing between dissolved and particulate metal ion concentrations in the presence of hydrous oxides. A case study on lead (II). Environ. Sci. Technol. 19(2): 141-146.

Gonzalez-Munoz, M.J., M.A. Rodríguez, S. Luque and J.R. Álvarez. 2006. Recovery of heavy metals from metal industry waste waters by chemical precipitation and nanofiltration. Desalination (Amsterdam) 200(1-3) : 742-744.

Groves, J.T., R.C. Haushalter, M. Nakamura, T.E. Nemo and B.J. Evans. 1981. High-valent iron-porphyrin complexes related to peroxidase and cytochrome P-450. J. Am. Chem. Soc. 103(10): 2884-2886.

Xueyuan, G. and L.J. Evans. 2008. Surface complexation modelling of Cd (II), Cu (II), Ni (II), Pb (II) and Zn (II) adsorption onto kaolinite. Geochim. Cosmochim. Acta 72(2): 267-276.

Guibal, E., Ch Roulph and P. Le Cloirec. 1992. Uranium biosorption by a filamentous fungus Mucor miehei pH effect on mechanisms and performances of uptake. Water Res. 26(8): 1139-1145.

Hall, J.L. 2002. Cellular mechanisms for heavy metal detoxification and tolerance. Journal of Experimental Botany 53(366): 1-11.

Hamlett, N.V., E.C. Landale, B.H. Davis and A.O. Summers. 1992. Roles of the Tn21 merT, merP and merC gene products in mercury resistance and mercury binding. J. Bacteriol. 174(20): 6377-6385.

Hartley, J., J.W.G. Cairney and A.A. Meharg. 1997. Do ectomycorrhizal fungi exhibit adaptive tolerance to potentially toxic metals in the environment? Plant Soil 189(2): 303-319.

Harvey, L.J. and H.J. McArdle. 2008. Biomarkers of copper status: A brief update. Brit. J. Nutr. 99(S3): S10-S13.

Hobman, J.L. and N.L. Brown. 1996. Overexpression of MerT, the mercuric ion transport protein of transposon Tn501 and genetic selection of mercury hypersensitivity mutations. Molecular and General Genetics MGG 250(1): 129-134.

Holum, J.R. 1979. Elements of General and Biological Chemistry An Introduction to the Molecular Basis of Life. Wiley & Sons, 565 pp.

Hosea, M., B. Greene, R. Mcpherson, M. Henzl, M.D. Alexander and D.W. Darnall. 1986. Accumulation of elemental gold on the alga Chlorella vulgaris. InorganicaChimica Acta 123(3): 161-165.

Hu, M.Z.C., J.M. Norman, B.D. Faison and M.E. Reeves. 1996. Biosorption of uranium by Pseudomonas aeruginosa strain CSU: Characterization and comparison studies. Biotechnol. Bioeng. 51(2): 237-247.

Huang, C.P., D. Westman, K. Quirk, J.P. Huang and A.L. Morehart. 1988a. Removal of cadmium (II) from dilute aqueous solutions by fungal biomass. Particulate Science and Technology 6(4): 405-419.

Huang, C.P., D. Westman, K. Quirk and J.P. Huang. 1988b. The removal of cadmium (II) from dilute aqueous solutions by fungal adsorbent. Water Science and Technology 20(11-12): 369-376.

Huang, C. and A.L. Morehart. 1990. The removal of Cu (II) from dilute aqueous solutions by *Saccharomyces cerevisiae*. Water Res. 24(4): 433-439.

Hudson, T.S., S. Subramanian and R.J. Allen. 1984. Determination of pantothenic acid, biotin and vitamin B12 in nutritional products. J. Assoc. Off. Anal. Chem. 67(5): 994-998.

Huisman, J.L., G. Schouten and C. Schultz. 2006. Biologically produced sulphide for purification of process streams, effluent treatment and recovery of metals in the metal and mining industry. Hydrometallurgy 83(1-4): 106-113.

Hüttermann, A., I. Arduini and D.L. Godbold. 1999. Metal pollution and forest decline. pp. 253-272. *In:* M.N.V. Prasad. (ed.). Heavy Metal Stress in Plants. Springer, Berlin, Heidelberg.

Iijima S. 1991. Helical microtubules of graphitic carbon. Nature 354(6348): 56-58.

Iyengar, G.V. 1998. Reevaluation of the trace element content in reference man. Radiat. Phys. Chem. 51(4-6): 545-560.

Järup, Lars. 2003. Hazards of heavy metal contamination. Br. Med. Bull. 68(1): 167-182.

Jentschke, G. and D.L. Godbold. 2000. Metal toxicity and ectomycorrhizas. Physiologia Plantarum 109(2): 107-116.

Ji, G. and S. Silver. 1992. Reduction of arsenate to arsenite by the ArsC protein of the arsenic resistance operon of *Staphylococcus aureus* plasmid pI258. Proc. Natl. Acad. Sci. 89(20): 9474-9478.

Ji, G., E.A. Garber, L.G. Armes, C.M. Chen, J.A. Fuchs and S. Silver. 1994. Arsenate reductase of *Staphylococcus aureus* plasmid pI258. Biochemistry 33(23): 7294-7299.

Jusoh, A., L. SuShiung and M.J.M.M. Noor. 2007. A simulation study of the removal efficiency of granular activated carbon on cadmium and lead. Desalination 206(1-3): 9-16.

Kabbashi, N.A., M.A. Atieh, A. Al-Mamun, M.E. Mirghami, M.D.Z. Alam and N. Yahya. 2009. Kinetic adsorption of application of carbon nanotubes for Pb(II) removal from aqueous solution. J. Environ. Sci. 21(4): 539-544.

Kanamaru, K., S. Kashiwagi and T.A. Mizuno. 1994. A copper-transporting P-type ATPase found in the thylakoid membrane of the cyanobacterium *Synechococcus* species PCC7942. Mol. Microbiol. 13(2): 369-377.

Kandah, M.I. and J.L. Meunier. 2007. Removal of nickel ions from water by multi-walled carbon nanotubes. J. Hazard. Mater. 146(1-2): 283-288.

Kang, K.C., S.S. Kim, J.W. Choi and S.H. Kwon. 2008. Sorption of Cu^{2+} and Cd^{2+} onto acid- and base-pretreated granular activated carbon and activated carbon fiber samples. J. Ind. Eng. Chem. 14(1): 131-135.

Kang, S.Y., J.U. Lee, S.H. Moon and K.W. Kim. 2004. Competitive adsorption characteristics of Co^{2+}, Ni^{2+} and Cr^{3+} by IRN-77 cation exchange resin in synthesized wastewater. Chemosphere 56(2): 141-147.

Kapoor, A. and T. Viraraghavan. 1996. Biosorption of heavy metals on fungal biomass. Proceedings of the 4th CSCE Environmental Engineering Speciality Conference, Edmonton, Alberta.

Kapoor, A. and T. Viraraghavan. 1997. Heavy metal biosorption sites in *Aspergillus niger*. Bioresour. Technol. 61(3): 221-227.

Kim, H.J., K. Baek, B.K. Kim and J.W. Yang. 2005. Humic substance-enhanced ultrafiltration for removal of cobalt. J. Hazard. Mater. 122(1-2): 31-36.

Kogej, A. and A. Pavko. 2001. Laboratory experiments of lead biosorption by self-immobilized *Rhizopus nigricans* pellets in the batch stirred tank reactor and the packed bed column. Chem. Biochem. Eng. Q. 15(2): 75-80.

Köhler, S.J., P. Cubillas, J.D. Rodríguez-Blanco, C. Bauer and M. Prieto. 2007. Removal of cadmium from wastewaters by aragonite shells and the influence of other divalent cations. Environ. Sci. Technol. 41(1): 112-118.

Kommel, L., E. Kimmari, M. Saarna and M. Viljus. 2013. Processing and properties of bulk ultrafine-grained pure niobium. J. Mater. Sci. 48(13): 4723-4729.

Kongsricharoern, N. and C. Polprasert. 1995. Electrochemical precipitation of chromium (Cr6+) from an electroplating wastewater. Water Sci. Technol. 31(9): 109.

Ku, Y. and I.L. Jung. 2001. Photocatalytic reduction of Cr (VI) in aqueous solutions by UV irradiation with the presence of titanium dioxide. Water Res. 35(1): 135-142.

Kuo, Chao-Yin and Han-Yu Lin. 2009. Adsorption of aqueous cadmium (II) onto modified multi-walled carbon nanotubes following microwave/chemical treatment. Desalination 249(2): 792-796.

Kurdi-Haidar, B., D. Heath, S. Aebi and S.B. Howell. 1998a. Biochemical characterization of the human arsenite-stimulated ATPase (hASNA-I). Journal of Biological Chemistry 273(35): 22173-22176.

Kurdi-Haidar, B., D.K. Hom, D.E. Flittner, D. Heath, L. Fink, P. Naredi, et al. 1998b. Dual cytoplasmic and nuclear distribution of the novel arsenite-stimulated human ATPase (hASNA-I). Journal of Cellular Biochemistry 71(1): 1-10.

Kuroda, M., S. Dey, O.I. Sanders and B.P. Rosen. 1997. Alternate energy coupling of ArsB, the membrane subunit of the Ars anion-translocating ATPase. J. Biol. Chem. 272(1): 326-331.

Kuyucak, N. 1990. Feasibility of biosorbents applications. *In*: B. Volesky. (ed.). Biosorption of Heavy Metals. Boca Raton: CRC Press, pp. 371-378.

Labanda, J., M.S. Khaidar and J. Llorens. 2009. Feasibility study on the recovery of chromium (III) by polymer enhanced ultrafiltration. Desalination 249(2): 577-581.

Landaburu-Aguirre, J., V. García, E. Pongrácz and R.L. Keiski. 2009. The removal of zinc from synthetic wastewaters by micellar-enhanced ultrafiltration: Statistical design of experiments. Desalination 240(1-3): 262-269.

Lane, T.W., M.A. Saito, G.N. George, I.J. Pickering, R.C. Prince and F.M. Morel. 2005. A cadmium enzyme from a marine diatom. Nature 435: 42.

Lee, D.C., C.J. Park, J.E. Yang, Y.H. Jeong and H.I. Rhee. 2000. Screening of hexavalent chromium biosorbent from marine algae. Appl. Microbiol. Biotechnol. 54(3): 445-448.

Levinson, Hillel S. and Inga Mahler. 1998. Phosphatase activity and lead resistance in Citrobacter freundii and Staphylococcus aureus. FEMS Microbiology Letters 161(1): 135-138.

Lewis, S., R.D. Handy, B. Cordi, Z. Billinghurst and M.H. Depledge. 1999. Stress proteins (HSP's): Methods of detection and their use as an environmental biomarker. Ecotoxicology 8(5): 351-368.

Li, J. and B.P. Rosen. 1998. Steric limitations in the interaction of the ATP binding domains of the ArsA ATPase. J. Biol. Chem. 273(12): 6796-6800.

Li, J., S. Liu and B.P. Rosen. 1996. Interaction of ATP binding sites in the ArsA ATPase, the catalytic subunit of the Ars pump. J. Biol. Chem. 271(41): 25247-25252.

Li, Y., F. Liu, B. Xia, Q. Du, P. Zhang, D. Wang, et al. 2010. Removal of copper from aqueous solution by carbon nanotube/calcium alginate composites. J. Hazard. Mater. 177(1-3): 876-880.

Liu, J. and B.P. Rosen. 1997. Ligand interactions of the ArsC arsenate reductase. J. Biol. Chem. 272(34): 21084-21089.

Liu, Z.R., L.M. Zhou, W.E.I. Peng, Z.E.N.G. Kai, C.X. Wen and H.H. Lan. 2008. Competitive adsorption of heavy metal ions on peat. Journal of China University of Mining and Technology 18(2): 255-260.

Lu, A., S. Zhong, J. Chen, J. Shi, J. Tang and X. Lu. 2006. Removal of Cr(VI) and Cr(III) from aqueous solutions and industrial wastewaters by natural clino-pyrrhotite. Environ. Sci. Technol. 40(9): 3064-3069.

Leuf, E.T., T. Prey and G.P. Kubicek. 1991. Biosorption of zinc by fungal mycelial wastes. Appl. Microbiol. Biotechnol. 34(5): 688-692.

Ma, J.F., P.R. Ryan and E. Delhaize. 2001. Aluminium tolerance in plants and the complexing role of organic acids. Trends Plant. Sci. 6(6): 273-278.

Maret, W. 2013. Zinc biochemistry: From a single zinc enzyme to a key element of life. Adv. Nutr. 4(1): 82-91.

Marschner, Horst. 1995. Mineral nutrition of higher plants. Institute of Plant Nutrition, University of Hohenheim.

Matlock, Matthew M., Brock S. Howerton and David A. Atwood. 2002a. Chemical precipitation of heavy metals from acid mine drainage. Water Res. 36(19): 4757-4764.

Matlock, M.M., B.S. Howerton, M.A. Van Aelstyn, F.L. Nordstrom and D.A. Atwood. 2002b. Advanced mercury removal from gold leachate solutions prior to gold and silver extraction: A field study from an active gold mine in Peru. Environ. Sci. Technol. 36(7): 1636-1639.

Matlock, Matthew M., Kevin R. Henke and David A. Atwood. 2002c. Effectiveness of commercial reagents for heavy metal removal from water with new insights for future chelate designs. Journal of Hazardous Materials 92(2): 129-142.

Mattuschka, B. and G. Straube. 1993. Biosorption of metals by a waste biomass. J. Chem. Technol. Biotechnol. 58(1): 57-63.

Megharaj, M., S. Avudainayagam and R. Naidu. 2003. Toxicity of hexavalent chromium and its reduction by bacteria isolated from soil contaminated with tannery waste. Curr. Microbiol. 47(1): 0051-0054.

Melo, J.S. and S.F. D'souza. 2004. Removal of chromium by mucilaginous seeds of *Ocimum basilicum*. Bioresour. Technol. 92(2): 151-155.

Mendel, Ralf R. 2005. Molybdenum: Biological activity and metabolism. Dalton Trans. 21: 3404-3409.

Merrin, J.S., R. Sheela, N. Saswathi, R.S. Prakasham and S.V. Ramakrishna. 1998. Biosorption of chromium VI using Rhizopus arrhizus. Indian J. Exp. Biol. 36(10): 1052-1055.

Mertz, Walter. 1981. The essential trace elements. Science 213(4514): 1332-1338.

Mirbagheri, S.A. and S.N. Hosseini. 2005. Pilot plant investigation on petrochemical wastewater treatment for the removal of copper and chromium with the objective of reuse. Desalination 171(1): 85-93.

Mohan, D. and S. Chander. 2006. Removal and recovery of metal ions from acid mine drainage using lignite—a low cost sorbent. J. Hazard. Mater. 137(3): 1545-1553.

Molinari, R., T. Poerio and P. Argurio. 2008. Selective separation of copper (II) and nickel (II) from aqueous media using the complexation-ultrafiltration process. Chemosphere 70(3): 341-348.

Montgomery, Hugh. 1930. The copper content and the minimal molecular weight of the hemocyanins of Busycon canaliculatum and of Loligopealei. Biol. Bull. Rev. 58(1): 18-27.

Motsi, T., N.A. Rowson and M.J.H. Simmons. 2009. Adsorption of heavy metals from acid mine drainage by natural zeolite. Int. J. Miner. Process. 92(1-2): 42-48.

Mudhoo, A., S.K. Sharma, V.K. Garg and C.H. Tseng. 2011. Arsenic: An overview of applications, health and environmental concerns and removal processes. Crit. Rev. Env. Sci. Tec. 41(5): 435-519.

Mullen, M.D., D.C. Wolf, T.J. Beveridge and G.W. Bailey. 1992. Sorption of heavy metals by the soil fungi Aspergillus niger and Mucor rouxii. Soil Biol. Biochem. 24(2): 129-135.

Muraleedharan, T.R. and C. Venkobachar. 1990. Mechanism of biosorption of copper (II) by Ganoderma lucidum. Biotechnol. Bioeng. 35(3): 320-325.

Murthy, Z.V.P. and Latesh B. Chaudhari. 2008. Application of nanofiltration for the rejection of nickel ions from aqueous solutions and estimation of membrane transport parameters. J. Hazard. Mater. 160(1): 70-77.

Muthukrishnan, M. and B.K. Guha. 2008. Effect of pH on rejection of hexavalent chromium by nanofiltration. Desalination 219(1-3): 171-178.

Neumann, D., O. Lichtenberger, D. Günther, K. Tschiersch and L. Nover. 1994. Heat-shock proteins induce heavy-metal tolerance in higher plants. Planta 194(3): 360-367.

Neumann, D., U. Zur Nieden, O. Lichtenberger and I. Leopold. 1995. How does Armeriamaritima tolerate high heavy metal concentrations?. J. Plant. Physiol. 146(5-6): 704-717.

Nguyen, C.M., S. Bang, J. Cho and K.W. Kim. 2009. Performance and mechanism of arsenic removal from water by a nanofiltration membrane. Desalination 245(1-3): 82-94.

Nieboer, E. and D.H.S. Richardson. 1980. The replacement of the nondescript term 'heavy metals' by a biologically and chemically significant classification of metal ions. Environmental Pollution Series B, Chemical and Physical 1(1): 3-26.

Nies, D.H. and S. Silver. 1995. Ion efflux systems involved in bacterial metal resistances. J. Ind. Microbiol. 14(2): 186-199.

Niu, H. and B. Volesky. 1999. Characteristics of gold biosorption from cyanide solution. J. Chem. Technol. Biotechnol. 74(8): 778-784.

Niu, H., X.S. Xu, J.H. Wang and J.H. Volesky. 1993. Removal of lead from aqueous solutions by *Penicillium* biomass. J. Chem. Technol. Biotechnol. 42(6): 785-787.

Niyogi, S., T. Emilia Abraham and S.V. Ramakrishna. 1998. Removal of chromium (VI) ions from industrial effluents by immobilized biomass of Rhizopus arrhizus. J. Sci. Ind. Res. 57(10-11): 809-816.

Nolan, K.R. 1983. Copper toxicity syndrome. J. Orthomol. Psych. 12(4): 270-282.

Oden, K.L., T.B. Gladysheva and B.P. Rosen. 1994. Arsenate reduction mediated by the plasmid-encoded ArsC protein is coupled to glutathione. Mol. Microbiol. 12(2): 301-306.

Ostroski, I.C., M.A. Barros, E.A. Silva, J.H. Dantas, P.A. Arroyo and O.C. Lima. 2009. A comparative study for the ion exchange of Fe (III) and Zn (II) on zeolite NaY. J. Hazard. Mater. 161(2-3): 1404-1412.

Özverdi A. and M. Erdem. 2006. Cu^{2+}, Cd^{2+} and Pb^{2+} adsorption from aqueous solutions by pyrite and synthetic iron sulphide. J. Hazard. Mater. 137(1): 626-632.

Papadopoulos, A., D. Fatta, K. Parperis, A. Mentzis, K.J. Haralambous and M. Loizidou. 2004. Nickel uptake from a wastewater stream produced in a metal finishing industry by combination of ion-exchange and precipitation methods. Sep. Purif. Technol. 39(3): 181-188.

Park, H.G., T.W. Kim, M.Y. Chae and I.K. Yoo. 2007. Activated carbon-containing alginate adsorbent for the simultaneous removal of heavy metals and toxic organics. Process. Biochem. 42(10): 1371-1377.

Parkin, Michael J. and I. Stuart Ross. 1986a. The regulation of Mn^{2+} and Cu^{2+} uptake in cells of the yeast *Candida utilis* grown in continuous culture. FEMS Microbiology Letters 37(1): 59-62.

Parkin, Michael J. and I. Stuart Ross. 1986b. The specific uptake of manganese in the yeast *Candida utilis*. Microbiology 132(8): 2155-2160.

Patlolla, A.K., C. Barnes, C. Yedjou, V.R. Velma and P.B. Tchounwou. 2009a. Oxidative stress, DNA damage and antioxidant enzyme activity induced by hexavalent chromium in Sprague-Dawley rats. Environ. Toxicol. 24(1): 66-73.

Patlolla, A.K., C. Barnes, D. Hackett and P.B. Tchounwou. 2009b. Potassium dichromate induced cytotoxicity, genotoxicity and oxidative stress in human liver carcinoma (HepG2) cells. Int. J. Environ. Res. Public Health 6(2): 643-653.

Pillay, K., E.M. Cukrowska and N.J. Coville. 2009. Multi-walled carbon nanotubes as adsorbents for the removal of parts per billion levels of hexavalent chromium from aqueous solution. J. Hazard. Mater. 166(2-3): 1067-1075.

Prasad, M.N.V. and K. Strzałka. 1999. Impact of heavy metals on photosynthesis. pp.117-138. *In*: M.N.V. Prasad. (ed.). Heavy Metal Stress in Plants. Springer, Berlin, Heidelberg.

Pribula, Alan. 1991. The elements, their origin, abundance and distribution (Cox, PA). J. Chem. Educ. 68(4): A112.

Qian, H., L. Sahlman, P.O. Eriksson, C. Hambraeus, U. Edlund and I. Sethson. 1998. NMR solution structure of the oxidized form of MerP, a mercuric ion binding protein involved in bacterial mercuric ion resistance. Biochemistry 37(26): 9316-9322.

Rai, U.N., M. Gupta, R.D. Tripathi and P. Chandra. 1998. Cadmium regulated nitrate reductase activity in *Hydrilla verticillata* (1. f.) Royle. Water Air Soil Pollut. 106(1-2): 171-177.

Rao, C.R.M. and G.S. Reddi. 2000. Platinum group metals (PGM); occurrence, use and recent trends in their determination. TrAC Trends in Analytical Chemistry 19(9): 565-586.

Rao, C.R.N., L. Iyengar and C. Venkobachar. 1993. Sorption of copper (II) from aqueous phase by waste biomass. J. Environ. Eng. 119(2): 369-377.

Rao, G.P., C. Lu and F. Su. 2007. Sorption of divalent metal ions from aqueous solution by carbon nanotubes: A review. Sep. Purif. Technol. 58(1): 224-231.

Remacle, J. 1990. The cell wall and metal binding. pp. 83-92. *In*: B. Volesky. (ed.). Biosorption of Heavy Metals. CRC Press, Boca Raton, Florida, USA.

Rensing, C., U. Kües, U. Stahl, D.H. Nies and B. Friedrich. 1992. Expression of bacterial mercuric ion reductase in *Saccharomyces cerevisiae*. J. Bacteriol. 174(4): 1288-1292.

Rensing, C., Y. Sun, B. Mitra and B.P. Rosen. 1998. Pb (II)-translocating P-type ATPases. J. Biol. Chem. 273(49): 32614-32617.

Reyes, I., M. Villarroel, M.C. Diez and R. Navia. 2009. Using lignimerin (a recovered organic material from Kraft cellulose mill wastewater) as sorbent for Cu and Zn retention from aqueous solutions. Bioresour. Technol. 100(20): 4676-4682.

Rosenstein, R., A. Peschel, B. Wieland and F. Götz. 1992. Expression and regulation of the antimonite, arsenite and arsenate resistance operon of *Staphylococcus xylosus* plasmid pSX267. Journal of Bacteriology 174(11): 3676-3683.

Rugh, C.L., H.D. Wilde, N.M. Stack, D.M. Thompson, A.O. Summers and R.B. Meagher. 1996. Mercuric ion reduction and resistance in transgenic *Arabidopsis thaliana* plants expressing a modified bacterial merA gene. Proc. Natl. Acad. Sci. 93(8): 3182-3187.

Rugh, C.L., J.F. Senecoff, R.B. Meagher and S.A. Merkle. 1998a. Development of transgenic yellow poplar for mercury phytoremediation. Nature Biotechnology 16(10): 925-928.

Rugh, C.L., G.M. Gragson, R.B. Meagher and S.A. Merkle. 1998b. Toxic mercury reduction and remediation using transgenic plants with a modified bacterial gene. Hortscience 33(4): 618-621.

Saeed, A., M. Iqbal and M.W. Akhtar. 2002. Application of biowaste materials for the sorption of heavy metals in contaminated aqueous medium. Pak. J. Sci. Ind. Res. 45(3): 206-211.

Sahlman, L., W. Wong and J. Powlowski. 1997. A mercuric ion uptake role for the integral inner membrane protein, MerC, involved in bacterial mercuric ion resistance. J. Biol. Chem. 272(47): 29518-29526.

Salt, D.E., N. Kato, U. Kramer, R.D. Smith and I. Raskin. 2000. The Role of root exudates in nickel hyperaccumulation and tolerance in accumulator and non accumulator species of Thlaspi. 189-200. *In*: N. Terry, G. Banuelos. (eds.). Phytoremediation of Contaminated Soil and Water. Lewis Publishers Inc, Boca Raton.

Schiering, N., W. Kabsch, M.J. Moore, M.D. Distefano, C.T. Walsh and E.F. Pai. 1991. Structure of the detoxification catalyst mercuric ion reductase from Bacillus sp. strain RC607. Nature 352(6331): 168-172.

Selifonova, O., R. Burlage and T. Barkay. 1993. Bioluminescent sensors for detection of bioavailable Hg (II) in the environment. Appl. Environ. Microbiol. 59(9): 3083-3090.

Sen, M. and M. Ghosh Dastidar. 2010. Adsorption-desorption studies on Cr (VI) using non-living fungal biomass. Asian J. Chem. 22(3): 2331.

Shahalam, A.M., A. Al-Harthy and A. Al-Zawhry. 2002. Feed water pretreatment in RO systems: Unit processes in the Middle East. Desalination 150(3): 235-245.

Shaw, B.P., S.K. Sahu and R.K. Mishra. 2004. Heavy metal induced oxidative damage in terrestrial plants. pp. 84-126. *In*: Heavy Metal Stress in Plants. Springer, Berlin, Heidelberg.

Shehata, S.A. and S.A. Badr. 1980. Growth response of Scenedesmus to different concentrations of copper, cadmium, nickel, zinc and lead. Environ. Int. 4(5-6): 431-434.

Sheng, G., S. Wang, J. Hu, Y. Lu, J. Li, Y. Dong and X. Wang. 2009. Adsorption of Pb (II) on diatomite as affected via aqueous solution chemistry and temperature. Colloids Surf. A Physicochem. Eng. Asp. 339(1-3): 159-166.

Siegel, S., P. Keller, M. Galun, H. Lehr, B. Siegel and E. Galun. 1986. Biosorption of lead and chromium by *Penicillium* preparations. Water Air Soil Pollut. 27(1-2): 69-75.

Siegel, S.M. and B.Z. Siegel. 1987. Fungal biosorption: A comparative study of metal uptake by *Penicillium* and *Cladosporium*. Metals Speciation, Separation and Recovery. Lewis Publishers Chelsea, MI, pp. 339-364.

Sigg, L., M. Sturn and D. Kistler. 1987. Vertical transport of heavy metals by settling particles in Lake Zurich. Limnol. Oceanogr. 32(1): 112-130.

Silver, S. 1996. Bacterial resistances to toxic metal ions-a review. Gene 179(1): 9-19.

Silver, S. and L.T. Phung. 1996. Bacterial heavy metal resistance: New surprises. Annu. Rev. Microbiol. 50(1): 753-789.

Smith, T., K. Pitts, J.A. McGarvey and A.O. Summers. 1998. Bacterial oxidation of mercury metal vapor, Hg (0). Appl. Environ. Microbiol. 64(4): 1328-1332.

Stadtman, T.C. 1990. Selenium biochemistry. Annu. Rev. Biochem. 59(1): 111-127.

Stern, B.R. 2010. Essentiality and toxicity in copper health risk assessment: Overview, update and regulatory considerations. J. Toxicol. Environ. 73(2-3): 114-127.

Stowers, C.C., B.M. Cox and B.A. Rodriguez. 2014. Development of an industrializable fermentation process for propionic acid production. J. Ind. Microbiol. Biotechnol. 41(5): 837-852.

Strandberg, G.W., I.I. Shumate, J.R. Jr. Parrott and D.A. McWhirter. 1980. Microbial uptake of uranium, cesium and radium. No. CONF-800814-12. Oak Ridge National Lab., TN (USA).

Strandberg, G.W., S.E. Shumatc and J.R. Parrot. 1981. Microbial cells as biosorbents for heavy metals: Accumulation of uranium by *Saccharomyces cerevisiae* and *Pseudomonas aeruginosa*. Appl. Environ. Microbiol. 41(1): 237-245.

Strage, J. and M.R. Macnair. 1991. Evidence for a role for the cell membrane in copper tolerance of Mimulus guttatus Fischer ex DC. New Phytologist 119(3): 383-388.

Sud, D., G. Mahajan and M.P. Kaur. 2008. Agricultural waste material as potential adsorbent for sequestering heavy metal ions from aqueous solutions: A review. Bioresour. Technol. 99(14): 6017-6027.

Sutton, D.J. and P.B. Tchounwou. 2007. Mercury induces the externalization of phosphatidylserine in human renal proximal tubule (HK-2) cells. Int. J. Environ. Res. Public Health 4(2): 138-144.

Sutton, D.J., P.B. Tchounwou, N. Ninashvili and E. Shen. 2002. Mercury induces cytotoxicity and transcriptionally activates stress genes in human liver carcinoma (HepG2) cells. Int. J. Mol. Sci. 3(9): 965-984.

Taffarel, S.R. and J. Rubio. 2009. On the removal of Mn^{2+} ions by adsorption onto natural and activated Chilean zeolites. Miner. Eng. 22(4): 336-343.

Tam, Paul C.F. 1995. Heavy metal tolerance by ectomycorrhizal fungi and metal amelioration by Pisolithustinctorius. Mycorrhiza 5(3): 181-187.

Tan T. and P. Cheng. 2003. Biosorption of metal ions with *Penicillium chrysogenum*. Appl. Biochem. Biotechnol. 104(2): 119-128.

Tchounwou, P.B., I. 1999. Development of public health advisories for arsenic in drinking water. Rev. Environ. Health 14(4): 211-230.

Tchounwou, P.B., A.B. Ishaque and J. Schneider. 2001. Cytotoxicity and transcriptional activation of stress genes in human liver carcinoma cells (HepG2) exposed to cadmium chloride. Mol. Cell Biochem. 222(1-2): 21-28.

Tchounwou, P.B., C.G. Yedjou, D.N. Foxx, A.B. Ishaque and E. Shen. 2004a. Lead-induced cytotoxicity and transcriptional activation of stress genes in human liver carcinoma (HepG 2) cells. Molecular and Cellular Biochemistry 255(1-2): 161-170.

Tchounwou, Paul B., Jose A. Centeno and Anita K. Patlolla. 2004b. Arsenic toxicity, mutagenesis and carcinogenesis: A health risk assessment and management approach. Molecular and Cellular Biochemistry 255(1-2): 47-55.

Tchounwou, P.B., C.G. Yedjou, A.K. Patlolla and D.J. Sutton. 2012. Heavy metal toxicity and the environment. Experientia Supplementum 101: 133-164.

Terfassa, B., J.A. Schachner, P. Traar, F. Belaj and N.C.M. Zanetti. 2014. Oxorhenium (V) complexes with naphtholate-oxazoline ligands in the catalytic epoxidation of olefins. Polyhedron 75: 141-145.

Tobin, J.M., B. L'homme and J.C. Roux. 1993. Immobilisation protocols and effects on cadmium uptake by *Rhizopus arrhizus* biosorbents. Biotechnol. Tech. 7(10): 739-744.

Townsley, C.C., I.S. Ross and A.S. Atkins. 1986. Copper removal from a simulated leach effluent using the filamentous fungus *Trichoderma viride*. Immobilisation of ions by biosorption. Chichester, UK: Ellis Harwood 159.

Trajanovska, S., L.B. Margaret and M. Bhave. 1997. Detection of heavy metal ion resistance genes in Gram-positive and Gram-negative bacteria isolated from a lead-contaminated site. Biodegradation 8(2): 113-124.

Tseng, T.S., S.S. Tzeng, C.H. Yeh, F.C. Cheng, Y.M. Chen and C.Y. Lin. 1993. The heat-shock response in rice seedlings: Isolation and expression of cDNAs that encode class I low-molecular-weight heat-shock proteins. Plant Cell Physiol. 34(1): 165-168.

Tsezos, M. and B. Volesky. 1982. The mechanism of uranium biosorption by *Rhizopus arrhizus*. Biotechnol. Bioeng. 24(2): 385-401.

Tsezos, Marios, Zoe Georgousis and Emmanouela Remoudaki. 1997. Mechanism of aluminum interference on uranium biosorption by *Rhizopus arrhizus*. Biotechnol. Bioeng. 55(1): 16-27.

Üçer, A., A. Uyanik and Ş.F. Aygün. 2006. Adsorption of Cu (II), Cd (II), Zn (II), Mn (II) and Fe (III) ions by tannic acid immobilised activated carbon. Separation and Purification Technology 47(3): 113-118.

Vala, A.K., N. Anand, P.N. Bhatt and H.V. Joshi. 2004. Tolerance and accumulation of hexavalent chromium by two seaweed associated aspergilli. Mar. Pollut. Bull. 48(9-10): 983-985.

Vala, A.K. 2010. Tolerance and removal of arsenic by a facultative marine fungus *Aspergillus candidus*. Bioresour. Technol. 101(7): 2565-2567.

Vala, A.K. and R.J. Patel. 2011. Biosorption of trivalent arsenic by facultative marine *Aspergillus niger*. pp. 459-464. *In*: A. Mason. (ed.). Bioremediation: Biotechnology, Engineering and Environmental Management. Nova Science Publishers, New York.

Vala, A.K. and V. Sutariya. 2012. Trivalent arsenic tolerance and accumulation in two facultative marine fungi. Jundishapur J. Microbiol. 5(4): 542-545.

Van Assche, F. and H. Clijsters. 1990. Effects of metals on enzyme activity in plants. Plant Cell Environ. 13(3): 195-206.

Van Tichelen, Katia K., Jan V. Colpaert and Jaco Vangronsveld. 2001. Ectomycorrhizal protection of *Pinus sylvestris* against copper toxicity. New Phytol. 150(1): 203-213.

Vasudevan, Padma, V. Padmavathy and S.C. Dhingra. 2003. Kinetics of biosorption of cadmium on Baker's yeast. Bioresour. Technol. 89(3): 281-287.

Viczec, S.A., K.B. Jensen and K.A. Francesconi. 2016. Arsenic-containing phosphatidylcholines: A new group of arsenolipids discovered in herring caviar. AngewandteChemie 128(17): 5345-5348.

Vieirna, R.H. and B. Volesky. 2000. Biosorption: A solution to pollution? Int. J. Microbiol. 3(1): 17-24.

Vierling, E. 1991. The roles of heat shock proteins in plants. Annu. Rev. Plant Physiol. Plant. Mol. Biol. 42(1): 579-620.

Volesky, B. and H.A. May-Phillips. 1995. Biosorption of heavy metals by *Saccharomyces cerevisiae*. Appl. Microbiol. Biotechnol. 42(5): 797-806.

Volesky, B., H. May and Z.R. Holan. 1993. Cadmium biosorption by *Saccharomyces cerevisiae*. Biotechnol. Bioeng. 41(8): 826-829.

Volesky, B., J. Weber and R.H.S.F. Vieira. 1999. Biosorption of Cd and Cu by different types of *Sargassum* biomass. Process Metallurgy. Vol. 9. Elsevier, pp. 473-482.

Volesky, B. 1994. Advances in biosorption of metals: Selection of biomass types. FEMS Microbiol. Rev. 14(4): 291-302.

Wainwright, S.J. and H.W. Woolhouse. 1977. Some physiological aspects of copper and zinc tolerance in *Agrostis tenuis* Sibth.: Cell elongation and membrane damage. J. Exp. Bot. 28(4): 1029-1036.

Wan Ngah, W.S. and M.A.K.M. Hanafiah. 2008. Removal of heavy metal ions from wastewater by chemically modified plant wastes as adsorbents: a review. Bioresour. Technol. 99(10): 3935-3948.

Wang, H., A. Zhou, F. Peng, H. Yu and J. Yang. 2007. Mechanism study on adsorption of acidified multiwalled carbon nanotubes to Pb (II). J. Colloid Interface Sci. 316(2): 277-283.

Wang, Suwei and Xianglin Shi. 2001. Molecular mechanisms of metal toxicity and carcinogenesis. Molecular and Cellular Biochemistry 222(1-2): 3-9.

White, C. and G.M. Gadd. 1987. The uptake and cellular distribution of zinc in *Saccharomyces cerevisiae*. Microbiology 133(3): 727-737.

Wilburn, D.R. 2012. Global exploration and production capacity for platinum-group metals from 1995 through 2015. US Department of the Interior, US Geological Survey.

Wollgiehn, R. and D. Neumann. 1999. Metal stress response and tolerance of cultured cells from *Silene vulgaris* and *Lycopersicon peruvianum*: Role of heat stress proteins. J. Plant Physiol. 154(4): 547-553.
World Health Organization. Trace elements in human nutrition and health. World Health Organization, 1996.
Wu, J. L.S. Tisa and B.P. Rosen. 1992. Membrane topology of the ArsB protein, the membrane subunit of an anion-translocating ATPase. J. Biol. Chem. 267(18): 12570-12576.
Wysocki, R., P. Bobrowicz and S. Ulaszewski. 1997. The *Saccharomyces cerevisiae* ACR3 gene encodes a putative membrane protein involved in arsenite transport. J. Biol. Chem. 272(48): 30061-30066.
Yanagisawa, H., Y. Matsumoto, M. Machida. 2010. Adsorption of Zn (II) and Cd (II) ions onto magnesium and activated carbon composite in aqueous solution. Appl. Surf. Sci. 256(6): 1619-1623.
Yedjou, C.G. and P.B. Tchounwou. 2006. Oxidative stress in human leukemia (HL-60), human liver carcinoma (HepG2) and human (Jurkat-T) cells exposed to arsenic trioxide. Metal ions in biology and medicine: proceedings of the... International Symposium on Metal Ions in Biology and Medicine held...= Les ions metalliquesenbiologie et enmedecine:... Symposium international sur les ions metalliques.... Vol. 9. NIH Public Access.
Yedjou, Clement G. and Paul B. Tchounwou. 2007a. *In vitro* cytotoxic and genotoxic effects of arsenic trioxide on human leukemia (HL-60) cells using the MTT and alkaline single cell gel electrophoresis (Comet) assays. Molecular and Cellular Biochemistry 301(1-2): 123-130.
Yedjou, Clement G. and Paul B. Tchounwou. 2007b. N-acetyl-l-cysteine affords protection against lead-induced cytotoxicity and oxidative stress in human liver carcinoma (HepG2) cells. International Journal of Environmental Research and Public Health 4(2): 132-137.
Ying, X. and F. Zhang. 2006. Experimental research on heavy metal wastewater treatment with dipropyl dithiophosphate. J. Hazard. Mater. 137(3): 1636-1642.
Young, S.B. 2013. Minerals, Metals and Sustainability: Meeting Future Material Needs, by William J. Rankin. Boca Raton, FL, USA: CRC Press, 2011, 440 pp. ISBN 9780415684590, paperback, $99.95. J. Ind. Ecol. 17(2): 334-335.
Zhou, T. and B.P. Rosen. 1997. Tryptophan fluorescence reports nucleotide-induced conformational changes in a domain of the ArsA ATPase. J. Biol. Chem. 272(32): 19731-19737.

10

Ketamine: Its Abuse and Effect on Aquatic Ecosystem

Rupak Thapa

INTRODUCTION

In today's world, nearly 70% of people are under some sorts of prescribed medication and it is of no surprise that lots of chemicals either from pharmaceuticals, hospitals or metabolic excreta ultimately go to water and have its influence on the aquatic ecosystem. Hundreds and thousands of the most frequently consumed medications include ibuprofen, naproxen, beta-blockers, antibiotics, antidepressants, and immunosuppressive drugs. With increasing population and urbanization, chemicals are being exponentially discharged in water every day. Our water life is home to various pharmaceutical products that come via different sources (Fig. 10.1).

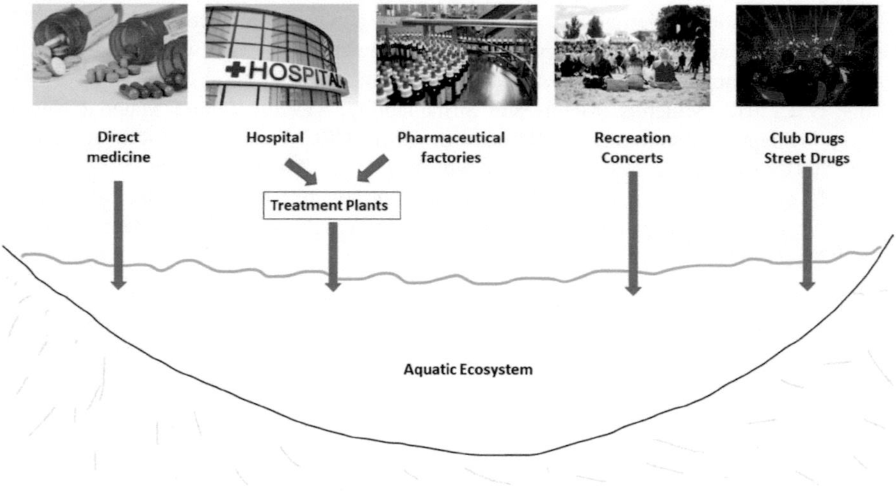

FIGURE 10.1 Sources of drug which ultimately goes to the aquatic environment.

One of the most abused drugs in various parts of the world is ketamine. Though it is a control substance in the United States of America and not easily available without

Department of Physiology, SIU School of Medicine, Life Science II, room 245, Southern Illinois University 1125 Lincoln Drive, Mail Code 6512, Carbondale, IL 62901.
E-mail: rupak.thapa@siu.edu

a prescription, it is legal and easily available in other countries. Popularly known by its nickname "Special K", it is misused in parties, clubs, streets, etc. and often found to be mixed with other drugs and alcohol. Ketamine is abused worldwide because it relieves pain and provides the heavenly pleasure of detaching from the self. These illicit drugs directly or indirectly mix with water and impact the aquatic life severely. Therefore, it is very important to understand the fate and side effects of ketamine and other illicit drug products' pollution in the aquatic environment as it not only affects aquatic life but also pose threat to human health.

History

History of ketamine dates to the 1950s, when phencyclidine (PCP) was used as an anesthetic drug synthesized by Parke Davis Company in 1956 (Maddox et al. 1965). PCP also goes by name "angel dust" and has mind-altering effects including hallucinations, confusions, and unpredictable behavior. PCP has been shown to have a drunken state effect in rats, delirium effects in dogs and catatonia state effects in monkeys. Though it proved to be a potent anesthetic medication, its use was discontinued in humans due to various side effects associated with it. In pursuit of better anesthetic with lesser side effects, Parke Davis discovered various analogues of phencyclidine which were experimented on rats, dogs, and monkeys. Among all the analogues, Ketamine became popular due to its fast-acting nature. Calvin Stevens first synthesized Ketamine commercially in 1962 at Parke Davis Company and it experimented in humans in 1965. Ketamine is a unique drug as it has hypnotic, analgesic and amnesic effects.

Ketamine not only provides relief from pain but also induces a trance-like heavenly state where users experience the pinnacle of relaxation and full-body buzz. Users have reported having experienced sensation of detachment from their body and floating in the universe, which is like the near-death experience "K-hole". Therefore, Ketamine also goes by the name "Special K", a very popular street and party drug. It was on the West Coast where its abuse was reported for the first time. Later, in the 1980s its abuse increased exponentially. Several forms of the drug became available in the market including injection, powder, crystals, capsule, tablets, and solutions. Eventually, Ketamine became a popular socio-recreational drug linked to club, party, and dance. It was found to be misused as an adulterant with ecstasy, alcohol, tobacco products and marijuana. Misuse came in light when it was used in gay dance during the 1990s in the UK where it was used under different pseudonyms like Special K, Super K, Green, Jet and Mean Green. As the drug was misused in recreational activities, the federal government in the USA classified Ketamine as a controlled substance in 1999.

Routes of Administration

The intravenous infusion and intramuscular injection are the most common methods of administering Ketamine. Depending on the situation, anesthesiologists provide Ketamine via various routes, for example, oral, intranasal, anal, etc. (Fig. 10.2) (Table 10.1).

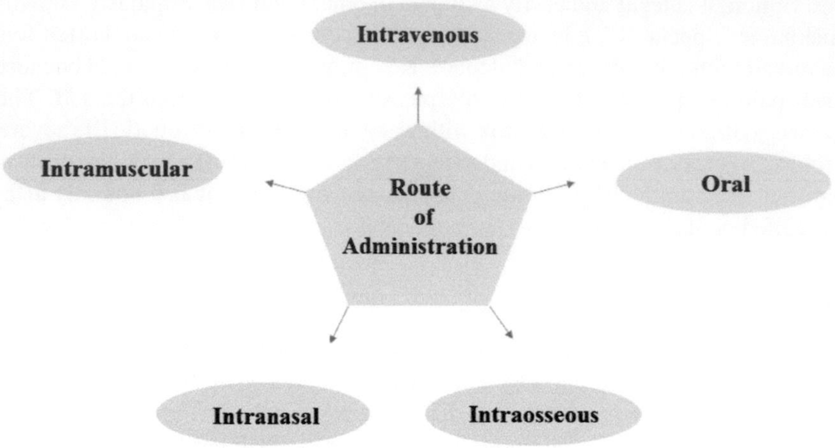

FIGURE 10.2 Common routes of administration.

TABLE 10.1 Ketamine dose.

S. No.	Route of Administration	Dose	Duration	Drug Effect
1	Intravenous (anesthesia)	2 mg/kg	60 sec	5-10 mins
2	Intramuscular (anesthesia)	10 mg/kg	3 to 4 mins	12-25 mins
3	Intravenous (analgesia)	0.2-0.75 mg/kg	–	–

Ketamine and its Environmental Occurence

As ketamine possesses the unique property of anesthetic, analgesic and hypnotic, this drug has been used and misused profoundly. Ketamine has been found to be detected in the environment in various parts of the world (Table 10.2). The conventional wastewater treatment was found to be ineffective against ketamine and its metabolites which possess a threat to the aquatic environment. The environmental occurrence of ketamine is mainly due to two major factors:

1. Veterinary, hospitals and medical purpose
2. Individual, recreational purpose

Environmental detection of ketamine poses a threat to aquatic life as well as animals that are under its direct or indirect influence. Long term use of ketamine is associated with central nervous system malfunction which causes behavioral change and addiction. The toxic effect of ketamine on animals is associated with an increase in oxidative stress and neurotoxicity. There have been several reports of ecotoxicological consequences for ketamine and its metabolites (Table 10.3).

TABLE 10.2 Ketamine detection in various parts of the world.

S. No.	Location	Amount Detected (Ketamine)	Sample Type	Reference
1	China (36 major rivers in Bohai sea and North Yellow sea region in September 2015)	0.05 to 4.5 ng/ltr	(3-5) 10 ltr water samples were mixed in one 60 ltr stainless steel barrel and finally, 0.5L mixed water was stored in polyethylene terephthalate bottles	(Wang et al. 2016)
2	Taipei Taiwan	Influent: 147 ng/ltr Effluent: 206-343 ng/ltr	Samples (1 ltr) collected from each site in amber glass bottles and stored in ice-packed coolers 11 river sampling sites, effluent from 5 hospitals	(Lin et al. 2010)
3	China (sewage treatment plant in Guangzhou)	64.9-673.7 mg/day/1000 people	24 hr composite wastewater samples were collected from a municipal sewage treatment plant.	(Zang et al. 2019)
4	Belgium (Antwerpen-Zuid, Brussel-Noord, and Deurne)	8-27 ng/ltr Ketamine DHNK not detected Norketamine detected	24 hr composite influent wastewater samples that were collected from 3 different WWTPs in Belgium	(Nuijs et al. 2013)
5	Columbia, Bogata, and Medellin	Bogata 3.7 mg/day/1000 people Medellin 1.5 mg/day/1000 people	24 hr composite influent wastewater (IWW) samples collected daily over 7 consecutive days in 2015	(Bijlsma et al. 2016)
6	Taipei area (Northern Taiwan) and Kaohsiung-Pingtung area (Southern Taiwan)	Hospital effluent: 10 μg/ltr Ketamine and Norketamine in similar concentration The ratio of Ketamine/Norketamine = 0.3-4.6 (ratio like that found in urine sample)	Grab sample (3 ltr) 20 river samples 13 hospital effluent 3 ltr collected in amber glass bottles and stored in ice pack coolers Sample vacuum filtered - 0.22 μm cellulose acetate membrane	(Li et al. 2017)

Contd.

TABLE 10.2 Contd.

S. No.	Location	Amount Detected (Ketamine)	Sample Type	Reference
7	17 cities in Italy	Ketamine mass loads progressively increased from 2010/13. In Milan rise from 1 to 1.5 g/day in 2008-2010 to 3.4-3.6 g/day in 2013/14	24 hr composite samples collected daily for at least 7 consecutive days (4 yrs monitory study)	(Castiglonli et al. 2019)
8	Hong Kong	1400-1600 mg/ day/1000 people	Hourly raw wastewater composite samples were collected at both inlet channels applying a time proportional sampling mode, 250 ml every 15 min	(Yin et al. 2019)
9	Canada	Untreated :40 ng/ltr Treated: 10 ng/ltr	24 hr composites (equal volume equal hours) untreated and treated samples	(Yargeau et al. 2014)
10	Beijing Shanghai Guangzhou Shenzhen	5-6 ng/ltr 13 ng/ltr 89 ng/ltr 500 ng/ltr	Nine domestic sewage treatment plants (STPs), four locations. Each STP was sampled for four days by collecting four consecutive 24 hr composite samples	(Khan et al. 2014)

TABLE 10.3 Effect of Ketamine on different animals.

S. No.	Animal	Ketamine Conc.	Time	Effect	Reference
1	Zebrafish	0.5-2mm	Lasting upto 20hr	Normal morphology Neurotoxic to motor neuron and possibly affects differentiating neurons	(Kanungo et al. 2011)
2	Medaka fish (*Oryzias latipes*)	Environment relevant exposure of 0.004 μM to 40 μM	One to 14 days	Ketamine induced oxidative stress disrupted the expression of acetylcholinesterase and p53 regulated apoptosis pathways	(Liao et al. 2018)

Contd.

TABLE 10.3 Contd.

S. No.	Animal	Ketamine Conc.	Time	Effect	Reference
3	Pregnant Rat	40 mg/kg	3 hr	Nerve injury The density of hippocampal nerve and dendritic spine changed Gene related to Wnt/β-catenin pathway were downregulated	(Zhang et al. 2019)
4	*C. elegans*	54.3 μg/ltr	60 hr	Vulva deformity	(Wang et al. 2019)
5	*Daphnia magna*	30.93 mg/ltr	48 hr	Reproductive toxicity Reduction of the number of total live offspring by 33.6-49.5-8%	(Li et al. 2017)
6	Rhesus monkey	20-50 mg/kg	24 hr 3 hr	Neural cell death in perinatal monkey Does not produce neuronal cell death	(Slikker et al. 2007)
7	*Stachybotrys chartarum* *Staphylococcus epidermidis* *Borrelia burgdorferi*	1:80 dilution 1:16 1:320	24 hr 48 hr 7 days	Prevent fungal growth	(Torres et al. 2018)
8	Bacterial community	1 mg/ltr	40 days	Shift in bacterial community	(Wang et al. 2018)

Phytodegradation

Irradiation byproducts are important chemical products which are formed with the help of light. Ketamine has been shown not to degrade in absence of light (dark condition) and is not vulnerable to microbial degradation and hydrolysis, whereas in the presence of light it has been shown that it is degraded to its byproducts which resembles human metabolites. The photochemical reactions in water can be divided into two groups, namely direct photolysis and the indirect photolysis (Fig. 10.3).

Direct photolysis is the formation of byproducts as a result of radiation absorption where sunlight and oxygen play an indispensable role. For direct photolysis to occur, substance must absorb light radiation above 280 nm as it is the radiation of sunlight at the surface of the Earth. When a molecule absorbs light, the molecule promotes electrons from the ground state to excited state. It procures both electronic and vibrational energy which induce photoionization, ultimately leading to bond breakage and formation of byproducts. Unlike direct photolysis, indirect photolysis occurs when dissolved molecules, called photosensitizers, absorb sunlight. Photosensitizers can be chromophoric dissolved organic matter, bicarbonate and nitrate. Both direct and indirect photolysis chemical reactions ultimately lead to the formation of demethylation of ketamine to form norketamine, followed by hydroxylation to form hydroxynorketamine (HNK). Similarly, HK (hydroxyketamine) is also formed from the hydroxylation of ketamine (Fig. 10.4).

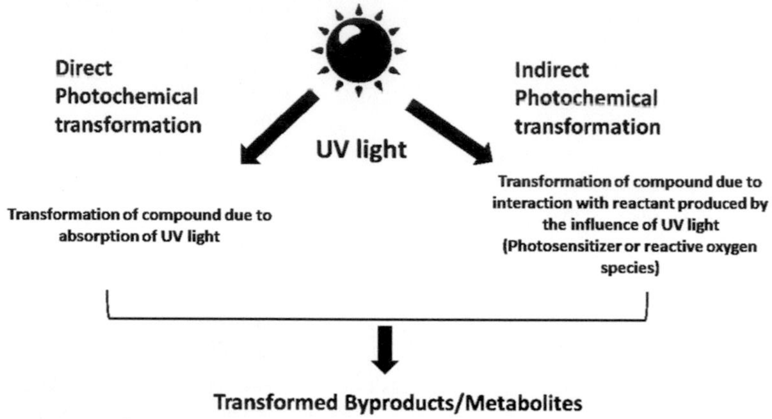

FIGURE 10.3 Direct and indirect photolysis.

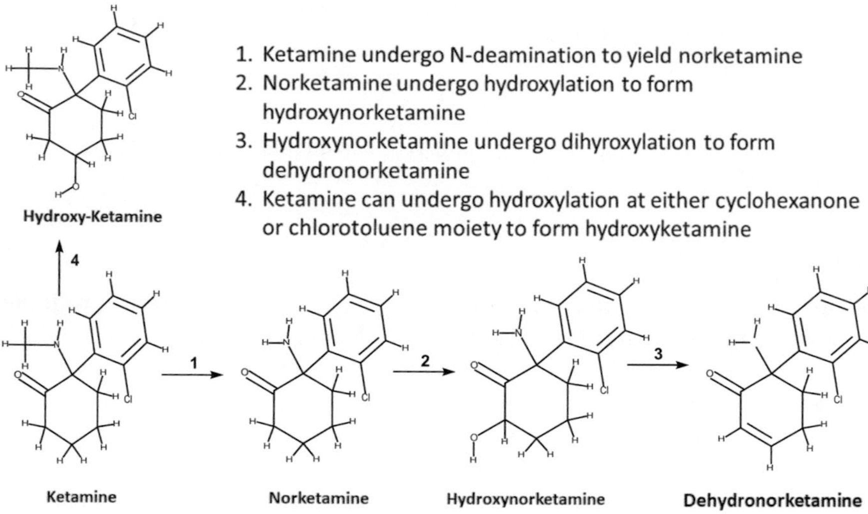
Hydroxy-Ketamine
1. Ketamine undergo N-deamination to yield norketamine
2. Norketamine undergo hydroxylation to form hydroxynorketamine
3. Hydroxynorketamine undergo dihyroxylation to form dehydronorketamine
4. Ketamine can undergo hydroxylation at either cyclohexanone or chlorotoluene moiety to form hydroxyketamine
Ketamine
Norketamine
Hydroxynorketamine
Dehydronorketamine

FIGURE 10.4 Ketamine conversion to different by-products.

Direct and indirect photolysis in water depends on various factors:

1. Structure of chemical substance
2. Electronic absorption spectrum of chemical substance

3. Amount of CDOM (Chromophoric dissolved organic matter) present in water
4. Number of radicals present in water
5. Wavelength of light

Environmental Fate

Due to its dissociative nature, Ketamine has been exploited in various recreational and medicinal sectors. Its illicit use has led Ketamine and its metabolites to the environment through various means.

Terrestrial Fate: Ketamine has a Log Koc value of 3.06. Koc measurement gives information regarding the mobility of substance in the soil. It is one of the important parameters for determining the environmental exposure level of chemicals. High Koc means it is strongly adsorbed with soil and does not move. Low Koc means it is mobile. Ketamine Koc value indicates it has low mobility in soil. pKa value of ketamine is 7.5, which indicates ketamine will exist in slightly cation form in the soil.

Aquatic Fate: Log Koc value of ketamine suggests ketamine be adsorbed with particles, suspended solids and residues. Ketamine exists as slightly cationic with ph values of 5 to 9. Volatilization also does not occur.

Atmospheric Fate: Ketamine has a vapor pressure of 52 nm Hg at 25°C, which suggests that it will prevail in both particulate and vapor phases. Ketamine is prone to direct photolysis in the presence of sunlight as it absorbs light at 290 nm.

Microbial Shift

Wang et al. in 2018 performed an experiment in the presence and absence of light and under the presence and absence of oxygen. High-throughput sequencing with Illumina Miseq was used to study the microbial population in the above mentioned experimental design, and the goal was to visualize the change in the population of the microbial community. In the following organisms, noticeable differences have been observed between control and ketamine-exposed samples (Wang et al. 2018):

a. *Methylophilaceae*
b. *Saprospiraceae*
c. *Xanthomonadaceae*
d. *WCHB1-69*
e. *Desulfobulbaceae*
f. *Porphyromonadaceae*
g. *Family X1*
h. *Peptococcaceae*
i. *Rhizobiaceae*
j. *Methylophilaceae*

Methamphetamine and Ketamine concentration was varied based on location of varied microbial population. Microorganisms play an indispensable role in the ecosystem and maintenance of steady microbial community is critical for ecosystem balance. *Methylophilacea,* a family within the order Methylophilales, consists of genera Methulobacillus, Methylovorus, and Methylotenera. They are obligate and restricted facultative methylotrophs which can use methanol as an exclusive source of carbon and energy. The presence of the methyl group in ketamine contributes to influential Methylophilaceae as a significant one for the demethylation pathway of ketamine breakdown. Saprospiraceae consists of genera Sprospira, Lewinella, Haliscomenobacter, Aureispira, Rubidimonas, and Portibacter. They are mostly aquatic and associated mostly with activated sludge. They seem to be involved in the breakdown of complex organic matter. WCHB1-69 also goes by name CFB group bacteria and its abundance is correlated with an increase in chlorine in the water. Saprospiraceae, Peptococcaceae, and Desulfobulbaceae are generally present scantly but when they are present profoundly, they make it difficult for other bacteria and algae to survive. These pose threats not only to the microbial community but also to the humans as they are associated with various ailments like inflammation. Therefore, microbial shift associated with biodegradation of chemical compounds like ketamine and methamphetamine confers serious threat as they disturb the aquatic ecosystem.

ACKNOWLEDGEMENTS

I would like to thank Southern Illinois University, Carbondale and my professor Dr. Brent Bany for constant guidance and support.

REFERENCES

Bijlsma, L., A. Botero-Coy, R. Rincón, G. Peñuela and F. Hernández. 2016. Estimation of illicit drug use in the main cities of Colombia by means of urban wastewater analysis. Sci. Total Environ. 565: 984-993.

Castiglioni, S., A. Borsotti, I. Senta and E. Zuccato. 2015. Wastewater analysis to monitor spatial and temporal patterns of use of two synthetic recreational drugs, ketamine and mephedrone, in Italy. Environ. Sci. Technol. 49(9): 5563-5570.

Kanungo, J., E. Cuevas, S. Ali and M. Paule. 2011. Ketamine induces motor neuron toxicity and alters neurogenic and proneural gene expression in zebrafish. J. Appl. Toxicol. 33(6): 410-417.

Khan, U., A. van Nuijs, J. Li, W. Maho, P. Du and K. Li, et al. 2014. Application of a sewage-based approach to assess the use of ten illicit drugs in four Chinese megacities. Sci. Total Environ. 487: 710-721.

Li, S., Y. Wang and A. Lin. 2017. Ecotoxicological effect of ketamine: Evidence of acute, chronic and photolysis toxicity to Daphnia magna. Ecotoxicol. Environ. Saf. 143: 173-179.

Liao, P., W. Yang, C. Yang, C. Lin, C. Hwang and P. Chen. 2018. Illicit drug ketamine induces adverse effects from behavioral alterations and oxidative stress to p53-regulated apoptosis in medaka fish under environmentally relevant exposures. Environ. Pollut. 237: 1062-1071.

Lin, A., X. Wang and C. Lin. 2010. Impact of wastewaters and hospital effluents on the occurrence of controlled substances in surface waters. Chemosphere 81(5): 562-570.

Maddox V.H., E.F. Godefroi and R.F. Parcell. 1965. The synthesis of phencyclidine and other 1-arylcyclohexylamines. J. Med. Chem. 8: 230-235.

Slikker, W., X. Zou, C. Hotchkiss, R. Divine, N. Sadovova, N. Twaddle, et al. 2007. Ketamine-Induced neuronal cell death in the perinatal rhesus monkey. Toxicol. Sci. 98(1): 145-158.

Torres, G., C. Hoehmann, J. Cuoco, K. Hitscherich, C. Pavia, M. Hadjiargyrou, et al. 2018. Ketamine intervention limits pathogen expansion *in vitro*. Pathog. Dis. 76(2): fty006.

Van Nuijs, A., A. Gheorghe, P. Jorens, K. Maudens, H. Neels and A. Covaci. 2013. Optimization, validation and the application of liquid chromatography-tandem mass spectrometry for the analysis of new drugs of abuse in wastewater. Drug. Test Anal. 6(7-8): 861-867.

Wang, D., Q. Zheng, X. Wang, J. Du, C. Tian, Z . Wang, et al. 2016. Illicit drugs and their metabolites in 36 rivers that drain into the Bohai Sea and north Yellow Sea, north China. Environ. Sci. Pollut. Res. 23(16): 16495-16503.

Wang, Z., Z. Xu and X. Li. 2018. Biodegradation of methamphetamine and ketamine in aquatic ecosystem and associated shift in bacterial community. J. Hazard. Mater. 359: 356-364.

Wang, Z., Z. Xu and X. Li. 2019. Impacts of methamphetamine and ketamine on C.elegans's physiological functions at environmentally relevant concentrations and eco-risk assessment in surface waters. J. Hazard. Mater. 363: 268-276.

Yargeau, V., B. Taylor, H. Li, A. Rodayan and C. Metcalfe. 2014. Analysis of drugs of abuse in wastewater from two Canadian cities. Sci. Total Environ. 487: 722-730.

Yin, X., C. Guo, Y. Teng and J. Xu. 2019. Development and application of the analytical method for illicit drugs and metabolites in fish tissues. Chemosphere 233: 532-541.

Zhang, X., J. Zhao, T. Chang, Q. Wang, W. Liu and L. Gao. 2020. Ketamine exerts neurotoxic effects on the offspring of pregnant rats via the Wnt/β-catenin pathway. Environ. Sci. Pollut. Res. 27: 305-314.

Zhang, X., R. Huang, P. Li, Y. Ren, J. Gao, J. Mueller, et al. 2019. Temporal profile of illicit drug consumption in Guangzhou, China monitored by wastewater-based epidemiology. Environ. Sci. Pollut. Res. 26(23): 23593-23602.

Index